T0326351

Electromagnetic Surface Waves

Electromagnetic Surface Waves
A Modern Perspective

John A. Polo, Jr.
Department of Physics and Technology
Edinboro University of Pennsylvania, Edinboro, PA, USA

Tom G. Mackay
School of Mathematics
University of Edinburgh, Edinburgh, UK

Akhlesh Lakhtakia
Department of Engineering Science and Mechanics
The Pennsylvania State University, University Park, PA, USA

AMSTERDAM • BOSTON • HEIDELBERG • LONDON • NEW YORK • OXFORD
PARIS • SAN DIEGO • SAN FRANCISCO • SINGAPORE • SYDNEY • TOKYO

ELSEVIER

Elsevier
225 Wyman Street, Waltham, MA 02451, USA
32 Jamestown Road, London NW1 7BY

First edition

Notices

Knowledge and best practice in this field are constantly changing. As new research and experience broaden our understanding, changes in research methods, professional practices, or medical treatment may become necessary.

Practitioners and researchers must always rely on their own experience and knowledge in evaluating and using any information, methods, compounds, or experiments described herein. In using such information or methods they should be mindful of their own safety and the safety of others, including parties for whom they have a professional responsibility.

To the fullest extent of the law, neither the Publisher nor the authors, contributors, or editors, assume any liability for any injury and/or damage to persons or property as a matter of products liability, negligence or otherwise, or from any use or operation of any methods, products, instructions, or ideas contained in the material herein.

Library of Congress Cataloging-in-Publication Data
A catalog record for this book is available from the Library of Congress

British Library Cataloguing-in-Publication Data
A catalogue record for this book is available from the British Library

ISBN: 978-0-12-397024-4

For information on all Elsevier publications
visit our website at store.elsevier.com

This book has been manufactured using Print On Demand technology. Each copy is produced to order and is limited to black ink. The online version of this book will show color figures where appropriate.

Working together
to grow libraries in
developing countries

www.elsevier.com • www.bookaid.org

To
His mother, Eunice Polo (1928–2002)
—*John A. Polo, Jr.*

His father, William Mackay (1927–2012).
—*Tom G. Mackay*

His students who have accompanied him on old and new paths
—*Akhlesh Lakhtakia*

Contents

Preface

Uniform plane waves are the simplest (nontrivial) solutions of the Maxwell equations in the frequency domain. The components of the field phasors and their spatial variation along a fixed axis are described using a Cartesian coordinate system. In the simplest case, the medium of propagation is a linear and homogeneous material that extends to infinity in all directions. The mathematical apparatus needed here is quite elementary.

Plane waves in air (approximated as vacuum) and isotropic dielectric materials are sufficiently simple to be taught in an undergraduate course using the approach of separation of variables. Plane waves in uniaxial dielectric materials are typically introduced in a first-year graduate course. Plane waves in bianisotropic materials are taught, if at all, in a specialized graduate course—with a convenient approach provided by the three-dimensional spatial Fourier transform.

Scattering of an incident electromagnetic field occurs at the planar boundary of two half-spaces occupied by dissimilar linear and homogeneous materials. If the incident field is a plane wave, analytical treatment of the reflected and refracted plane waves does not require additional skills beyond the ability to satisfy standard boundary conditions, in exactly the same way as in the undergraduate electromagnetics course. The representation of each plane wave—whether incident, reflected, or refracted—involves just one material and is obtained after pretending that this material fills all space.

An electromagnetic *surface* wave, in sharp contrast, is a creature of two materials. The surface wave straddles the planar interface of two half-spaces each occupied by a different material; remove the interface by making the two partnering materials identical, and the surface wave vanishes. Moreover, there is no guarantee that, at a specific frequency, a chosen pair of partnering materials will necessarily support the existence of a surface wave. Thus, a surface wave is far more delicate than a plane wave.

Electromagnetic plane waves have been known in one form or another for several millenniums, but the history of electromagnetic surface waves spans slightly more than a century. Initial progress was sporadic and slow, but the discovery of optical means of launching surface plasmon-polariton waves reported in 1959 led to a quickened pace during the 1970s. At optical frequencies, a surface plasmon-polariton wave is guided by the interface of a metal and a dielectric material (which could be air). Thousands of research papers are published every year on applications of this type of surface wave nowadays, many of the authors having only a beginner's knowledge of electromagnetics.

Inevitably, attention began to be paid during the last four decades to other types of electromagnetic surface waves as well. The complexity of partnering materials has increased from isotropic to anisotropic and even to bianisotropic in one sense, and from homogeneous to periodically nonhomogeneous (in the direction normal to the interface) in another sense. These new types of surface waves have not had the vogue experienced by surface plasmon-polariton waves yet, but the spectacular growth of nanotechnologies during the last two decades promises a very bright future for all types of surface waves.

The attraction of surface waves stems from most of the energy of a surface wave being confined to the close vicinity of the interface of the two partnering materials. Any change in the composition of either partnering material in that vicinity could alter—even eliminate—the surface wave. Conversely, by engineering the composition close to the interface, a surface wave can be allowed to propagate. At optical frequencies, the close vicinity is often less than a few micrometers. Nanotechnologies allow engineering of composition with a precision of a few nanometers, enabling hitherto unrealizable geometries for field confinement.

From a modern perspective, it does not matter if a partnering material is homogeneous, isotropic, and dielectric or if it is bianisotropic and periodically nonhomogeneous in the direction normal to the interface. A 4×4 matrix formalism can be devised to solve a dispersion equation and then determine the field phasors on each side of the interface. The same formalism can be used to analyze and design a practical prism-coupled configuration for launching a surface wave, and an extension allows treatment of the practical grating-coupled configuration as well. In this way, simple and complex surface plasmon-polariton waves, Zenneck waves, Fano waves, Dyakonov waves, Tamm waves, and Dyakonov-Tamm waves can all be treated in a unified fashion.

Bianisotropic materials offer a far richer palette of electromagnetic phenomenons than the simpler dielectric ones. A permittivity dyadic, a permeability dyadic, and two magnetoelectric dyadics are needed to characterize a linear bianisotropic material, whereas a linear dielectric material requires just the permittivity dyadic. Attention has been focused in this book only on electromagnetic surface waves guided by the interface of two materials both of which can be described using only permittivity dyadics. The research literature on electromagnetic surface waves involving bianisotropic materials is presently scant, but we expect it to grow in the next two decades to incorporate significant quantums of theoretical and experimental results.

One aim of this monograph is to expose the mathematical unity of diverse surface-wave phenomenons. The other aim is to familiarize researchers with the characteristics of these phenomenons when the partnering materials are dielectric (isotropic or anisotropic) and periodically nonhomogeneous. Researchers have found during the past 5 years that periodic nonhomogeneity is the cause of the existence of multiple surface-wave modes at a specified frequency and propagating in a particular direction in the interface plane—but with different phase speed, attenuation rate, and field profiles. This multiplicity will surely enhance applications of surface waves in optical sensing, imaging, communication, and harvesting of solar energy.

This monograph was written to collate and organize research results that have become available chiefly during the last decade—in order to inspire further theoretical research, galvanize extensive experimental efforts, and seed application-oriented research. Accordingly, the monograph is directed chiefly toward doctoral students and post-doctoral scholars. Familiarity with the frequency-domain Maxwell equations and matrix manipulations is assumed. Some knowledge of dyadics would be helpful, although the basics provided in an appendix ought to suffice.

We express our deep gratitude to Muhammad Faryad (Pennsylvania State University) for tireless assistance with calculations, diagrams, and graphs. His cheerful, positive, and helpful attitude exceeded the norms of collegiality. We also thank him for reading a draft version and suggesting changes.

We thank Jeremy W. Galusha and Michael H. Bartl (University of Utah) for an image of a natural diamond-based photonic crystal on a scanning electron microscope, Yi-Jun Jen (National Taipei University of Technology) for a cross-sectional image of a sculptured nematic thin film on a scanning electron microscope, Sandia National Laboratory for an image on a scanning electron microscope of a photonic crystal called a woodpile, and Osamu Takayama (Institut de Ciències Fotòniques, Barcelona) for two photographs related to the first observation of a Dyakonov wave in a laboratory.

AL thanks Andreas Otto (Universität Düsseldorf) and H. Angus Macleod (University of Arizona) for discussions on the history of optical means to excite simple surface plasmon-polariton waves. Also, AL thanks his doctoral students Stephen E. Swiontek and Drew P. Pulsifer for photographs of equipment, cross-sectional images of sculptured thin films on scanning electron microscopes, and graphs. JAP and TGM join AL in thanking both Stephen and Drew for reading a draft version and suggesting changes.

AL is grateful to the Trustees of the Pennsylvania State University for granting a sabbatical leave of absence in Fall 2012, the Charles Godfrey Binder Endowment at the same university for ongoing support of his research activities, and the US National Science Foundation for supporting his research on surface multiplasmonics during the last 2 years.

At Elsevier, Lisa Tickner, Tracey Miller, and Erin Hill-Parks in succession herded the three academic cats that we are. Largely as a result of their frequent prodding, we wrote fewer papers during the last 2 years—but we still managed to push the publication deadline by a month. A longer delay would have helped us reduce errors in the monograph, but please do not blame the editors for any errors that you may find in the monograph—instead, reprimand AL by e-mail, for he is ready to shoulder the blame. Plus, he was the delayer-in-chief.

John A. Polo, Jr.
Edinboro, PA, USA

Tom G. Mackay
Edinburgh, Scotland, United Kingdom

Akhlesh Lakhtakia
University Park, PA, USA

January 2013

List of Acronyms and Principal Symbols

Acronyms

AED	angular existence domain
CCD	charge-coupled device
CLC	cholesteric liquid crystal
CTF	columnar thin film
FDTD	finite-difference time-domain
LRSPP	long-range surface-plasmon-polariton
NPV	negative phase velocity
PVD	physical vapor deposition
RIU	refractive index unit(s)
RCWA	rigorous coupled-wave approach
SEM	scanning electron microscope
SNTF	sculptured nematic thin film
SP	surface plasmon
SPP	surface plasmon-polariton
SPR	surface-plasmon resonance
SRSPP	short-range surface-plasmon-polariton
STF	sculptured thin film

Scalars

$A_{L,R}$	circular absorptances
$A_{s,p}$	linear absorptances
c_0	speed of light in free space
$C_{1,2,3,4}$	four amplitudes for eigenmodal representation of field phasors
f_ℓ	volume fraction of constituent ℓ
h	handedness parameter
i	$\sqrt{-1}$
k_0	free-space wavenumber
$k_{x,y}^{(m,n)}$	wavenumber of Floquet harmonic of order (m,n)
q	wavenumber of surface wave (canonical boundary-value problem)

$r_{\mathrm{LL,LR,RL,RR}}$	circular reflection coefficients
$r_{\mathrm{ss,sp,ps,pp}}$	linear reflection coefficients
$R_{\mathrm{LL,LR,RL,RR}}$	circular reflectances
$R_{\mathrm{ss,sp,ps,pp}}$	linear reflectances
t	time
$t_{\mathrm{LL,LR,RL,RR}}$	circular transmission coefficients
$t_{\mathrm{ss,sp,ps,pp}}$	linear transmission coefficients
$T_{\mathrm{LL,LR,RL,RR}}$	circular transmittances
$T_{\mathrm{ss,sp,ps,pp}}$	linear transmittances
v_p	phase speed
x, y, z	Cartesian coordinates
$\alpha_{1,2,3,4}$	four eigenvalues of $\left[\underline{\underline{A}}\right]$
$\beta^{\mathcal{A}}_{s,p}$	decay constants for s- and p-polarized surface waves in material $\mathcal{A}$
$\beta^{\mathcal{A}}_{1,2}$	decay constants for surface waves in material $\mathcal{A}$
$\beta^{\mathcal{B}}_{3,4}$	decay constants for surface waves in material $\mathcal{B}$
χ	tilt angle
χ_v	vapor flux angle
δ	biaxiality
$\delta_{\alpha\beta}$	Kronecker delta
Δn	birefringence
ϵ	permittivity
ϵ_0	permittivity of free space
η_0	intrinsic impedance of free space
$\gamma^{\mathcal{A}}$	angle between the xz plane and some significant plane of material $\mathcal{A}$
$\gamma^{\mathcal{B}}$	angle between the xz plane and some significant plane of material $\mathcal{B}$
λ_0	free-space wavelength
κ	wavenumber in prism-coupled configuration
μ	permeability
μ_0	permeability of free space
ω	angular frequency
Ω	structural half-period
ψ	surface-wave propagation angle with respect to x axis in the interface plane
ρ	sensitivity
$\tilde{\rho}_e$	externally impressed electric charge density (time domain)
ρ_{RI}	refractive-index sensitivity
$\sigma_{1,2,3,4}$	four eigenvalues of $\left[\underline{\underline{Q}}\right]$
$\tau^{\mathcal{A,B}}$	phase angles for refractive-index modulation
θ_{inc}	angle of incidence for prism-coupled and grating-coupled configurations

3-Vectors

$\tilde{B}, B$	primitive magnetic field (time, frequency domain)
$\tilde{D}, D$	induction electric field (time, frequency domain)
$\tilde{E}, E$	primitive electric field (time, frequency domain)
$e(z)$	auxiliary electric field phasor
$\tilde{H}, H$	induction magnetic field (time, frequency domain)
$h(z)$	auxiliary magnetic field phasor
$\tilde{J}_e, J_e$	externally impressed electric current density (time, frequency domain)
k	wave vector
$\tilde{M}$	magnetization (time domain)
P	time-averaged Poynting vector
$\tilde{P}$	polarization (time domain)
r	position
u_s	unit vector in the interface plane normal to u_{prop}
u_{prop}	unit vector in direction of surface-wave propagation in the interface plane
u_x, u_y, u_z	unit vectors parallel to Cartesian axes

3 × 3 Dyadics

$\underline{\underline{I}}$	identity dyadic
$\underline{\underline{U}}$	shape dyadic
$\underline{\underline{0}}$	null dyadic
$\underline{\underline{\epsilon}}$	permittivity constitutive dyadic
$\underline{\underline{\zeta}}$	magnetoelectric constitutive dyadic
$\underline{\underline{\mu}}$	permeability constitutive dyadic
$\underline{\underline{\xi}}$	magnetoelectric constitutive dyadic

4-Vectors

$[f(z)]$	auxiliary phasor vector comprising x and y components of $e(z)$ and $h(z)$
$[v]^{(1,2,3,4)}$	four eigenvectors of $\begin{cases} [P] \text{ for homogeneous materials} \\ [A] \text{ for periodically nonhomogeneous materials} \end{cases}$

4 × 4 Matrixes

$[\underline{\underline{0}}]$	null matrix
$[\underline{\underline{A}}]$	matrix related to $[\underline{\underline{Q}}]$ by Eq. (3.57)
$[\underline{\underline{G}}]$	diagonal matrix containing eigenvalues of $[\underline{\underline{P}}]$
$[\underline{\underline{I}}]$	identity matrix
$[\underline{\underline{J}}]$	matrix with every element being 1
$[\underline{\underline{P}}]$	matrix in the matrix ordinary differential equation (3.22)
$[\underline{\underline{P}}(z)]$	matrix in the matrix ordinary differential equation (3.49)
$[\underline{\underline{Q}}]$	matrix characterizing electromagnetic response of one structural period of a periodically nonhomogeneous material, per Eq. (3.53)
$[\underline{\underline{V}}]$	diagonalizing matrix for $[\underline{\underline{P}}]$

RCWA quantities

$[\vdots]$	column vector
$[\vdots]$	matrix

Operators and functions

$\{\cdot\}^*$	complex conjugate
$\{\cdot\}^\dagger$	conjugate transpose
$\det\{\cdot\}$	determinant
$\text{Im}\{\cdot\}$	imaginary part
$\{\cdot\}^{-1}$	inverse
$\text{Re}\{\cdot\}$	real part
$\{\cdot\}^T$	transpose
$\mathcal{U}(\cdot)$	unit step function

1 Surface Waves

1.1 Introduction

The planar interface of two dissimilar materials plays a vital role in many optical phenomenons. In elementary optics, for instance, the material interface appears to be responsible for reflection and refraction: light approaches the interface along a straight path, and then abruptly breaks up into two parts, both of which continue along straight paths in two different directions. Of course, the entirety of the refracting material is involved in what is actually seen, but that realization requires understanding of a conceptual framework based on the fundamental laws of electromagnetism.

One can, however, discern a more gradual change at the interface when total internal reflection occurs. Thus Isaac Newton realized that the incident wave, which seems not to enter the refracting material, actually penetrates that material with an amplitude that decays exponentially with distance from the interface on the microscopic scale [1]. The penetrating wave, known as an evanescent wave, was once mainly a theoretical curiosity. In recent years, however, evanescent waves have been used in newly developing technologies such as near-field spectroscopy [2].

Another phenomenon, the electromagnetic surface wave, is even more intimately tied to the interface. Such a wave travels in a direction parallel to the interface but, on either side of the interface, its amplitude is minuscule after a certain distance from the interface. This localization is a propitious quality that is exploited, for example, in some extremely sensitive bio/chemical sensors [3,4]. The electromagnetic surface wave is the focus of this book.

The study of electromagnetic surface waves began in the first decade of the 20th century. Yet, nearly a century later, a single type of wave, the surface-plasmon-polariton (SPP) wave, dominates the technoscientific scene, at least at optical frequencies. The concentration on the SPP wave has resulted in wonderful technological developments with the creation of extremely sensitive chemical sensors and biochemical sensors, and improvements in this mature technology continue to this day. Even in this highly developed application, the two partnering materials which meet at the interface may be simple: one is a typical metal—a plasmonic material at optical frequencies—and the other is a homogeneous, isotropic, dielectric material. The SPP wave then is labeled as *simple*, and its electric and magnetic fields have simple spatial profiles, as discussed in Chapter 2.

Electromagnetic Surface Waves. http://dx.doi.org/10.1016/B978-0-12-397024-4.00001-3

However, we now know that, in partnership with a plasmonic material, a more complex polarizable material can also support the propagation of SPP waves. The surface waves guided by such interfaces can have some technoscientifically interesting properties which easily could be exploited in the near future; very significantly, multiple SPP-wave modes for a specified range of directions of propagation and for a specified range of angular frequencies are possible. The electric and magnetic fields of some of these SPP-wave modes have complicated spatial profiles, as discussed in Chapter 6.

Furthermore, while the interface of a plasmonic material and a polarizable material supports an SPP wave, a variety of other types of surface waves can be supported by the interface of two polarizable materials. Since polarizable materials such as dielectric materials are less dissipative, in general, than plasmonic materials such as metals, the advantage of these materials for long-range propagation of surface waves is apparent. Yet, as we shall see, surface waves guided by the interface of two dielectric materials can exhibit other intriguing properties.

Research on electromagnetic surface waves of new types progressed at a slow rate for a few decades, but is now advancing at an accelerated rate. What is responsible for this activity? As always, electromagnetics is a fertile playground for the theorist. After years of research, exploring surface-wave propagation under many different possible conditions, a solid theoretical foundation has been established which supports current work. As with research on simple SPP-wave propagation, theory has led the way in the discovery of new types of electromagnetic surface waves.

The contribution of modern nanotechnology to the development of research on electromagnetic surface waves is significant. Although nature provides ample materials and situations that motivate explorations purely out of scientific curiosity, there is nothing like the prospects of new and useful devices to spur on a frenzy of research activity. Recent developments in nanotechnology provide a host of new possibilities. A variety of materials with desirable complexities have been developed. In particular, materials with an engineered nanoscale structure have provided a platform for investigation and produced some of the most interesting results. While it was not possible to contemplate such materials when the notion of surface waves supported only by dielectric materials emerged, their production is now nearly routine. With the current ability to design and fabricate a vast variety of materials, it is possible to foresee some interesting applications; many more uses will surely take us by surprise.

1.2   A Brief History

The notion of an electromagnetic surface wave made a significant appearance in 1907 when Zenneck [5] authored a theoretical paper exploring the possibility of a wave guided by the interface of the atmosphere and either earth or a large body of water. His focus was on radio waves, a region of the electromagnetic spectrum far from the optical regime in which we are particularly interested when nanomaterials are to be used to guide surface waves. Nonetheless, the principles involved are the same, owing to the scale invariance of the Maxwell postulates. Although the Zenneck wave, as it is now called [6], was further investigated, also theoretically, by such notables as Sommerfeld [7] and Bouwkamp [8], its practical existence remains controversial [9,10].

During the late 1940s, certain electronic phenomenons observed in metals exposed to electric fields were resolved in terms of plasma oscillations [11,12]. Antecedent literature exists [13], and the electronic-plasma oscillations were also related to the oscillations of ionic plasmas in vacuum tubes [14]. The quantum of plasma oscillations is a quasiparticle dubbed the *plasmon* [15]. A few years later, the energy losses of electrons impinging on a metal film were explained in terms of electronic-plasma oscillations occurring at the film's surfaces [16]. Naturally, the quantum of these oscillations came to be called the *surface plasmon* (SP).

A train of SPs traveling along a metal/vacuum interface can be treated classically as an SP wave. When vacuum is replaced by a dielectric material, the quasiparticles are called surface plasmon-polaritons, each with a plasmonic component in the metal and a polaritonic component in the dielectric material. Just as a plasmon is the quantum of plasma oscillations, a *polariton* is the quantum of polarization [17] in dielectric matter [18]. At a frequency where the imaginary part of the permittivity of the metal is small enough in magnitude to be ignored in comparison to the real part, the SPP wave is called a Fano wave [19].

SPP waves cannot be excited by shining light directly at either a dielectric film lying atop a metal film or a metal film lying atop a dielectric film. In 1959, T. Turbadar published a seminal paper [20] showing that the reflectance of a parallel-polarized plane wave from a thin aluminum film deposited on a glass prism exhibited a sharp dip as the angle of incidence exceeded the critical angle for a glass/air interface. A proper choice of the thickness of the metal film could transform the reflectance low into a reflectance null even [21]. No relation to the excitation of an SP wave was drawn by Turbadar, but was provided in 1968 by Otto [22] and separately by Kretschmann and Raether [23]. Thus emerged the Turbadar-Otto and the Turbadar-Kretschmann-Raether configurations as practical methods to excite SPP waves optically using evanescent waves generated by total internal reflection. When an SPP wave is excited in this manner, energy is absorbed from the evanescent wave, thereby resulting in a reduction in the intensity of reflected light, a process known as attenuated total reflection or frustrated total reflection [24]. SPP waves can also be excited using a surface-relief grating, a waveguide, or even an optical fiber [4]. Indeed, as early as 1902, Wood [25,26] had observed dark lines in the pattern of light reflected from a surface-relief grating of metal, which we now attribute to the excitation of SPP waves. No matter the method, the phenomenon of a light wave coupling to an SPP wave is known as surface-plasmon resonance (SPR).

From the 1970s, there has been a continual development of instruments exploiting SPR. At first, metal films and films adhering to the surface of the metal were explored. In the early 1980s, Nylander *et al.* [27] were among the first to develop sensing using attenuated total reflection. The Swedish company Phamacia Biosensors AB[1] was formed in 1984 in order to commercially produce an SPR instrument. Today there is a wide choice of SPR instruments, chiefly SPR sensors, available from many companies [3,4].

---

[1] This company was renamed Biacore AB in 1996 and was acquired by GE Healthcare in 2006.

Over the last three decades, the propagation of SPP waves guided by the interface of a metal and more complex dielectric materials has been investigated. At first, the focus was on a purely theoretical understanding of SPP waves guided by metal/anisotropic-dielectric interfaces. With the growing interest in liquid crystals during the 1980s, a few researchers [28–31] investigated the use of SPR for characterizing liquid crystals. Furthermore, in some spectral regimes, the metal may be replaced by an alloy [32] or a semiconductor [33].

Opposite signs of the real parts of the permittivity scalars of the two isotropic partnering materials—and analogous conditions if one or both partnering materials are anisotropic—are essential to SPP-wave propagation. However, the interface of two homogeneous dielectric materials of which at least one is anisotropic may support surface-wave propagation of another type, even though the real parts of all components of the permittivity dyadics of both materials are positive. Although research began in the 1970s [34], interest in surface waves guided by the interface of two dielectric materials began to take off after Dyakonov in 1988 [35] explored the propagation of a surface wave guided by the interface of a uniaxial dielectric material and an isotropic dielectric material. Since then, the scope of the term *Dyakonov wave* has expanded to include surface waves guided by the interface of two homogeneous dielectric materials, at least one of which is anisotropic [36]; even bianisotropic partnering materials are admissible [37–39].[2] Due to the complicated nature of the field expressions in an anisotropic material, theoretical research on Dyakonov waves guided by the interface of various homogeneous dielectric materials under different conditions continues to this day. The narrow range of directions for Dyakonov-wave propagation makes experimental work difficult, so much so that the first observation of a Dyakonov wave was made only in 2009 [41], more than two decades after its theoretical introduction.

The recent explosion of nanotechnology has opened a whole new realm of possibilities for both SPP waves and Dyakonov waves within the past 10 years. Now, materials can be designed and manufactured with specific anisotropic nature as well as nonhomogeneities. For example, columnar thin films (CTFs) [42], formed by physical vapor deposition (PVD) of a bulk material after conversion to a collimated vapor, comprise straight nanowires or columns. CTFs are effectively homogeneous under optical illumination. By changing either the evaporated material or deposition conditions, the anisotropy of the generally biaxial CTF can be engineered. If the substrate upon which the film is deposited is rocked and/or rotated during the deposition process, the nanowires can be sculptured into various shapes to create sculptured thin films (STFs) [42]. Under the right conditions, it is possible to create a nonhomogeneous material with optical properties that vary continuously along the direction perpendicular to a planar interface. By the turn of the new millennium, understanding and characterization of CTFs and STFs was sufficiently advanced to allow consideration of these materials for designing interfaces to support surface-wave propagation.

---

[2]Parenthetically, SPP waves guided by the planar interface of a metal and a dielectric material which moves at constant velocity relative to the metal have also been considered [40]. In effect, the planar interface here is of a metal and a bianisotropic material.

Beginning in 2007 [43–45], theoretical research has characterized the properties of SPP and Dyakonov waves supported by metal/CTF and CTF/isotropic-dielectric material interfaces, respectively. The theory for surface-wave propagation guided by interfaces formed with CTFs is not fundamentally different from that describing propagation guided by interfaces formed with naturally occurring biaxial crystals. However, the optical properties of CTFs are selectable over a continuous range, thereby making possible the design of interfaces for surface-wave propagation. Additionally, CTFs—and STFs in general—are porous materials, a property which has potential for technological exploitation [42].

The controllable nonhomogeneity and anisotropy of STFs is creating new avenues of surface-wave research and development which are inaccessible with crystalline materials. A search, in 2007 [46], for surface waves guided by the interface of an isotropic, homogeneous, dielectric material and an STF that is periodically nonhomogeneous in the direction normal to the interface gave positive results. The theoretically discovered wave was named a Dyakonov-Tamm wave. The first part of the name of this wave is appropriate because the surface wave is an extension of the Dyakonov wave, which is guided by the interface of two dielectric materials of which at least one is anisotropic. Furthermore, due to the periodicity of the STF, the new surface wave is an optical analog of the electron wave guided by the surface of a metal proposed by Tamm in 1932 [47], thereby making *Tamm* an appropriate modifier.

The first Dyakonov-Tamm wave investigated is guided by the interface of an isotropic, homogeneous, dielectric material and a chiral STF, which is composed of nanohelices oriented perpendicular to the interface. Since then, calculations have shown that an isotropic, homogeneous, dielectric material partnered with either (i) an STF having a periodically nonhomogeneous morphology [48] or (ii) a periodically layered material [49] can also support surface-wave propagation. Furthermore, calculations [50–53] for an interface of two chiral STFs with different orientations and/or constitutions also predict the possibility of Dyakonov-Tamm waves. Periodically ordered liquid crystals could be employed for surface-wave propagation, as well. STFs and liquid crystals each have certain advantages with respect to various applications. Liquid crystals may be controlled dynamically with an applied electric field or respond to physical conditions such as temperature and pressure, while PVD allows for an almost unlimited variety of unidirectional nonhomogeneity when fabricating STFs.

The interface of a periodically nonhomogeneous dielectric material and a metal supports SPP waves with a very interesting property. In 2008, calculations for both the interface of a metal and a periodically nonhomogeneous sculptured nematic thin film (SNTF) [54,55] and the interface of a metal and a chiral STF [56] showed that multiple SPP-wave modes can be guided by a single interface along a specific direction. All of the SPP-wave modes for a single interface have different field profiles, attenuation rates, and phase speeds, but all of them have the same frequency. In 2009, experiments verified the existence of multiple SPP-wave modes guided by both a metal/SNTF interface [57] and a metal/chiral-STF interface [58,59].

Other periodically nonhomogeneous, dielectric materials have also received theoretical attention. Rugate filters [60], which are constructed with isotropic materials and are mostly designed to possess a sinusoidally varying refractive index, reject light in a narrow frequency band and are routinely manufactured. Interfaces of rugate filters with both homogeneous dielectric materials [61] and other rugate filters [62] in various configurations, all support multiple surface waves called Tamm waves. The interface of a rugate filter and a metal can guide several SPP waves [63]; if the imaginary part of the permittivity scalar of the metal can be ignored, those SPP waves may be called Fano waves [61].

The Reusch pile is a stack of layers of an anisotropic dielectric material with an incremental rotation from one layer to the next about an axis normal to the layers [64,65]. The interface of a Reusch pile and an isotropic dielectric material supports a single Dyakonov-Tamm wave mode [49], similarly to an interface of an isotropic dielectric material and a chiral STF with a continuously rotating permittivity dyadic.

The chiral STF, the periodically nonhomogeneous SNTF, the rugate filter, and the Reusch pile may be considered as examples of one-dimensional (1-D) photonic crystals. A few researchers have pursued the possibility of surface-wave propagation at interfaces formed with 2-D [66] and 3-D [67] photonic crystals. These materials lie well within the current technical reach of modern nanotechnology.

Homogeneous materials that support the propagation of plane waves with negative phase velocity (NPV) have been a very popular research topic in recent years [68], and have not escaped the scrutiny of investigation for surface-wave propagation [69,70]. Although realization of composite NPV materials which can be considered homogeneous and exhibit low dissipation in the optical frequency range awaits further development [71–73], theoretical work predicts that interfaces incorporating an NPV material will support the propagation of surface waves with some interesting characteristics [74–76].

1.3 Simple SPP Wave

We start a more technical look at surface-wave propagation with the simple SPP wave, these days a technologically important wave, as a foundation for further discussion. The qualifier *simple*, here, implies several properties of the partnering materials forming the interface which guides the SPP wave. When the materials are isotropic, homogeneous, linear, non-magnetic, and achiral [17], the characteristics of the SPP wave are quite simple and straightforward.

1.3.1 Canonical Boundary-Value Problem

Let the interface of the two partnering materials, for purposes of this discussion, be the plane $z = 0$, with a metal filling the half-space $z < 0$, and a dielectric material filling the half-space $z > 0$. This is the geometry of what may be called the canonical problem of SPP-wave propagation. Parenthetically, the material occupying the half-space $z < 0$ is most commonly, but not necessarily, a metal.

A fundamental property of a surface wave is its localization to the interface. Thus, it must be guided along some direction which is parallel to the interface and have an amplitude which decreases with distance far away from the interface. Such a wave may seem quite distinct from a plane wave, an eigenmodal solution of the frequency-domain Maxwell postulates in a bulk material, which extends indefinitely in all directions with electric and magnetic field phasors described as [17]

$$\left.\begin{array}{l} \underline{E}(\underline{r}) = \underline{\mathcal{E}} \exp(i\underline{k} \cdot \underline{r}) \\ \underline{H}(\underline{r}) = \underline{\mathcal{H}} \exp(i\underline{k} \cdot \underline{r}) \end{array}\right\}, \tag{1.1}$$

where $\underline{\mathcal{E}}$ and $\underline{\mathcal{H}}$ are amplitude vectors with complex-valued components; $\underline{k}$ is the wave vector which can also have complex-valued components; $\underline{r} = x\underline{u}_x + y\underline{u}_y + z\underline{u}_z$ is the position vector with $\underline{u}_x$, $\underline{u}_y$, and $\underline{u}_z$ being the Cartesian unit vectors; and an $\exp(-i\omega t)$ dependence on time t is implicit with $i = \sqrt{-1}$ and ω as the angular frequency.

However, the required surface-wave characteristics may be achieved with a slight modification of the plane wave described by Eq. (1.1); thus,

$$\underline{E}(\underline{r}) = \begin{cases} \underline{\mathcal{E}}_{met} \exp(i\underline{k}_{met} \cdot \underline{r}), & z < 0, \\ \underline{\mathcal{E}}_{diel} \exp(i\underline{k}_{diel} \cdot \underline{r}), & z > 0, \end{cases} \tag{1.2}$$

and

$$\underline{H}(\underline{r}) = \begin{cases} \underline{\mathcal{H}}_{met} \exp(i\underline{k}_{met} \cdot \underline{r}), & z < 0, \\ \underline{\mathcal{H}}_{diel} \exp(i\underline{k}_{diel} \cdot \underline{r}), & z > 0, \end{cases} \tag{1.3}$$

with the subscripts $_{met}$ and $_{diel}$ denoting the metal and the dielectric material, respectively. If the component of the wave vector perpendicular to the interface is complex valued with a non-zero imaginary part, the amplitude of the wave either decays or grows with distance from the interface. Of course, we must check that, in making a component of the wave vector complex valued, we obtain a legitimate result which satisfies the Maxwell curl postulates; i.e.

$$\left.\begin{array}{l} \underline{k}_{met} \times \underline{\mathcal{E}}_{met} = \omega\mu_0\underline{\mathcal{H}}_{met} \\ \underline{k}_{met} \times \underline{\mathcal{H}}_{met} = -\omega\epsilon_{met}\underline{\mathcal{E}}_{met} \end{array}\right\}, \quad z < 0, \tag{1.4}$$

and

$$\left.\begin{array}{l} \underline{k}_{diel} \times \underline{\mathcal{E}}_{diel} = \omega\mu_0\underline{\mathcal{H}}_{diel} \\ \underline{k}_{diel} \times \underline{\mathcal{H}}_{diel} = -\omega\epsilon_{diel}\underline{\mathcal{E}}_{diel} \end{array}\right\}, \quad z > 0, \tag{1.5}$$

where ϵ_{met} and ϵ_{diel} are the permittivity scalars of the metal and the dielectric material, respectively, and are functions of the angular frequency ω. Typically, $\text{Re}\{\epsilon_{met}\} < 0$, $\text{Re}\{\epsilon_{diel}\} > 0$, and $\text{Im}\{\epsilon_{diel}\} = 0$ are assumed for analysis of SPP-wave propagation. The permeability of free space (i.e. vacuum) is denoted by $\mu_0 = 4\pi \times 10^{-7}$ H m^{-1}, and the permittivity of free space by $\epsilon_0 = 8.854 \times 10^{-12}$ F m^{-1}.

Although the component of the wave vector along the direction of propagation parallel to the plane $z = 0$ *may* be complex valued, it must have a non-zero real part in

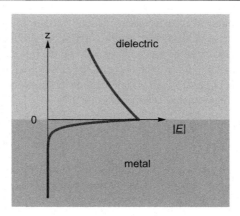

Figure 1.1 Typical variation of the amplitude of the electric field phasor of an SPP wave as a function of distance from the interface $z = 0$, when both partnering materials are homogeneous, isotropic, non-magnetic, and achiral.

order to describe propagation. Thus, we write the wave vector in the metal side of the interface as[3]

$$\underline{k}_{met} = q\underline{u}_{prop} - \alpha_{met}\underline{u}_z, \tag{1.6}$$

and the wave vector in the dielectric side of the interface as

$$\underline{k}_{diel} = q\underline{u}_{prop} + \alpha_{diel}\underline{u}_z, \tag{1.7}$$

where q, α_{met}, and α_{diel} are complex valued. The unit vectors $\underline{u}_{prop}$ and $\underline{u}_z$ are parallel to the propagation direction and normal to the interface, respectively; $\underline{u}_{prop} \cdot \underline{u}_z = 0$. Both the real and imaginary parts of q are positive, so that the direction of propagation is also the direction of attenuation. The phase speed of the SPP wave is denoted by $v_p = \omega/\text{Re}\{q\}$.

The wave is localized to the interface and decays as $z \to \pm\infty$, provided the twin conditions

$$\left.\begin{array}{l} \text{Im}\,\{\alpha_{met}\} > 0 \\ \text{Im}\,\{\alpha_{diel}\} > 0 \end{array}\right\} \tag{1.8}$$

are satisfied. Figure 1.1 illustrates a typical exponentially decaying behavior of SPP field magnitudes on both sides of the interface $z = 0$. Although the spatial profile of the

---

[3]In Chapters 1, 2, and 6, the half-space $z < 0$ in the canonical boundary-value problem for SPP-wave propagation is occupied by a metal, and the spatial dependence of the field phasors is taken to be $\exp[i(q\underline{u}_{prop} - \alpha_{met}\underline{u}_z) \cdot \underline{r}]$ with $\text{Im}\{\alpha_{met}\} > 0$. In Chapters 4, 5, and 7, sometimes the half-space $z < 0$ in the canonical boundary-value problem for surface-wave propagation is occupied by an isotropic homogeneous dielectric material. In consonance with Chapter 3, we then adopt the spatial dependence $\exp[i(q\underline{u}_{prop} + \alpha^{B}\underline{u}_z) \cdot \underline{r}]$ with $\text{Im}\{\alpha^{B}\} < 0$. Both representations are identical, as can be ascertained by setting $\alpha_{met} = -\alpha^{B}$.

electromagnetic field depends on the angular frequency and the partnering materials, typically in the visible regime [77],

   i. the penetration depth $\Delta_{met} = 1/\text{Im}\,\{\alpha_{met}\}$ into the metal is $\sim$25 nm,
   ii. the penetration depth $\Delta_{diel} = 1/\text{Im}\,\{\alpha_{diel}\}$ into the dielectric material is $\sim$250 nm,
   iii. the propagation length $\Delta_{prop} = 1/\text{Im}\,\{q\}$ of the SPP wave in the direction parallel to $\underline{u}_{prop}$ is less than 400 µm, and
   iv. the SPP wavelength $2\pi/\text{Re}\{q\} \sim 500$ nm in the same direction.

Although the wave vectors in the two half-spaces are different, their components parallel to the unit vector $\underline{u}_{prop}$ are the same. This is a necessary condition in order to satisfy the standard boundary conditions simultaneously everywhere along the interface [17,78]. Substitution of Eqs. (1.2), (1.3), (1.6), and (1.7) into Eqs. (1.4) and (1.5) leads to the constraint

$$q^2 = k_0^2(\epsilon_{diel}/\epsilon_0) - \alpha_{diel}^2 = k_0^2(\epsilon_{met}/\epsilon_0) - \alpha_{met}^2 \qquad (1.9)$$

on the two wave vectors, where $k_0 = \omega\sqrt{\epsilon_0\mu_0} = 2\pi/\lambda_0$ is the wavenumber in free space and λ_0 is the wavelength in free space.

After splitting the description of the surface wave into two parts, one each for the two half-spaces, we must ensure that the components of the electric and magnetic fields parallel to the interface are continuous across the interface [17,78]. Enforcement of these continuity conditions leads to the requirement

$$\alpha_{diel}\epsilon_{met} = -\alpha_{met}\epsilon_{diel}; \qquad (1.10)$$

furthermore, the SPP wave must be p polarized, i.e. the magnetic field phasor is orthogonal to both $\underline{u}_{prop}$ and $\underline{u}_z$, whereas the electric field phasor lies wholly in the plane formed by $\underline{u}_{prop}$ and $\underline{u}_z$. Of course, this characteristic is displayed only if the partnering dielectric material is both homogeneous and linear. If that material is either nonhomogeneous in the direction perpendicular to the interface [63] or is nonlinear [79], both s-polarized and p-polarized SPP waves are possible. The electric field phasor of an s-polarized wave is orthogonal to both $\underline{u}_{prop}$ and $\underline{u}_z$, whereas its magnetic field phasor lies wholly in the plane formed by $\underline{u}_{prop}$ and $\underline{u}_z$.

What are the implications of the analysis so far? In order to extract the essence of the simple SPP wave, let us temporarily neglect the imaginary part of the permittivity of the dielectric material, it being typically very small compared to the real part, and let us also ignore $\text{Im}\,\{\epsilon_{met}\}$ for simplicity. Then, Eq. (1.10) yields

$$\frac{\text{Im}\,\{\alpha_{diel}\}}{\text{Im}\,\{\alpha_{met}\}} = -\frac{\epsilon_{diel}}{\epsilon_{met}}, \qquad (1.11)$$

and in conjunction with Eq. (1.8), requires that the permittivity scalars of the two partnering materials have opposite signs. Metals are traditionally used as one of the two partnering materials to form the interface for SPP-wave propagation since commonly $\text{Re}\{\epsilon_{met}\} < 0$ at optical frequencies; that requirement can be fulfilled in different spectral regimes by semiconductors as well [32,33]. In contrast, generally $\epsilon_{\text{diel}} \approx \text{Re}\{\epsilon_{\text{diel}}\} > 0$ in the same frequency range. Furthermore, due to Eq. (1.9), purely real permittivity scalars lead to the conclusion that $q > k_0\sqrt{\epsilon_{\text{diel}}/\epsilon_0}$. Thus, the SPP wavelength $2\pi/q$ is less than the wavelength $2\pi/k_0\sqrt{\epsilon_{\text{diel}}/\epsilon_0}$ of a plane wave in the partnering dielectric material in the bulk, and as a consequence the phase speed of the SPP is also less than the phase speed of the plane wave in the partnering dielectric material in the bulk. The foregoing conclusions do not have to be modified significantly if the magnitudes of the imaginary parts are not ignored.

So far, our discussion has been rather abstract: propagation of a surface wave guided by the interface of two semi-infinite materials, with an unknown agent exciting the surface wave. The solution of this canonical problem is vital to the understanding of surface-wave propagation, but the canonical problem is unimplementable in practice.

1.3.2   Practical Configurations

In practice, only materials of finite extent may be used, and the SPP wave must be excited in some manner. The mismatch in wavelength of the SPP wave and the plane wave in the bulk dielectric partnering material prevents excitation of the SPP wave by direct coupling to a beam of light. Therefore, four practical configurations have been devised to excite SPP waves: two types of prism-coupled configuration, as well as grating-coupled and waveguide-coupled configurations. Although all four configurations are equivalent to each other for the excitation of SPP waves, each is suitable for different applications; furthermore, when careful interpretation of experimental or calculated data from any of the configurations is needed, resort must be made to the solution of the canonical problem.

1.3.2.1   Prism-Coupled Configurations

Two configurations, known as the Turbadar-Kretschmann-Raether [23] and Turbadar-Otto [22] configurations, use the evanescent wave generated by total internal reflection to excite the SPP wave. An evanescent wave, like the SPP wave, has a wavenumber greater than that of a plane wave in the partnering dielectric material in the bulk.

Turbadar-Kretschmann-Raether configuration. This configuration, by far the more popular of the two prism-coupled configurations, employs a prism made of a refractive index n_{prism} that exceeds the refractive index $n_{\text{diel}} = \sqrt{\epsilon_{\text{diel}}/\epsilon_0}$ (assumed real and positive) of the partnering dielectric material. A thin metal film is supposed to be deposited on (or otherwise placed in intimate contact with) the base of the prism with the partnering dielectric material forming an interface on the opposite side of the metal film, as shown in Figure 1.2a. When a beam of light strikes the prism/metal interface at an angle θ_{inc} greater than the critical angle for total internal reflection from an interface of the prism and the partnering dielectric material, an evanescent wave is formed which is able to pass through the suitably thin metal film and into the partnering dielectric material. At a certain value of θ_{inc}, the component of the evanescent wave's wave vector parallel to the interface is close in magnitude to the real part of q that satisfies Eqs. (1.9) and (1.10), and the SPP wave is excited.

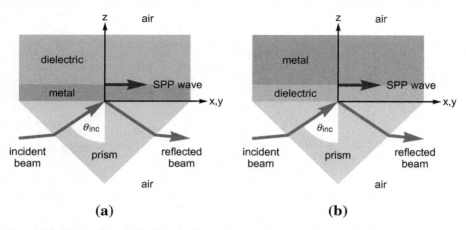

Figure 1.2 Schematics of (a) Turbadar-Kretschmann-Raether and (b) Turbadar-Otto configurations for exciting an SPP wave using a prism.

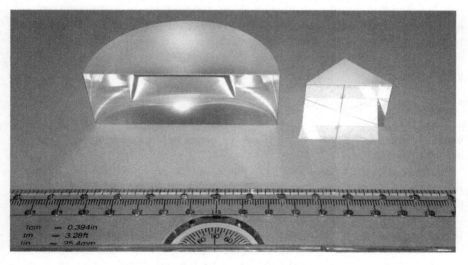

Figure 1.3 Semi-circular and triangular prisms.

Figure 1.3 is a photograph of semi-circular and triangular prisms made of borosilicate glass, which has a refractive index of 1.51509 at $\lambda_0 = 632.8$ nm. Containing about 10% boric oxide, this type of crown glass is resistant to chemical and environmental damage, besides exhibiting very low dispersion over the visible spectral regime. In practice, the base of the prism is put in contact with a borosilicate-glass slide on which the metal film is deposited. A very thin layer of an index-matching liquid is used to stick the slide to the prism, thereby eliminating a discontinuity-causing layer of air between the slide and the prism.

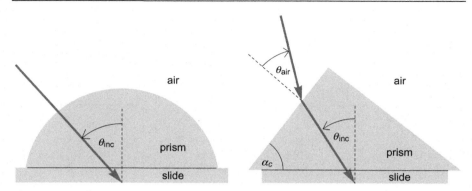

Figure 1.4 Angle of incidence θ_{inc} in the Turbadar-Kretschmann-Raether configuration with semi-circular and triangular prisms.

Triangular prisms are used more commonly than semi-circular prisms. This is because the cross-section of a semi-circular prism is not a complete semi-circle but must be reduced by the thickness of the slide on which the metal film is deposited. The reduction involves an extra manufacturing step. Although the measurement of the angle θ_{inc} is greatly simplified, as shown in Figure 1.4, only slides of a specific thickness must be used. No such limitation on the slide thickness exists with a triangular prism, but θ_{inc} has to be inferred—after measuring the angle θ_{air} shown in Figure 1.4—by using the formula

$$\theta_{inc} = \alpha_c - \sin^{-1}\left(\frac{\sin\theta_{air}}{n_{prism}}\right), \tag{1.12}$$

where α_c is the corner angle of the triangular prism and n_{prism} is the refractive index of the prism material.

Turbadar-Otto configuration. This configuration is similar to the Turbadar-Kretschmann-Raether configuration with the positions of the metal and partnering dielectric material reversed, as shown in Figure 1.2b. Light striking the interface of the prism and the partnering dielectric material at an angle greater than the critical angle for these two materials creates an evanescent wave. If the partnering dielectric material is sufficiently thin, then the evanescent wave may reach that material's interface with the metal with a significant amplitude. Just as for the Turbadar-Kretschmann-Raether configuration, at a certain angle of incidence θ_{inc} at the interface of the prism and the partnering dielectric material, the component of the wave vector parallel to the interface of the evanescent wave is close in magnitude to the real part of q that satisfies Eqs. (1.9) and (1.10), and the SPP wave is excited. In order for the Turbadar-Otto configuration to work efficiently, the thickness of the partnering dielectric material must be $\sim$200 nm, which is quite a low value. This requirement has limited the popularity of the Turbadar-Otto configuration for many applications, particularly for sensing chemicals and biochemicals when the partnering dielectric material is a liquid containing the analyte. The Turbadar-Kretschmann-Raether configuration—with a thin metal film rather than a thin dielectric film—is much more convenient for those applications.

1.3.2.2   Grating-Coupled Configuration

It is also possible to use a surface-relief grating for exciting SPP waves [3], thereby avoiding the use of a prism. Both partnering materials are of finite thickness but their interface is now the undulating surface $z = g(x, y)$ instead of the plane $z = 0$. The undulating surface can be singly periodic with $g(x, y) \equiv g(x) = g(x \pm L_x)$, say, where L_x is the period along the x axis, as schematically illustrated in Figure 1.5. Alternatively, the surface $z = g(x, y)$ can be doubly periodic with $g(x, y) = g(x \pm L_x, y) = g(x, y \pm L_y)$, where L_x and L_y are the periods along the x and y axes, respectively.

Suppose that a p-polarized plane wave is incident on a doubly periodic grating from, say, free space on the other side of the partnering dielectric material, the wave vector of the incident plane wave having the projection $k_0(\underline{u}_x \cos \psi + \underline{u}_y \sin \psi) \sin \theta_{\text{inc}}$ in the xy plane. The field phasor excited in the metal can be written for $z \geqslant \max_{x,y} g(x, y)$ as the sum of Floquet harmonics [80]—also called Bloch waves [81]—whose spatial variations are stated as

$$\exp \left(i \underline{k}_{\text{met}}^{(m,n,\pm)} \cdot \underline{r} \right), \tag{1.13}$$

where

$$\underline{k}_{\text{met}}^{(m,n,\pm)} = \underline{\kappa}^{(m,n)} \mp \alpha_{\text{met}}^{(m,n)} \underline{u}_z, \tag{1.14}$$

$$\underline{\kappa}^{(m,n)} = k_x^{(m)} \underline{u}_x + k_y^{(n)} \underline{u}_y, \tag{1.15}$$

$$k_x^{(m)} = k_0 \sin \theta_{\text{inc}} \cos \psi + m(2\pi / L_x), \quad m \in \{0, \pm 1, \pm 2, ...\}, \tag{1.16}$$

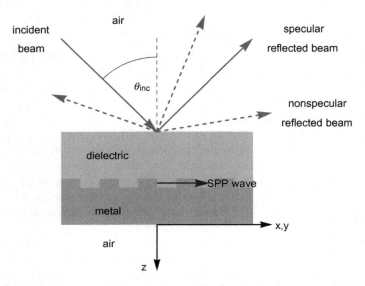

Figure 1.5 Schematic of the grating-coupled configuration for exciting an SPP wave. The surface-relief grating shown is singly periodic.

$$k_y^{(n)} = k_0 \sin\theta_{\text{inc}} \sin\psi + n(2\pi/L_y), \quad n \in \{0, \pm 1, \pm 2, ...\}, \tag{1.17}$$

$$\underline{k}_{\text{met}}^{(m,n,\pm)} \cdot \underline{k}_{\text{met}}^{(m,n,\pm)} = k_0^2 \epsilon_{\text{met}}/\epsilon_0. \tag{1.18}$$

Likewise, for $z \leqslant \min_{x,y} g(x, y)$ in the partnering dielectric material, the field phasor is the sum of Floquet harmonics with spatial variations

$$\exp(i\underline{k}_{\text{diel}}^{(m,n,\pm)} \cdot \underline{r}), \tag{1.19}$$

where

$$\underline{k}_{\text{diel}}^{(m,n,\pm)} = \underline{\kappa}^{(m,n)} \pm \alpha_{\text{diel}}^{(m,n)} \underline{u}_z \tag{1.20}$$

and

$$\underline{k}_{\text{diel}}^{(m,n,\pm)} \cdot \underline{k}_{\text{diel}}^{(m,n,\pm)} = k_0^2 (\epsilon_{\text{diel}}/\epsilon_0). \tag{1.21}$$

For a singly periodic grating, the index $n \in \{0\}$.

Provided the incidence conditions are such that (i) the magnitude of $\underline{\kappa}^{(\tilde{m},\tilde{n})}$ for some $(\tilde{m}, \tilde{n})$ is close to the real part of q that satisfies Eqs. (1.9) and (1.10), and (ii) both $\alpha_{\text{met}}^{(\tilde{m},\tilde{n})}$ and $\alpha_{\text{diel}}^{(\tilde{m},\tilde{n})}$ have positive imaginary parts, an SPP wave supported by the periodically undulating interface $z = g(x, y)$ will be excited by the incident p-polarized plane wave. This form of coupling is especially suited for enhanced harvesting of solar energy in photovoltaic cells [82–84].

Instead of a simple grating, a compound grating may be used [85]. Each period of a compound grating comprises two or more simple gratings, each several periods long. A compound grating can help excite SPP waves over a range of free-space wavelengths and for several different incidence conditions. However, the excitation of a specific SPP wave may be less efficient with compound grating than with a simple grating [85,86].

1.3.2.3 Waveguide-Coupled Configurations

For applications in integrated optics, excitation of SPP waves is attempted using a waveguide-coupled configuration called end-fire configuration. The plane $x = 0$ separates a waveguide ($x < 0$) whose axis is the x axis from the metal/dielectric structure ($x > 0$), as shown in Figure 1.6a. The cross-sectional dimensions of both the waveguide—often a parallel-plate waveguide but can have another cross section—and the metal/dielectric structure must be sufficiently large for the fields to decay significantly before reaching the outer limits of the waveguide and the metal/dielectric structure far from the central axis. A normal mode of the waveguide is excited with an axial variation $\exp(i\kappa_{wg}x)$ such that κ_{wg} is close to the solution q of Eqs. (1.9) and (1.10); of course, the magnetic field phasor of the waveguide mode must be oriented substantially along the y axis. This widely used configuration was devised by Stegeman et al. [87] in 1983.

In a different waveguide-coupled configuration, a thin metal overlayer is placed on part of a dielectric waveguide and the dielectric partnering material is put on top of the overlayer, as shown in Figure 1.6b. An SPP wave is excited when the normal mode in the bulk of the dielectric waveguide is phase-matched to the SPP wave. This configuration is attractive in some sensor systems because of its ruggedness and small size [3].

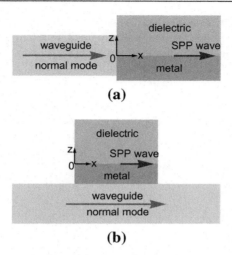

Figure 1.6 Schematics of two waveguide-coupled configurations for exciting an SPP wave.

1.4 Dielectric Materials

The simple SPP wave, as we have seen in Section 1.3, is guided by the interface of two isotropic, homogeneous materials, of which one is dielectric. The use of more complex materials not only expands the range of possibilities for SPP waves, but allows for the propagation of other types of surface waves.

1.4.1 Solid Crystals

Nature provides us with many anisotropic dielectric materials in the form of crystals, and these were the first materials considered to form interfaces supporting Dyakonov waves. A crystal is identified as belonging to one of seven systems [88].

A crystal that belongs to the cubic system is characterized just like an isotropic dielectric material by the constitutive relations

$$\left. \begin{array}{l} \underline{D}(\underline{r}) = \epsilon \underline{E}(\underline{r}) \\ \underline{B}(\underline{r}) = \mu_0 \underline{H}(\underline{r}) \end{array} \right\}, \tag{1.22}$$

containing the permittivity scalar ϵ. Although this quantity is complex valued in general, it is real valued in the absence of dissipation (which is often assumed in crystal optics). Furthermore, ϵ is a function of the angular frequency ω. Here, $\underline{D}$ is the electric displacement phasor and $\underline{B}$ is the magnetic displacement phasor.

In order to accommodate the directional properties of anisotropic dielectric crystals, constitutive relations of the form

$$\left. \begin{array}{l} \underline{D}(\underline{r}) = \underline{\underline{\epsilon}} \cdot \underline{E}(\underline{r}) \\ \underline{B}(\underline{r}) = \mu_0 \underline{H}(\underline{r}) \end{array} \right\} \tag{1.23}$$

are required, with the permittivity 3×3 dyadic $\underline{\underline{\epsilon}}$ being a function of ω.[4] For a uniaxial crystal—which belongs to either the tetragonal, or the trigonal or the hexagonal system—the permittivity dyadic has the form [17,88]

$$\underline{\underline{\epsilon}} = \epsilon_{11}(\underline{u}_1\underline{u}_1 + \underline{u}_2\underline{u}_2) + \epsilon_{33}\underline{u}_3\underline{u}_3, \tag{1.24}$$

where $\underline{u}_1$, $\underline{u}_2$, and $\underline{u}_3$ is a right-handed triad of three mutually orthogonal unit vectors: $\underline{u}_1 \times \underline{u}_2 = \underline{u}_3$ *et cyclicum*. For biaxial dielectric crystals, the form of $\underline{\underline{\epsilon}}$ depends upon the crystal system [17,88]:

$$\underline{\underline{\epsilon}} = \epsilon_{11}\underline{u}_1\underline{u}_1 + \epsilon_{22}\underline{u}_2\underline{u}_2 + \epsilon_{33}\underline{u}_3\underline{u}_3 \tag{1.25}$$

for an orthorhombic crystal,

$$\underline{\underline{\epsilon}} = \epsilon_{11}\underline{u}_1\underline{u}_1 + \epsilon_{22}\underline{u}_2\underline{u}_2 + \epsilon_{33}\underline{u}_3\underline{u}_3 + \epsilon_{12}(\underline{u}_1\underline{u}_2 + \underline{u}_2\underline{u}_1) \tag{1.26}$$

for a monoclinic crystal, and

$$\underline{\underline{\epsilon}} = \epsilon_{11}\underline{u}_1\underline{u}_1 + \epsilon_{22}\underline{u}_2\underline{u}_2 + \epsilon_{33}\underline{u}_3\underline{u}_3 \\ + \epsilon_{12}(\underline{u}_1\underline{u}_2 + \underline{u}_2\underline{u}_1) + \epsilon_{13}(\underline{u}_1\underline{u}_3 + \underline{u}_3\underline{u}_1) + \epsilon_{23}(\underline{u}_2\underline{u}_3 + \underline{u}_3\underline{u}_2) \tag{1.27}$$

for a triclinic crystal. More general forms of $\underline{\underline{\epsilon}}$ arise when, for example, gyrotropic effects are taken into account [17]. Furthermore, in the most general case of dielectric anisotropy, the dyadic $\underline{\underline{\epsilon}}$ is specified by nine complex-valued components; i.e.

$$\underline{\underline{\epsilon}} = \sum_{\ell=1}^{3} \sum_{m=1}^{3} \epsilon_{\ell m}\underline{u}_\ell\underline{u}_m. \tag{1.28}$$

1.4.2   Particulate Composite Materials

Crystals are homogeneous continuums at macroscopic length scales, i.e. over distances much larger than the dimensions of the unit cell. The degree of anisotropy of naturally occurring crystals tends to be quite small, while practical applications in many instances will require a larger degree of anisotropy. Materials with a larger degree of anisotropy can be obtained as particulate composite materials containing electrically small particles [89]. Electrically small particles are required to have linear dimensions much smaller than the wavelength(s) inside both the host material and the bulk material of which the component particles are made [90,91], thereby ensuring that resonances do not arise within the component particles [92]. Such a particulate composite material can be considered as an effectively homogeneous material with a permittivity dyadic $\underline{\underline{\epsilon}}$ that can be predicted using homogenization theories [73,93].

1.4.3   Nanoengineered Materials

Several types of surface waves treated in this book require materials that can be considered as periodically nonhomogeneous continuums at subwavelength scales. Such

---

[4]See Appendix A for a functional introduction to dyadics.

materials are not readily found in nature, but nanotechnology provides a means for fabricating material continuums with high anisotropy and periodic nonhomogeneity. By adjusting the composition and structure of materials on the nanoscale, it is possible to manipulate both anisotropy and periodicity in a controlled manner, thus allowing for the design of materials with specific properties for particular uses. Even though nanostructured materials are composed of discrete entities at the nanoscale, they appear at optical frequencies as continuously varying materials. Let us look at some of the available nanostructured materials, which are currently driving innovative work on surface-wave propagation, and their properties.

1.4.3.1   Columnar Thin Films

When a well-collimated vapor is directed toward a planar substrate at a suitable temperature and pressure in a low-pressure chamber, the evaporated material self-aggregates into parallel columns, or nanowires, thereby producing a CTF [42, Chapter 7], as shown in Figures 1.7 and 1.8. With this PVD process, it is currently possible, in low-pressure chambers commonly used for academic research, to obtain nanowires tilted at an angle $\chi \in [20°, 90°]$ to the substrate plane by proper selection of the acute angle χ_v between the average direction of the vapor and the substrate plane. Lower values of χ are expected to be achievable through lower values of χ_v in low-pressure chambers with a larger working volume. In general, $\chi > \chi_v$ except when $\chi_v = 90°$, in which case $\chi = 90°$. Empirically determined relationships between the tilt angle χ and the vapor flux angle χ_v are available for a few materials [94], but these relationships are most likely dependent on the particular apparatus used to produce the CTF. The nanowire diameter is roughly between 10 and 100 nm, depending on the deposited material and conditions. With structural features at such a small length scale, the CTF appears as a homogeneous continuum at optical frequencies. The thickness of the CTF is limited to about 10 μm with commonly used inorganic materials. At greater thickness, the columnar morphology begins to deteriorate. Polymeric materials may be grown several times thicker but the columnar diameter is also proportionally larger [95].

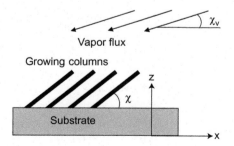

Figure 1.7 Schematic of the growth of a CTF. When a well-collimated vapor is incident on the substrate at an average angle $\chi_v \in (0°, 90°]$ with respect to the substrate plane, nanowires grow tilted at an angle $\chi \geqslant \chi_v$ such that $\chi \in (0°, 90°]$. The xz plane is the morphologically significant plane of the CTF.

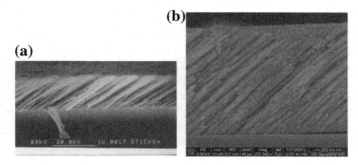

Figure 1.8 Cross-sectional images of two CTFs on a scanning electron microscope. The CTFs were fabricated by evaporating (a) magnesium fluoride and (b) chalcogenide glass of nominal composition $Ge_{28}Sb_{12}Se_{60}$.

Columnar growth is inherently anisotropic. If the vapor is taken to travel in the xz plane and the substrate plane is the xy plane, the cross-sectional dimensions of the nanowires along the x and the y axes are different. Thus, it is appropriate to model a CTF as an orthorhombic dielectric continuum with

$$\underline{\underline{\epsilon}}_{CTF} = \underline{\underline{S}}_y(\chi) \cdot (\epsilon_a \underline{u}_z \underline{u}_z + \epsilon_b \underline{u}_x \underline{u}_x + \epsilon_c \underline{u}_y \underline{u}_y) \cdot \underline{\underline{S}}_y^{-1}(\chi) \qquad (1.29)$$

as its permittivity dyadic. The real symmetric dyadic

$$\underline{\underline{S}}_y(\chi) = \underline{u}_y \underline{u}_y + (\underline{u}_x \underline{u}_x + \underline{u}_z \underline{u}_z) \cos \chi + (\underline{u}_z \underline{u}_x - \underline{u}_x \underline{u}_z) \sin \chi \qquad (1.30)$$

in Eq. (1.29) captures the tilt of the nanowires with respect to the substrate plane in the morphologically significant xz plane. The three principal permittivity scalars of a CTF are denoted by ϵ_a, ϵ_b, and ϵ_c.

Accordingly, a CTF is a birefringent material [17]: along any fixed direction, two different plane waves with different phase speeds, attenuation rates, and polarization states can travel in this material. With the assumption of negligible dissipation (i.e. $\epsilon_{a,b,c}$ are real and positive), as shown in Appendix B, the two optic ray axes (also called the crystallographic axes) of a CTF are parallel to the vectors

$$\underline{a}_{\pm} = \underline{\underline{S}}_y(\chi) \cdot \left\{ \underline{u}_x \sqrt{\frac{\epsilon_b - \epsilon_c}{\epsilon_b - \epsilon_a}} \pm \underline{u}_z \sqrt{\frac{\epsilon_c - \epsilon_a}{\epsilon_b - \epsilon_a}} \right\}, \qquad (1.31)$$

and the two optic axes are parallel to the vectors

$$\underline{b}_{\pm} = \underline{\underline{S}}_y(\chi) \cdot \left\{ \pm \underline{u}_x \sqrt{\frac{(\epsilon_b - \epsilon_c)\epsilon_a}{(\epsilon_b - \epsilon_a)\epsilon_c}} + \underline{u}_z \sqrt{\frac{(\epsilon_c - \epsilon_a)\epsilon_b}{(\epsilon_b - \epsilon_c)\epsilon_c}} \right\}, \qquad (1.32)$$

where it has been assumed that either $\epsilon_b > \epsilon_c > \epsilon_a$ or $\epsilon_b < \epsilon_c < \epsilon_a$. Both optic ray axes and both optic axes lie in the xz plane, which is the morphologically significant plane of the CTF. The two plane waves have the same phase speed if they are traveling along an optic axis, but both have the same rate of flow of energy if they are traveling along an optic ray axis.

Decreasing χ_v from 90° increases the porosity and exacerbates anisotropy. Thus, the choice of χ_v during fabrication crucially affects the optical response characteristics of a CTF. The ability to choose constitutive parameters $\epsilon_{a,b,c}$ and χ over a continuous range should allow a greater degree of flexibility in designing interfaces for surface-wave propagation. This is especially attractive, since CTFs may be fabricated from a wide variety of materials with different bulk properties [96].

The porosity of CTFs may be of use in practical applications of surface waves— for example, in optical sensing. The optical properties of a CTF can be altered by embedding a material in the porous structure [97,98].

1.4.3.2   Sculptured Thin Films

Rocking and/or rotating the substrate during PVD results in curved nanowires and thereby a sculptured thin film (STF) is fabricated [42, Chapter 1]. Like CTFs, an STF is a porous material. At a given distance z from the substrate, the STF has a local orthorhombic anisotropy. With smooth, continuous motion of the substrate, the permittivity dyadic of the STF likewise varies smoothly with distance from the substrate. Conversely, the orientation of the substrate relative to the oncoming vapor can be changed abruptly during deposition of the film to create a discontinuous change in the permittivity dyadic. Thus, in addition to partnering an STF with a second STF made by evaporating another material to form an interface for surface-wave propagation, it is also possible to create a constitutive change within an STF by abruptly changing the motion of the substrate during deposition to create an internal interface [53].

Rocking the substrate about the y axis produces nanowires which can be described by a curve in the xz plane [99]. A film thus produced is referred to as a sculptured nematic thin film (SNTF). Computer-controlled rocking during deposition can produce nearly an unlimited number of nanowire shapes. Some common shapes of nanowires are chevronic, sigmoid, and slanted sigmoid. A cross-sectional image on a scanning electron microscope (SEM) of an SNTF with chevronic morphology is shown in Figure 1.9.

The permittivity dyadic of an SNTF is a straightforward modification of Eq. (1.29); thus,

$$\underline{\underline{\epsilon}}_{\mathrm{SNTF}}(z) = \underline{\underline{S}}_y[\chi(z)] \cdot [\epsilon_a(z)\underline{u}_z\underline{u}_z + \epsilon_b(z)\underline{u}_x\underline{u}_x + \epsilon_c(z)\underline{u}_y\underline{u}_y] \cdot \underline{\underline{S}}_y^{-1}[\chi(z)], \quad (1.33)$$

where $\chi(z)$ and $\epsilon_{a,b,c}(z)$ depend on $\chi_v(z)$. If $\chi_v(z)$ is piecewise constant, the SNTF is piecewise homogeneous. If $\chi_v(z)$ is periodic, so is $\underline{\underline{\epsilon}}_{\mathrm{SNTF}}(z)$ [57]. Nonhomogeneity along the z axis can also be effected by changing the material to be evaporated.

Rotating the substrate about an axis perpendicular to the substrate plane produces a chiral STF [100], which is composed of helical nanowires oriented perpendicular to the substrate plane [101]. In addition to choice of material, two geometrical parameters define the chiral STF: the tilt angle and the pitch. As shown in Figure 1.10, the tilt angle χ is the angle between a tangent to the helical nanowire and the substrate plane and is determined by the vapor flux angle χ_v, which is kept fixed during deposition. The pitch, designated by 2Ω, is the repeat distance along the z axis. Together, the deposition rate,

Figure 1.9 Cross-sectional image of an SNTF of chevronic morphology. The SNTF was fabricated by evaporating chalcogenide glass of nominal composition $Ge_{28}Sb_{12}Se_{60}$.

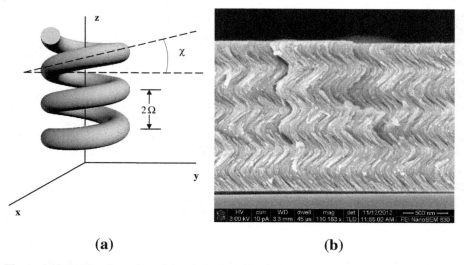

       (a)   **(b)**

Figure 1.10 (a) The tilt angle and the pitch of a helix. (b) Cross-sectional image of a chiral STF on a scanning electron microscope. The chiral STF was fabricated by evaporating chalcogenide glass of nominal composition $Ge_{28}Sb_{12}Se_{60}$.

the rotation rate of the substrate, and the vapor flux angle determine the pitch of the chiral STF. Incidentally, the sense of rotation of the substrate determines whether the chiral STF is right handed or left handed. Figure 1.10 also shows the cross-sectional image of a chiral STF on a scanning electron microscope.

On any xy plane, the chiral STF is locally orthorhombic. The principal values of the permittivity dyadic $\underline{\underline{\epsilon}}_{\text{chiral STF}}(z)$ remain the same for all z, but the principal axes of that dyadic rotate smoothly about the z axis with changing z. Accordingly, the permittivity dyadic of a chiral STF is given by [42, Chapter 7]

$$\underline{\underline{\epsilon}}_{\text{chiral STF}}(z) = \underline{\underline{S}}_z(z) \cdot \underline{\underline{S}}_y(\chi) \cdot (\epsilon_a \underline{u}_z \underline{u}_z + \epsilon_b \underline{u}_x \underline{u}_x + \epsilon_c \underline{u}_y \underline{u}_y) \cdot \underline{\underline{S}}_y^{-1}(\chi) \cdot \underline{\underline{S}}_z^{-1}(z),$$

(1.34)

where χ is a fixed angle just the same as for a CTF in Section 1.4.3.1 and the real symmetric dyadic

$$\underline{\underline{S}}_z(z) = \underline{u}_z \underline{u}_z + (\underline{u}_x \underline{u}_x + \underline{u}_y \underline{u}_y) \cos\left(h\frac{\pi z}{\Omega}\right) + (\underline{u}_y \underline{u}_x - \underline{u}_x \underline{u}_y) \sin\left(h\frac{\pi z}{\Omega}\right)$$

(1.35)

indicates rotation about the z axis with period 2Ω. The structural handedness parameter h appearing in the definition of the rotation dyadic $\underline{\underline{S}}_z(z)$ is equal to either $+1$ for right handedness or -1 for left handedness. Setting $h = 0$ reduces a chiral STF to a CTF. The structural handedness of the chiral STF is crucial for circular-polarization filters [102, 103].

Rocking and rotation of the substrate can be done concurrently or sequentially [104]. The evaporating material can also be changed during deposition. Multiple materials can be simultaneously deposited [105]. Thus, the permittivity dyadic of a general STF may be expressed as

$$\underline{\underline{\epsilon}}_{\text{STF}}(z) = \underline{\underline{S}}(z) \cdot [\epsilon_a(z)\underline{u}_z\underline{u}_z + \epsilon_b(z)\underline{u}_x\underline{u}_x + \epsilon_c(z)\underline{u}_y\underline{u}_y] \cdot \underline{\underline{S}}^{-1}(z),$$

(1.36)

where $\underline{\underline{S}}(z)$ is a real symmetric dyadic with all three eigenvalues positive. If the STF is periodically nonhomogeneous along the z axis with period 2Ω, the constraint

$$\underline{\underline{\epsilon}}_{\text{STF}}(z \pm 2\Omega) = \underline{\underline{\epsilon}}_{\text{STF}}(z)$$

(1.37)

is satisfied.

1.4.3.3 Photonic Crystals

Photonic crystals are manufactured periodic assemblies of discrete nanostructured components which are usually composed of differing isotropic dielectric materials [106, 107]. They must be periodic at least in one direction [108]. If a photonic crystal is periodic in more than one direction, there is no reason for those directions to be mutually orthogonal, although the first conceptualization of a three-dimensional photonic crystal did have orthorhombic symmetry [109]. The propagation of light through photonic crystals is analogous to the motion of electrons through natural crystals, with the periodicity of the crystal affecting the relationship between the angular frequency and the wavenumber. Just as a natural crystal may exhibit band gaps—which are ranges of angular frequencies wherein electrons are unable to travel freely through the crystal—a photonic crystal may exhibit band gaps wherein the propagation of light is forbidden.

A one-dimensional photonic crystal usually consists of layers of dielectric materials which can be simply deposited sequentially. They are, thus, the most easily constructed photonic crystals, and have been fabricated and studied extensively from the mid-20th century. Current technology routinely employs these photonic crystals as mirrors and spectral filters in optical devices such as lasers and spectrometers [110].

The quarter-wave stack is an example of one-dimensional photonic crystals. It usually comprises alternating layers of two different isotropic dielectric materials, the product of the refractive index and the thickness of each layer being a quarter of the wavelength in free space [60]. Reflections from the layers interfere constructively, and with a sufficient number of unit cells, the quarter-wave stack acts as a mirror. Bragg mirrors—sometimes called distributed Bragg reflectors—are similar devices. Chiral STFs—and chiral liquid crystals (Section 1.4.4)—can also be considered as one-dimensional photonic crystals.

The construction of higher-dimensional photonic crystals, such as those illustrated in Figures 1.11 and 1.12, can be much more difficult, but researchers have shown great ingenuity in devising fabrication methods. These methods, many born from the microelectronics industry [111], form a vast field of study and detailed description is beyond the scope of this book. Among the wide variety of methods used to construct higher-dimensional photonic crystals are: layer-by-layer construction using lithographic and etching techniques [112], ion-beam milling [113], micromanipulation of microspheres [114], self-assembly of microspheres [115], interference lithography [116] (also known as holographic lithography), and hybrid techniques [117]. Researchers have expended significant effort in creating photonic crystals with large and complete band gaps, a

Figure 1.11 SEM image of a photonic crystal called the woodpile, made of silicon. Courtesy: Sandia National Laboratory.

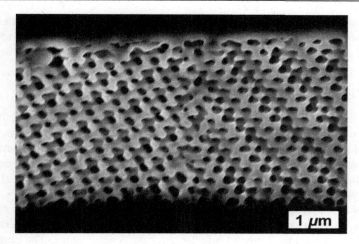

Figure 1.12 SEM image of a diamond-based photonic crystal found in scales of the beetle *Lamprocyphus augustus*. Courtesy: Jeremy W. Galusha and Michael H. Bartl, University of Utah.

complete band gap being one that exists for all directions of propagation. Many different structures and combinations of materials have been considered. In some cases, swapping the roles of the constituent dielectric materials can dramatically improve performance. For instance, a face-centered cubic lattice of electrically small dielectric spheres in air does not have a complete band gap. This is the structure of opals. A complete band gap is exhibited, however, by a face-centered cubic lattice of electrically small air spheres in a dielectric material, this being the inverse-opal structure [118].

1.4.3.4   Rugate Filters

Whereas the unit cells of one-dimensional photonic crystals usually comprise two homogeneous layers, a rugate filter is commonly conceptualized as being made of an isotropic dielectric material whose refractive index n_{rug} varies sinusoidally along one direction [60]; thus,

$$n_{rug}(z) = \sqrt{\frac{\epsilon_{rug}(z)}{\epsilon_0}} = \left(\frac{n_b + n_a}{2}\right) + \left(\frac{n_b - n_a}{2}\right) \sin\left(\frac{\pi z}{\Omega}\right), \qquad (1.38)$$

where 2Ω is the period, $\mathrm{Re}\{n_{rug}(z)\} > 0$ and dissipation is sufficiently small to be ignored. The high-reflectance spectral regime (i.e. the band gap) of a rugate filter is narrower than of a quarter-wave stack, which is often advantageous in some situations. Another advantage of rugate filters is that reflection bands at integer divisors of the central free-space wavelength of the band gap are greatly suppressed.

Because of its simplicity, the rugate filter has been extensively investigated and refined [119,120]. Several fabrication techniques—including PVD methods such as thermal evaporation [121] and reactive sputtering [122], chemical vapor deposition

[123], and electrochemical etching [124]—have been devised for rugate filters. Periodically nonhomogeneous SNTFs [54,125] can be considered akin to rugate filters.

1.4.4 Liquid Crystals

The study of liquid crystals is a vast field [126,127]. It is only possible to cover here some essentials necessary for the understanding of surface-wave propagation.

The molecules of a liquid crystal do not exhibit positional order in three-dimensional space, but some orientational order is still exhibited. Depending on the temperature and pressure, the orientational order can be short range or long range. In the optical regime, a liquid crystal can be considered as a material continuum, not unlike STFs. Indeed, many types of liquid crystals have optical properties similar to STFs.

Nematic liquid crystals are composed of aciculate molecules which are randomly positioned but their time-averaged orientations have a net average. The permittivity dyadic of a nematic liquid crystal is similar to that of a CTF, except that $\epsilon_c = \epsilon_a \neq \epsilon_b$ and $\chi = 0°$ in Eq. (1.29). Thus, these types of liquid crystals are uniaxial dielectric continuums, although biaxial analogs ($\epsilon_c \neq \epsilon_a \neq \epsilon_b$) have also been fabricated [128,129].

The molecules of a smectic liquid crystal are arranged in parallel layers, and thus have positional order along one direction. The long axes of the molecules are oriented perpendicular to the layers in smectic-A liquid crystals, but at an angle to that direction in smectic-C liquid crystals. With the assumption of the layers being parallel to the plane $z = 0$, Eq. (1.29) is applicable with the tilt angle $\chi = 90°$ for smectic-A liquid crystals and $\chi \in (0°, 90°)$ for smectic-C liquid crystals. Smectic liquid crystals may be either uniaxial ($\epsilon_c = \epsilon_a \neq \epsilon_b$) or biaxial ($\epsilon_c \neq \epsilon_a \neq \epsilon_b$).

Liquid crystals also exist in chiral phases, in which the molecules are inclined at a particular angle with respect to a direction known as the director, but twist azimuthally as a function of distance along the director. Both chiral nematic and chiral smectic liquid crystals exist, the permittivity dyadic being given by Eq. (1.34) with the z axis being the director. Chiral nematic liquid crystals are also called cholesteric liquid crystals (CLCs).

Although the molecules of liquid crystals are usually aciculate [126,127], some liquid crystals have molecules of other shapes, e.g. disks [130,131] and bananas [132]. As the orientational order depends on temperature, the anisotropy of liquid crystals changes with temperature. Some liquid crystals can even be isotropic when the temperature is in a certain range.

There are some particularly notable differences between STFs and liquid crystals. As the molecules of a liquid crystal can respond to various external stimuli such as temperature, pressure, and electric field, liquid crystals are easily tunable. In contrast, STFs made of non-polymeric materials are essentially unaltered by such external stimuli. On the other hand, it is possible to change the deposition conditions for STFs so that constitutive parameters become accessible in wide ranges. Various methods can be used to change the structural period of chiral liquid crystals, but the degree of control over this parameter is much less than for chiral STFs.

1.4.5   Reusch Piles

A cholesteric liquid crystal comprises aciculate molecules dispersed on a stack of closely spaced parallel planes. The molecules lying on a specific plane have an average orientation. This orientation progressively changes from plane to plane, the ensemble of orientations describing a helix in the thickness direction. The helix can be either left handed or right handed. When circularly polarized light of free-space wavelength in a certain spectral regime falls normally on a CLC of sufficiently large thickness, the reflectance is high if the handednesses of the CLC and the incident light are the same, but is low otherwise. The spectral regime is called the circular Bragg regime and the phenomenon is called the circular Bragg phenomenon [126,127]. This phenomenon, which is also displayed by chiral STFs, can also be seen for obliquely incident light [42].

Remarkably, two decades prior to the discovery of CLCs in cholesteryl benzoate [133,134], a coarse version of the cholesteric structure had been deduced by Reusch [64], who had been inspired by mica. The Reusch pile, as this cholesteric structure is called, is a stack of layers of an anisotropic dielectric material with an incremental rotation from one layer to the next about an axis normal to the layers.

Suppose a Reusch pile comprises $N_{\ell yr}$ layers, each of thickness $d_{\ell yr}$. The permittivity dyadic of the ℓth layer occupying the region $(\ell - 1)d_{\ell yr} < z < \ell d_{\ell yr}$ in the Reusch pile is given by

$$\underline{\underline{\epsilon}}_{\text{Reusch}}(z) = \underline{\underline{S}}_\ell \cdot \underline{\underline{S}}_y(\chi) \cdot (\epsilon_a \underline{u}_z \underline{u}_z + \epsilon_b \underline{u}_x \underline{u}_x + \epsilon_c \underline{u}_y \underline{u}_y) \cdot \underline{\underline{S}}_y^{-1}(\chi) \cdot \underline{\underline{S}}_\ell^{-1},$$
$$(\ell - 1)d_{\ell yr} < z < \ell d_{\ell yr}, \quad \ell \in [1, N_{\ell yr}]. \tag{1.39}$$

The dyadic

$$\underline{\underline{S}}_\ell = (\underline{u}_x \underline{u}_x + \underline{u}_y \underline{u}_y) \cos[h(\ell - 1)\pi/Q]$$
$$+ (\underline{u}_y \underline{u}_x - \underline{u}_x \underline{u}_y) \sin[h(\ell - 1)\pi/Q] + \underline{u}_z \underline{u}_z, \tag{1.40}$$

which indicates rotation of the ℓth layer about the z axis by an angle $h(\ell - 1)\pi/Q$ with respect to the layer labeled $\ell = 1$, is clearly a spatially uniform version of the continuous dyadic $\underline{\underline{S}}_z(z)$ of Eq. (1.35). The parameter $Q > 1$ is an integer, and the number of structural periods in the Reusch pile is the largest integer less than or equal to $N_{\ell yr}/2Q$. As $Q \to \infty$ while $2Qd_{\ell yr}$ remains fixed, the permittivity dyadic of a Reusch pile approaches that of a chiral STF.

A Reusch pile with $Q = 2$ does not discriminate between normally incident left-circularly polarized and right-circularly polarized light. Being periodically nonhomogeneous along the z axis, such a Reusch pile functions like a Bragg mirror but it does not exhibit the circular Bragg phenomenon [65]. Therefore, Reusch piles with $Q = 2$ are called equichiral.

A Reusch pile with $Q > 2$ is not structurally chiral like CLCs and chiral STFs. Instead, it is ambichiral [65]. It can preferentially reflect normally incident left-circularly polarized light in certain spectral regimes, but reflect normally incident right-circularly polarized light in other spectral regimes. Thus, several circular Bragg regimes are possible, of which only one survives in the limit $Q \to \infty$ [65,135].

1.5   Negative-Phase-Velocity Materials

In an isotropic, achiral, homogeneous material characterized by the constitutive relations

$$\left.\begin{array}{l} \underline{D}(\underline{r}) = \epsilon \underline{E}(\underline{r}) \\ \underline{B}(\underline{r}) = \mu \underline{H}(\underline{r}) \end{array}\right\},\tag{1.41}$$

where both the permittivity scalar ϵ and the permeability scalar μ are functions of the angular frequency ω, electromagnetic plane waves can be either uniform or nonuniform [78, Chapter 2]. When the planes of constant phase coincide with the planes of constant amplitude, the plane wave is classified as uniform; otherwise, it is classified as nonuniform.

Ordinarily, the condition

$$\mathrm{Re}\{\epsilon\}\mathrm{Im}\{\mu\} + \mathrm{Re}\{\mu\}\mathrm{Im}\{\epsilon\} > 0\tag{1.42}$$

holds, indicating that the direction of energy flow and attenuation, as quantitated through the time-averaged Poynting vector, of a uniform plane wave is the same as the direction of the phase velocity of a uniform plane wave. If, however, the condition [136]

$$\mathrm{Re}\{\epsilon\}\mathrm{Im}\{\mu\} + \mathrm{Re}\{\mu\}\mathrm{Im}\{\epsilon\} < 0\tag{1.43}$$

is true, the two directions are opposed to each other. While the opposition of the time-averaged Poynting vector and the phase velocity had been sporadically investigated during the 20th century [137,138], the actual fabrication of a specimen displaying a direct consequence of Eq. (1.43) emerged only in 2001 [139].[5] Many strange properties are associated with the NPV condition, including negative refraction (but not reversal of the law of Ibn Sahl [42, p. 126]), reversal of both the Doppler effect and Cerenkov radiation, and negative radiation pressure. However, *nonuniform* plane waves in isotropic, non-magnetic materials such as silver [140,141] can also exhibit negative phase velocity. Anisotropic dielectric materials [142] and bianisotropic materials [143,144] can also allow the propagation of plane waves with negative phase velocity.

Naturally occurring materials which support *uniform* plane waves with negative phase velocity are not readily available [145]. But nowadays, composite materials exhibiting the NPV characteristic at microwave frequencies are being routinely made [139,146,147]. These composite materials consisted of arrays of straight wires and split-ring resonators. In the microwave regime, the components of these materials were much smaller than the wavelength(s) in the component materials, thus making the composite materials appear as homogeneous continuums. Extension to optical frequencies has also occurred, but there are substantial dissipative losses associated with propagation at these frequencies [71,72].

Although the physical realization of negative phase velocity remains in doubt in some quarters [148], the consensus in the research community is that the rapid pace of development in nanotechnology shall lead to materials that exhibit negative phase velocity with low loss at optical frequencies. With this anticipation, researchers have produced many theoretical results relevant to surface-wave propagation.

[5]Materials satisfying Eq. (1.43) are often—and infelicitously—called *left-handed materials*, even though they are achiral [17].

1.6 Bianisotropic Materials

The constitutive relations of a linear bianisotropic material are as follows:

$$\left. \begin{aligned} \underline{D}(\underline{r}) &= \underline{\underline{\epsilon}}(\underline{r}) \cdot \underline{E}(\underline{r}) + \underline{\underline{\xi}}(\underline{r}) \cdot \underline{H}(\underline{r}) \\ \underline{B}(\underline{r}) &= \underline{\underline{\zeta}}(\underline{r}) \cdot \underline{E}(\underline{r}) + \underline{\underline{\mu}}(\underline{r}) \cdot \underline{H}(\underline{r}) \end{aligned} \right\}. \tag{1.44}$$

The dielectric properties of the material are captured by the permittivity dyadic $\underline{\underline{\epsilon}}(\underline{r})$, the magnetic properties by the permeability dyadic $\underline{\underline{\mu}}(\underline{r})$, and the magnetoelectric properties jointly by the dyadics $\underline{\underline{\xi}}(\underline{r})$ and $\underline{\underline{\zeta}}(\underline{r})$. All four constitutive dyadics are functions of the angular frequency ω. If the bianisotropic material were to be homogeneous, the constitutive dyadics would be independent of the position $\underline{r}$.

1.7 Taxonomy of Electromagnetic Surface Waves

As researchers tackle surface-wave propagation involving more and more complex and even exotic partnering materials, the types of surface waves have increased in number. Let us try to bring order to this large collection of surface waves by presenting a taxonomy of electromagnetic surface waves. Only the broad classes of surface waves are named in this section, while particular *species* of surface waves are denoted by the partnering materials forming the interface. The chief characteristics of each type of surface waves are also described. Table 1.1 lists the types and chief characteristics.

The range of directions along which a surface wave of a certain species may propagate in the interface plane is known as the angular existence domain (AED). As AEDs are generally symmetric with respect to certain rotations about the axis normal to the interface plane, two conventions for specifying AEDs are possible. The AED may refer to either (i) a single continuous interval of directions of propagation in the interface plane or (ii) the union of all intervals of directions of propagation in the interface plane. We adopt the former convention, as the narrowness of the individual continuous range of directions of propagation is the primary interest of the experimentalist and a thorny issue that must be dealt with if Dyakonov waves are to be exploited. Accordingly, for a given interface, there may be more than one AED, in which case the *total* AED refers to the union of the individual AEDs.

1.7.1 SPP Waves

An SPP wave is one that propagates guided by the interface of a metal and a dielectric material. In the optical regime the metal is isotropic with a permittivity which has a negative real part, whereas the dielectric material is isotropic with a permittivity which has a positive real part, both partnering materials being linear and homogeneous. A semiconductor may replace the metal, but application is then restricted to the far-infrared regime wherein the permittivity has a negative real part. With both materials linear and isotropic, the SPP wave can propagate along any direction parallel to the interface, the SPP wave has to be p polarized, and only one SPP-wave mode exists at a

Table 1.1 Types of surface waves and their characterization.

Name	Partnering material 1	Partnering material 2	Number of modes	Characteristics
SPP wave	metal	Homogeneous, isotropic dielectric	1	• 360° total AED • Excitable only by p-polarized plane waves • Not excitable by direct illumination • Commonly applied for biochemical sensing
		Homogeneous, anisotropic dielectric	1	• Large total AED $\leqslant 360°$ • Excitable by s- and p-polarized plane waves
		Periodically nonhomogeneous, isotropic dielectric	$\geqslant 1$	• Excitable by s- and p-polarized plane waves • Very high phase speed possible • Excitable by direct illumination
		Periodically nonhomogeneous, anisotropic dielectric	$\geqslant 1$	• Excitable by s- and p-polarized plane waves • Very high phase speed possible • Excitable by direct illumination
Dyakonov wave	Homogeneous, anisotropic dielectric	Homogeneous, isotropic dielectric	1	• Small total AED • Experimentally demonstrated
		Homogeneous, anisotropic dielectric	1	• Small total AED
		Homogeneous, dielectric-magnetic	1	• Moderate total AED
		Homogeneous, NPV material	$\geqslant 1$	• Large total AED • Tightly bound to the interface

(continued)

Table 1.1 Continued.

Name	Partnering material 1	Partnering material 2	Number of modes	Characteristics
Tamm wave	Periodically nonhomogeneous, isotropic dielectric	Homogeneous, isotropic dielectric	$\geqslant 1$	• 360° total AED • Excitable by s- and p-polarized plane waves • Experimentally demonstrated
		Periodically nonhomogeneous, isotropic dielectric	$\geqslant 1$	• 360° total AED • Excitable by s- and p-polarized plane waves • Very high phase speed possible
Dyakonov–Tamm wave	Periodically nonhomogeneous, anisotropic dielectric	Homogeneous, isotropic dielectric	$\geqslant 1$	• Moderate to large total AED
		Periodically nonhomogeneous, anisotropic dielectric	$\geqslant 1$	• Large total AED
	Homogeneous, anisotropic dielectric	Periodically nonhomogeneous, isotropic dielectric		Not yet studied

given frequency. The phase speed of an SPP wave is less than the speed $c_0 = 1/\sqrt{\epsilon_0\mu_0}$ of light in free space, with the assumption that the partnering dielectric material is nondissipative and its refractive index exceeds unity.

Replacement of the isotropic dielectric material with an anisotropic dielectric material—specified by a permittivity dyadic all of whose eigenvalues have positive real parts and all of whose eigenvectors have Cartesian components with zero imaginary parts—produces some modifications in the characteristics of SPP waves [149]. Thus, the phase speed, the rate of attenuation in the direction of propagation, and the decay rate away from the interface depend on the direction in which the SPP wave is propagating. Although the total AED is generally large, it may be less than 360° [45]. When the isotropic partnering dielectric material is replaced by an anisotropic part-nering dielectric material, the SPP wave may not be classified simply as p polarized [150] because the field phasors on the dielectric side of the interface contain at least one extraordinary component [78], Section 6.2]; also, an SPP wave may then be excited by either an s- or a p-polarized incident plane wave in the prism-coupled configurations.

An isotropic dielectric material which is periodically nonhomogeneous normal to the interface may also be partnered with a metal to support SPP-wave propagation. In this case, the interface may support multiple SPP-wave modes for propagation along any particular direction in the xy plane. These modes are either s polarized or p polarized, and require excitation by a plane wave of the same polarization state. In special circumstances, an SPP-wave mode may even have an arbitrary polariza-tion state [151,152]. Researchers have focused on various manufactured periodically nonhomogeneous dielectric materials, including rugate filters [61,153] and periodi-cally multilayered materials [154–156]. SPP waves with phase speeds exceeding c_0 are possible [157–159] and have been experimentally observed [160].

Much research activity has dealt with SPP waves guided by the interface of a metal and a dielectric material which is both *anisotropic* and whose constitutive properties vary periodically in the direction normal to the interface. Such interfaces may guide multiple SPP-wave modes along a specific direction in the xy plane. SPP-wave modes guided by the interface of a metal and either a periodically nonhomogeneous SNTF [54,57] or a chiral STF [56,58] have been characterized. With identical forms for the permittivity dyadics, analogous liquid crystals could replace the STFs. A polarization state cannot be assigned to these SPP waves [55,161].

1.7.2 Dyakonov Waves

A Dyakonov wave is a surface wave whose propagation is guided by the interface of two homogeneous dielectric materials, at least one of which is anisotropic. The partnering materials may even be identical (and anisotropic), but they must be differently oriented so that all eigenvectors of their permittivity dyadic do not coincide.

As dielectric materials are generally less dissipative than metals, Dyakonov waves should have much larger propagation distances than SPP waves. When dissipation is small enough to be ignored, the propagation distances of Dyakonov waves are infinite in theory. Unfortunately, Dyakonov waves are notorious for having a very small AED [36], often only a very small fraction of a degree.

After Dyakonov's report on surface-wave propagation guided by the interface of an isotropic dielectric material and a nondissipative uniaxial dielectric material whose optic axis is parallel to the interface [35], surface waves were theoretically shown to be possible even for other orientations of the uniaxial dielectric material's optic axis [162]. Other combinations of dielectric materials supporting Dyakonov waves include isotropic/biaxial [162,163], uniaxial/uniaxial [164,165], and biaxial/biaxial [166,167] ones. The first experimental observation [41] of Dyakonov waves was made for the isotropic/biaxial combination.

Because of the complexity of dealing with anisotropic materials, many studies on Dyakonov waves are limited to specific orientations of these materials. Suitable anisotropic materials may be naturally occurring crystals [36,166], but recent research has focused on manufactured materials because of their controllable properties. For example, CTFs are well suited for examining Dyakonov-wave propagation [43] because χ and $\epsilon_{a,b,c}$ in Eq. (1.29) can almost be tailored to order during fabrication. Some researchers have looked at photonic crystals [168] as a means of obtaining strong anisotropy in order to alleviate the problem of the small AED of Dyakonov waves observed with naturally occurring materials.

One strategy to increase the AED of Dyakonov waves is to incorporate magnetic properties into one or both partnering materials [69,70,169,170]. So far, theoretical studies show only a modest increase in the AED by the use of simple dielectric-magnetic materials [169]. However, the use of an isotropic NPV material can produce AEDs on the order of tens of degrees and Dyakonov waves that are tightly bound to the interface, and may even yield multiple Dyakonov-wave modes for propagation in a specified direction in the xy plane [69,170]. If NPV materials become a practical reality in the optical regime, they might offer considerable advantages over traditional materials for making interfaces for Dyakonov-wave propagation. One issue that must not be overlooked, though, is that a large degree of dissipation is a common trait of NPV materials [171].

Theoretical researchers have also examined the impact of gyrotropy [78, Chapter 7] on Dyakonov-wave propagation [38,170,172]. Surface-wave propagation guided by the interfaces of two bianisotropic materials has also been considered [37–39], but none of these efforts deals with practical implementation.

1.7.3   Tamm Waves

In 1932, Tamm theoretically determined the influence of the surface on electron states near the boundary of a finite material [47]. An understanding of the optical analog, a surface wave guided by the interface of two isotropic dielectric materials, at least one of which is periodically nonhomogeneous in the direction normal to the interface, has a long history [173]. The existence of this type of surface wave, called a Tamm wave, has been experimentally validated [174–176] and even exploited for optical biosensing [177]. The periodically nonhomogeneous dielectric material can be piecewise homogeneous [173–175,177–179], or it can be continuously nonhomogeneous such as a rugate filter [62,180]. As both partnering materials are isotropic, the total AED is 360°.

With proper selection of the two partnering materials, several Tamm-wave modes can be obtained for a given interface, some s polarized and the others p polarized [62, 180].

Usually, the ratio of c_0 to the phase speed of a Tamm wave is higher than the smallest refractive index in both partnering materials. But sometimes the phase speed of a Tamm wave can transcend that restriction, provided that both partnering materials are periodically nonhomogeneous [62].

The theoretical incorporation of NPV materials for exciting Tamm waves has been carried out [179, 181], but it cannot currently be implemented advantageously due to high losses in NPV materials [171].

1.7.4 Dyakonov-Tamm Waves

Dyakonov-Tamm waves are at the top of the scale in terms of the complexity of materials used to support surface-wave propagation. Combining aspects of both Dyakonov waves and Tamm waves, a Dyakonov-Tamm wave is guided by the interface of two dielectric materials, of which at least one is both anisotropic and periodically nonhomogeneous in the direction normal to the interface. As with Tamm waves, the periodic nonhomogeneity may be either piecewise homogeneous or continuous. Research on Dyakonov-Tamm waves has focused mainly on the use of STFs as the periodically nonhomogeneous material [46, 48, 50, 51, 53, 182]. Since liquid crystals can also have similar permittivity dyadics, they can be used to support Dyakonov-Tamm waves as well. The guidance of Dyakonov-Tamm waves by the interface of an isotropic dielectric material and a Reusch pile has also been theoretically established [49].

Early research [46, 48] indicated that the interface of a homogeneous isotropic dielectric material and a periodically nonhomogeneous anisotropic dielectric material supports only a single Dyakonov-Tamm wave, but that conclusion was later revised [182] to admit a multiplicity of Dyakonov-Tamm-wave modes by such interfaces. The spatial profiles of the fields are complicated enough that a polarization state cannot be assigned even in the homogeneous, isotropic partnering material. The total AED can be quite large, on the order of tens of degrees [46] typically, and can even be the maximum 360° possible [48]. The large total AED, however, seems to come at the price of a reduced range of allowable values of the refractive index of the isotropic partnering material, compared to Dyakonov waves.

When two periodically varying anisotropic dielectric materials are partnered, the interface may support multiple Dyakonov-Tamm-wave modes in some [50–52, 183] but not all [53] cases.

1.7.5 Emerging Types of Surface Waves

In this chapter, and in much of this book, certain assumptions regarding constitutive parameters have been made. All materials are passive [17, Section 1.7.2.2], and an $\exp(-i\omega t)$ dependence on time is implicit. A metal is a non-magnetic material characterized by only a permittivity scalar whose real part is negative and whose imaginary part is positive. All three eigenvalues of the permittivity dyadic of a purely dielectric material have positive real parts and either zero or positive imaginary parts.

An isotropic achiral NPV material is characterized by a permittivity scalar and a permeability scalar. The imaginary parts of these constitutive parameters are either zero or positive, as the NPV material is passive. Either the permittivity scalar or the permeability scalar or both must have a negative real part, in accordance with Eq. (1.43).

The boundary between one type of surface wave and another may not always be clear, when materials of other types are considered. For example, suppose that a surface wave is supported by the interface of an isotropic dielectric material and an anisotropic dielectric material whose permittivity dyadic has some eigenvalues with positive real parts and others with negative real parts [184,185]. Should this surface wave be called a Dyakonov wave, since one partnering material is anisotropic dielectric? Or, should it be called an SPP wave because either one (or even two, but not all three) of the eigenvalues of the permittivity dyadic of the anisotropic partnering material has a negative real part? Such and other taxonomic ambiguities will arise in research and are to be welcomed rather than dreaded.

The taxonomy of surface waves in this book is based mainly on the linear dielectric (and metallic) properties of the partnering materials. The classification of surface waves becomes more complicated on considering other relevant constitutive properties. In addition to magnetic properties captured by a permeability scalar or dyadic, what about gyrotropy [172,186], nonlinearity [66,187], and electro-optic properties [49,188,189]? Already, researchers have theoretically examined surface-wave propagation guided by interfaces formed with materials having these properties. As nanotechnology develops to encompass and reliably deliver a wider range of complex materials, the boundaries between the various types of surface waves enumerated in this book are bound to become fuzzier, and new types of surface waves will be named.

1.8   Applications

1.8.1   SPP Waves

Out of the various types of surface waves named in Section 1.7, the SPP wave is the only one which is being widely used in practical applications. As the characteristics of this wave are acutely sensitive to the permittivity of the partnering dielectric material, SPR is the basis for some extremely sensitive chemical and biochemical sensors, which can detect changes in the refractive index of an isotropic partnering material as low as 3×10^{-7} [4]. Although some initial uses for the sensing of gases and the characterization of thin films appeared in the 1980s, the real strength of SPR-based techniques is in the sensing of biologically relevant molecules.

The SPP wave, being confined to a very narrow region about the interface, is very sensitive to the immediate interfacial environment. Analyte molecules which find their way close to the interface alter the refractive index of the partnering dielectric material, and thus change the phase speed—among other characteristics—of the SPP wave. As a result, the conditions of the coupling of external light to the SPP wave are altered.

A prism-coupled configuration is very suitable for sensing analytes, with the Turbadar-Kretschmann-Raether configuration being far more convenient than the Turbadar-Otto configuration. In the angular interrogation approach, the free-space wavelength of the light source is fixed, and the angle at which the external light couples maximally to the SPP wave shifts due to the presence of the analyte in the partnering dielectric material. In the frequency interrogation approach, a light source of variable free-space wavelength is oriented at a fixed angle, but the wavelength which maximally couples to the SPP wave shifts. A significant drop in the intensity of the reflected light indicates coupling as the SPP wave is excited [3] and dissipates energy [55]. The angle of incidence must be greater than the critical angle for the interface of the prism material and the partnering dielectric material (which could be infiltrated by the analyte).

It is possible to sense a very wide range of chemical and biochemical species by exploiting SPR. Researchers have optimized protocols for sensing almost everything from quite small molecules of biological provenance to viruses and bacteria to pesticides and toxins to explosives [4]. Even a method of performing immunoassays on whole blood, an optically complex material, has been developed [190]. SPR sensors are deployed for such diverse uses as monitoring pollutants in the environment, pathological analyses in medical laboratories, purity and compositional analyses in the pharmaceutical industry, and basic research in analytical chemistry.

SPR sensors have some advantages over other types of sensors. They do not require markers to be attached to the analyte molecules, unlike fluorescence sensors, for which reason they are referred to as label-free sensors. SPR sensors are miniature and can be spatially multiplexed for sensing multiple analytes simultaneously and rapidly [191]. In addition to simple detection of molecules, SPR sensors are able to monitor reaction kinetics in real time [192].

Although most SPR sensors use a prism-coupled configuration, the grating-coupled configuration has also been investigated for sensing [193], because of their compactness. Grating-coupled sensors can be fabricated on silicon substrates, which allows easy integration with electronic circuitry in integrated sensors. Moreover, arrays of multiple grating-coupled SPR sensors are inexpensive to fabricate as they can be nano-imprinted [111]. A big disadvantage, however, is noise that comes about from the traversal of light through the analyte before and after interaction with the two partnering materials. Avenues for improvements continue to be explored [194,195].

Research on employing SPP waves for microscopy began nearly as early as research on SPR sensors, the earliest two demonstrations of SPP microscopy occurring in 1987 [196] and 1988 [197]. Subsequently, several techniques to take advantage of SPP waves were developed for various types of microscopic imaging. Some techniques allow observation of the intensity of the SPP wave itself [198] as it is guided by a patterned two-dimensional interface, which is useful [199] in the development of various planar structures toward the realization of plasmonic circuits. Optical elements to manipulate SPP waves are being designed and fabricated [200] for eventual use in two-dimensional microscopes. Many researchers have focused on three-dimensional SPR microscopes with subwavelength resolution [201]. Some researchers favor a scanning technique [202], while others are pursuing wide-field microscopy to obtain the image all at once [203]. In either case, a significant advantage of SPR microscopy is that, because of the extreme sensitivity of the SPP wave to changes in the region close to the interface, it is capable of obtaining images of objects with very low contrast, without the use of dyes or markers. SPR microscopy is, thus, very useful for monitoring living organisms with lateral resolution less than a micrometer and resolution along the z axis (i.e. the line of sight) on the order of nanometers. It is also common to combine SPR microscopy with fluorescence techniques [204] to advantage.

For the last three decades, research to bring down the cost of photovoltaic solar cells has gained a huge momentum and many techniques to increase the efficiency of light harvesting by solar cells have been investigated. Among other methods [205,206], the use of plasmonic structures to enhance the absorption of light by these solar cells is being intensively studied. An earlier idea of periodic texturing of the metallic backing layer of a thin-film solar cell to help excite SPP waves [207,208] has been revived during the last 5 years [84,209,210].

Applications of SPP waves are on the horizon in the area of communications [211,212]. Optical fibers have revolutionized communications in the past few decades with higher bandwidth and higher transmission rates than afforded by electrical currents along wires. Plasmonics—that is, the use of SPP waves for transmitting signals and signal processing—promises to bring the advantages realized with optical fibers to the nanoscale and more. While optical fibers offer high speed, miniaturization is limited by diffraction to half a wavelength, which is large in comparison to the nanoscale devices of today. Plasmonic waveguides of much smaller dimensions are capable of carrying signals between components on a chip, as well as between chips [213]. Recent research may soon make active plasmonic devices [214] and on-chip sources of SPP waves [215] practical. A transition to photonic/plasmonic communication technology may be possible in the not-too-distant future.

Plasmonics is currently a very popular research topic. Many researchers envision SPP waves guided by subwavelength-scale structures as a means of creating dense, high-speed circuitry. Some proposed structures [216] maximize the field intensity in the dielectric material and minimize the intensity in the metal, thereby reducing dissipation. Perhaps, surface waves guided by dielectric/dielectric interfaces may reduce dissipation even more, thereby promoting long-range communications.

1.8.2 Other Surface Waves

The technological exploitation of surface waves other than SPP waves in the optical regime is in its infancy. Tamm waves, which were observed more than three decades ago [174], have only recently been exploited for optical biosensing [177]. The first experimental observation of Dyakonov waves occurred just a few years ago [41]. The Dyakonov-Tamm wave [46] has yet to be observed experimentally.

Although reports of practical applications of non-SPP surface waves are scant, exciting proposals for potential uses abound. The introduction of a commercial SPR sensor, in 1990, occurred over 30 years after the discovery [20] of optical means to excite SPP waves. Forecasting scientific and technological development is always risky, but it seems that applications of non-SPP waves are poised to appear soon in abundance, especially considering the current rapid progress in nanotechnology [217,218].

Generally, Dyakonov, Tamm, and Dyakonov-Tamm waves are guided by interfaces formed with dielectric materials only. The dielectric materials of interest here typically exhibit dissipation many orders of magnitude smaller than that exhibited by metals. Thus, these waves have much less attenuation—ergo, much longer propagation distances—than SPP waves. This suggests possible applications in communications. Many of the non-SPP surface waves also have a much higher sensitivity to the direction of propagation and/or the dielectric properties of an isotropic partnering material than the more robust SPP wave. Perhaps the next generation of sensors could use these properties for greater sensitivity.

As both partnering materials needed to guide a Dyakonov wave, a Tamm wave, or a Dyakonov-Tamm wave are only slightly dissipative at most, direct exploitation of these waves for harvesting solar energy is highly unlikely. In contrast, as SPP-wave propagation requires one partnering material to be metallic, SPP waves are currently very attractive for solar-cell research.

1.8.3   STFs for Optical Sensing

STFs may be used to support the propagation of many types of surface waves, and their use could lead to several sensing modalities. Furthermore, STFs may be used in conjunction with nanostructured materials which possess exotic macroscopic optical properties. Infiltrating a material into the STF voids may offer an easy way to extend the range of STF properties, and thereby create new materials for surface-wave propagation.

SPR-based sensing of biomolecules requires the use of recognition molecules bound to the metal surface, to which analyte molecules in turn bind. The voids within STFs may offer a stable environment for recognition molecules, and at the same time protect the metal surface for use in the field. In dirty environments, STFs might filter out large particulate matter from the immediate vicinity of the interface, allowing more reliable detection. The different rates at which various analytes diffuse into the voids and different orientations which they take may make differentiation between more than one analyte possible. Multiple modes for some types of surface waves may both enhance the ability to distinguish between several analytes with a single interface, and make error-free sensing a reality.

2 Surface-Plasmon-Polariton Waves I

2.1 Introduction

For those researchers who focus on surface waves, the emergence of surface waves of many new types during the past few decades has been both exciting and stimulating. The uninitiated, however, may feel disoriented amidst the vast array of phenomenons tabulated in Table 1.1. As it may be difficult to see the forest for the trees, let us take a detailed look at the simple surface-plasmon-polariton waves supported by the interface of two homogeneous and isotropic materials, each characterized by a permittivity scalar. Let us also note here that the study and application of simple SPP waves does not need much help from nanotechnology, and, indeed, flourished for decades without it.

Surface waves of different types have several common features in their descriptions, analyses, and excitation strategies. So, in this chapter, we lay down a foundation, a landmark if you will, from which one might navigate to the general theory of surface-wave propagation in Chapter 3.

Let us begin this chapter with the canonical problem of an SPP wave guided by the interface of two semi-infinitely thick materials. As the partnering materials are taken to be isotropic, homogeneous, linear, non-magnetic, and achiral [17], the characteristics of the SPP wave are quite simple and straightforward. We label these SPP waves as *simple* SPP waves, regardless of the method adopted for their excitation.

A look at practical methods of exciting simple SPP waves guided by interfaces of materials of finite thickness follows. This chapter closes with a discussion of SPP waves when one of the two partnering materials is a nonlinear dielectric material.

2.2 Canonical Boundary-Value Problem

The canonical boundary-value problem is the propagation of an SPP wave guided by the planar interface formed by two materials filling adjoining half-spaces. There are a few reasons for studying such a clearly unphysical situation, which can only be studied theoretically. With semi-infinitely thick partnering materials, there is only one interface. The interpretation of a localized wave traveling parallel to the interface and straddling that interface is quite clear; it *is* a surface wave.

Let us consider an interface formed by two materials with finite thickness, a necessary condition in experiment and application. There are necessarily at least three

Electromagnetic Surface Waves. http://dx.doi.org/10.1016/B978-0-12-397024-4.00002-5

interfaces: the interface of the two materials of interest, and the interface of each material with the outside world. What is the effect of the two additional surfaces? That depends on the exact circumstances. If the materials are sufficiently thick, the outer surfaces may have no noticeable affect on the surface wave. For moderate thicknesses, the surface wave may retain its essential characteristics with only a slight perturbation. When the materials are even thinner, waveguide modes with energy spread over one or both partnering materials are possible. It is not always easy to distinguish a true surface wave from a waveguide mode, either in theory or experiment.

The solution of the canonical problem can serve as a guide, either for the interpretation of both theoretical and experimental results or to direct the design of an interface for a particular application.

2.2.1   Geometry

Let the half-space $z < 0$ be filled with a material having permittivity ϵ_{met}, and the half-space $z > 0$ be filled with a material having permittivity ϵ_{diel}. Both materials are non-magnetic and achiral [17]. Furthermore, both ϵ_{diel} and ϵ_{met} are implicit functions of the angular frequency ω, it being assumed that all fields have an $\exp(-i\omega t)$ dependence on time t. As in Section 1.3.1, the subscripts $_{diel}$ and $_{met}$ stand for *dielectric* and *metal*, respectively, but that distinction is not necessary until Section 2.2.9.

With the interface of the two materials as the plane $z = 0$ and both partnering materials being isotropic, a surface wave may travel in any direction in the xy plane. There is no loss of generality in orienting one of the axes along the direction of propagation. But subsequent chapters consider anisotropic materials and then, especially in experimental settings, a more general orientation of the x and y axes may be useful. Thus, with an eye to the future, let us assume that the surface wave is propagating at an angle ψ relative to the x axis. In addition to $\underline{u}_z$, two more unit vectors, both parallel to the interface, are useful:

$$\underline{u}_{prop} = \underline{u}_x \cos\psi + \underline{u}_y \sin\psi \tag{2.1}$$

is along the propagation direction, while

$$\underline{u}_s = -\underline{u}_x \sin\psi + \underline{u}_y \cos\psi \tag{2.2}$$

is perpendicular to both $\underline{u}_{prop}$ and $\underline{u}_z$. These three unit vectors form a right-handed triad:

$$\underline{u}_{prop} \times \underline{u}_s = \underline{u}_z, \quad \underline{u}_z \times \underline{u}_{prop} = \underline{u}_s, \quad \underline{u}_s \times \underline{u}_z = \underline{u}_{prop}. \tag{2.3}$$

Figure 2.1 is a schematic of the canonical problem.

2.2.2   Field Representation

With the geometry established, the next step is to represent the fields in each half-space. In the Cartesian coordinate system, plane waves are the simplest solutions of the Maxwell postulates [17]. The electric and magnetic fields of a plane wave are written as

$$\left.\begin{array}{l} \underline{\tilde{E}}(r, t) = \mathrm{Re}\{\underline{\mathcal{E}} \exp[i(\underline{k} \cdot \underline{r} - \omega t)]\} \\ \underline{\tilde{H}}(r, t) = \mathrm{Re}\{\underline{\mathcal{H}} \exp[i(\underline{k} \cdot \underline{r} - \omega t)]\} \end{array}\right\}, \tag{2.4}$$

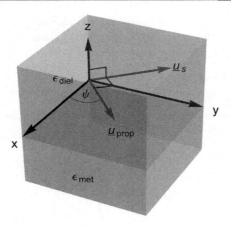

Figure 2.1 Schematic of the canonical boundary-value problem for the propagation of a surface wave guided by the planar interface of two dissimilar materials with permittivities ϵ_{met} and ϵ_{diel}.

where $\underline{\mathcal{E}}$ and $\underline{\mathcal{H}}$ are amplitude vectors with complex-valued components and $\underline{k}$ is the wave vector which can also have complex-valued components. The quantities

$$\left.\begin{array}{l} \underline{E}(\underline{r}) = \underline{\mathcal{E}} \exp{(i\underline{k} \cdot \underline{r})} \\ \underline{H}(\underline{r}) = \underline{\mathcal{H}} \exp{(i\underline{k} \cdot \underline{r})} \end{array}\right\} \tag{2.5}$$

are phasors corresponding to the electric and magnetic fields of a plane wave.

In the two half-spaces of Figure 2.1, these phasors are given by

$$\underline{E}(\underline{r}) = \begin{cases} \underline{\mathcal{E}}_{met} \exp{(i\underline{k}_{met} \cdot \underline{r})}, & z < 0, \\ \underline{\mathcal{E}}_{diel} \exp{(i\underline{k}_{diel} \cdot \underline{r})}, & z > 0, \end{cases} \tag{2.6}$$

and

$$\underline{H}(\underline{r}) = \begin{cases} \underline{\mathcal{H}}_{met} \exp{(i\underline{k}_{met} \cdot \underline{r})}, & z < 0, \\ \underline{\mathcal{H}}_{diel} \exp{(i\underline{k}_{diel} \cdot \underline{r})}, & z > 0. \end{cases} \tag{2.7}$$

Substitution of Eqs. (2.4) in the four source-free Maxwell postulates yields

$$\left.\begin{array}{l} \underline{k}_{met} \times \underline{\mathcal{E}}_{met} = \omega\mu_0\underline{\mathcal{H}}_{met} \\ \underline{k}_{met} \times \underline{\mathcal{H}}_{met} = -\omega\epsilon_{met}\underline{\mathcal{E}}_{met} \\ \underline{k}_{met} \cdot \underline{\mathcal{E}}_{met} = 0 \\ \underline{k}_{met} \cdot \underline{\mathcal{H}}_{met} = 0 \end{array}\right\}, \quad z < 0, \tag{2.8}$$

and

$$\left.\begin{array}{l} \underline{k}_{diel} \times \underline{\mathcal{E}}_{diel} = \omega\mu_0\underline{\mathcal{H}}_{diel} \\ \underline{k}_{diel} \times \underline{\mathcal{H}}_{diel} = -\omega\epsilon_{diel}\underline{\mathcal{E}}_{diel} \\ \underline{k}_{diel} \cdot \underline{\mathcal{E}}_{diel} = 0 \\ \underline{k}_{diel} \cdot \underline{\mathcal{H}}_{diel} = 0 \end{array}\right\}, \quad z > 0. \tag{2.9}$$

In order for the fields to represent a surface wave bound to the interface $z = 0$, the z-directed components of the wave vectors $\underline{k}_{met}$ and $\underline{k}_{diel}$ must be complex-valued with non-zero imaginary parts. Thus, we express the wave vectors as

$$\left.\begin{aligned} \underline{k}_{met} &= q\underline{u}_{prop} - \alpha_{met}\underline{u}_z \\ \underline{k}_{diel} &= q\underline{u}_{prop} + \alpha_{diel}\underline{u}_z \end{aligned}\right\}, \tag{2.10}$$

where q, α_{met}, and α_{diel} are complex-valued in general. In anticipation of satisfying the standard boundary conditions

$$\left.\begin{aligned} \underline{u}_x \cdot \underline{E}(x, y, 0-) &= \underline{u}_x \cdot \underline{E}(x, y, 0+) \\ \underline{u}_y \cdot \underline{E}(x, y, 0-) &= \underline{u}_y \cdot \underline{E}(x, y, 0+) \\ \underline{u}_x \cdot \underline{H}(x, y, 0-) &= \underline{u}_x \cdot \underline{H}(x, y, 0+) \\ \underline{u}_y \cdot \underline{H}(x, y, 0-) &= \underline{u}_y \cdot \underline{H}(x, y, 0+) \end{aligned}\right\}, \tag{2.11}$$

which can be conveniently recast as

$$\left.\begin{aligned} \underline{u}_{prop} \cdot \underline{E}(x, y, 0-) &= \underline{u}_{prop} \cdot \underline{E}(x, y, 0+) \\ \underline{u}_s \cdot \underline{E}(x, y, 0-) &= \underline{u}_s \cdot \underline{E}(x, y, 0+) \\ \underline{u}_{prop} \cdot \underline{H}(x, y, 0-) &= \underline{u}_{prop} \cdot \underline{H}(x, y, 0+) \\ \underline{u}_s \cdot \underline{H}(x, y, 0-) &= \underline{u}_s \cdot \underline{H}(x, y, 0+) \end{aligned}\right\} \tag{2.12}$$

at the interface $z = 0$, the wave vectors for both half-spaces have the same component $q\underline{u}_{prop}$ parallel to the interface. Finally, in order to describe a surface wave whose fields decreases in amplitude with increasing distance from the interface $z = 0$, the following conditions must be satisfied:

$$\left.\begin{aligned} \mathrm{Im}\{\alpha_{met}\} &> 0 \\ \mathrm{Im}\{\alpha_{diel}\} &> 0 \end{aligned}\right\}. \tag{2.13}$$

Eliminating $\underline{\mathcal{H}}_{met}$ from Eqs. $(2.8)_{1,2}$ and $\underline{\mathcal{H}}_{diel}$ from Eqs. $(2.9)_{1,2}$, we obtain

$$\left.\begin{aligned} (\underline{k}_{met} \cdot \underline{\mathcal{E}}_{met})\underline{k}_{met} - (\underline{k}_{met} \cdot \underline{k}_{met})\underline{\mathcal{E}}_{met} &= -k_0^2(\epsilon_{met}/\epsilon_0)\underline{\mathcal{E}}_{met} \\ (\underline{k}_{diel} \cdot \underline{\mathcal{E}}_{diel})\underline{k}_{diel} - (\underline{k}_{diel} \cdot \underline{k}_{diel})\underline{\mathcal{E}}_{diel} &= -k_0^2(\epsilon_{diel}/\epsilon_0)\underline{\mathcal{E}}_{diel} \end{aligned}\right\}, \tag{2.14}$$

where the free-space wavenumber $k_0 = \omega\sqrt{\epsilon_0\mu_0}$. The first terms on the left sides are zero by virtue of Eqs. $(2.8)_3$ and $(2.9)_3$, thereby yielding

$$\left.\begin{aligned} \underline{k}_{met} \cdot \underline{k}_{met} &= k_0^2\epsilon_{met}/\epsilon_0 \\ \underline{k}_{diel} \cdot \underline{k}_{diel} &= k_0^2\epsilon_{diel}/\epsilon_0 \end{aligned}\right\}. \tag{2.15}$$

Substituting the expressions for $\underline{k}_{met}$ and $\underline{k}_{diel}$ from Eqs. (2.10) in Eqs. (2.15), we finally get

$$\left.\begin{aligned} \alpha_{met} &= \sqrt{k_0^2(\epsilon_{met}/\epsilon_0) - q^2} \\ \alpha_{diel} &= \sqrt{k_0^2(\epsilon_{diel}/\epsilon_0) - q^2} \end{aligned}\right\}. \tag{2.16}$$

The square roots in Eqs. (2.16) must be taken so that Eqs. (2.13) are satisfied.

2.2.3   Linear Polarization States

It is possible to decompose the fields into two linear polarization states. A linear polarization state is one for which $\text{Re}\{\mathcal{E}\} \times \text{Im}\{\mathcal{E}\} = 0$, and likewise for $\mathcal{H}$ [78]. Thus, for a linear polarization state, a unique direction in three-dimensional real space may be associated with the complex-valued vector $\mathcal{E}$, and likewise for $\mathcal{H}$. A convenient choice of decomposition is:

* one polarization state with $\mathcal{E}$ parallel to $\underline{u}_s$ (s-polarization state), and
* the other polarization state with $\mathcal{H}$ parallel to $\underline{u}_s$ (p-polarization state).

Clearly, for the two linear polarization states, $\mathcal{E}$ and $\mathcal{H}$ satisfy the Maxwell divergence postulates represented in Eqs. (2.8)$_{3,4}$ and (2.9)$_{3,4}$.

After choosing an amplitude for $\mathcal{E}$ in the case of the s-polarized component, and an amplitude for $\mathcal{H}$ in the case of the p-polarized component, the Maxwell curl postulates represented in Eqs. (2.8) and (2.9) then provide the corresponding $\mathcal{H}$ and $\mathcal{E}$, respectively. The net result may be expressed as the amplitude vectors

$$\left.\begin{aligned}
\mathcal{E}_{met} &= b_s \underline{u}_s + b_p \left(\frac{\alpha_{met}\underline{u}_{prop} + q\underline{u}_z}{k_0 n_{met}} \right) \\
\mathcal{H}_{met} &= \frac{n_{met}}{\eta_0} \left[b_s \left(\frac{\alpha_{met}\underline{u}_{prop} + q\underline{u}_z}{k_0 n_{met}} \right) - b_p \underline{u}_s \right]
\end{aligned}\right\} \tag{2.17}$$

and

$$\left.\begin{aligned}
\mathcal{E}_{diel} &= c_s \underline{u}_s + c_p \left(\frac{-\alpha_{diel}\underline{u}_{prop} + q\underline{u}_z}{k_0 n_{diel}} \right) \\
\mathcal{H}_{diel} &= \frac{n_{diel}}{\eta_0} \left[c_s \left(\frac{-\alpha_{diel}\underline{u}_{prop} + q\underline{u}_z}{k_0 n_{diel}} \right) - c_p \underline{u}_s \right]
\end{aligned}\right\}. \tag{2.18}$$

Here, b_s and c_s are the amplitudes of the s-polarized component of the electric field phasor, and b_p and c_p are the amplitudes of the p-polarized component of the same phasor; the intrinsic impedance of free space is denoted by

$$\eta_0 = \sqrt{\mu_0/\epsilon_0}; \tag{2.19}$$

and the refractive indexes

$$\left.\begin{aligned}
n_{met} &= \sqrt{\epsilon_{met}/\epsilon_0} \\
n_{diel} &= \sqrt{\epsilon_{diel}/\epsilon_0}
\end{aligned}\right\} \tag{2.20}$$

have non-negative imaginary parts for passive materials.

2.2.4   Boundary Conditions

Equations (2.17) and (2.18) represent valid solutions of the Maxwell postulates on both sides of the interface $z = 0$ but additional conditions must be satisfied to completely determine the characteristics of the surface wave. The imposition of these conditions affects both the wave vectors and the complex amplitudes of the field phasors in both materials.

The fields of the surface wave must decay as $z \to \pm\infty$; hence, the wave vectors defined in Eqs. (2.10) must satisfy Eqs. (2.13). Care must be taken in the remainder of the canonical problem to assure that the imaginary parts of both α_{met} and α_{diel} are of the correct signs, in order to obtain a wave which decays with distance from the interface.

The boundary conditions at the interface are stated as Eqs. (2.12). Of these four conditions, Eqs. (2.12)$_{1,2}$ on the continuity of the components of $\underline{E}$ parallel to the interface $z = 0$ require that

$$b_s = c_s \tag{2.21}$$

and

$$b_p \frac{\alpha_{met}}{n_{met}} = -c_p \frac{\alpha_{diel}}{n_{diel}}, \tag{2.22}$$

and Eqs. (2.12)$_{3,4}$ on $\underline{H}$ require that

$$b_s \alpha_{met} = -c_s \alpha_{diel} \tag{2.23}$$

and

$$b_p n_{met} = c_p n_{diel}. \tag{2.24}$$

Equations (2.21) and (2.23) together require that

$$\alpha_{met} = -\alpha_{diel}, \tag{2.25}$$

while Eqs. (2.22) and (2.24) yield the condition

$$\alpha_{diel} \epsilon_{met} = -\alpha_{met} \epsilon_{diel}. \tag{2.26}$$

2.2.5 Amplitude Vectors

By virtue of Eqs. (2.16), Eq. (2.25) demands that $\epsilon_{met} = \epsilon_{diel}$, i.e. the materials on either side of the interface $z = 0$ are electromagnetically indistinguishable from each other at the angular frequency of interest. As the latter equation is a condition of the s-polarized surface wave only, we must set

$$b_s = c_s = 0, \tag{2.27}$$

thereby concluding that the surface wave cannot have s-polarized components in either partnering material.

Thus, we see that the surface wave guided by the interface of two linear, homogeneous, isotropic, non-magnetic, achiral materials must be p polarized, with

$$c_p = b_p \frac{n_{met}}{n_{diel}} \tag{2.28}$$

following from Eqs. (2.22) and (2.26). Equation (2.26) is the dispersion equation of the surface wave whose amplitude vectors are

$$
\left.\begin{aligned}
\underline{\mathcal{E}}_{met} &= b_p \left(\frac{\alpha_{met}\underline{u}_{prop} + q\underline{u}_z}{k_0 n_{met}} \right) \\
\underline{\mathcal{E}}_{diel} &= b_p \frac{n_{met}}{n_{diel}} \left(\frac{-\alpha_{diel}\underline{u}_{prop} + q\underline{u}_z}{k_0 n_{diel}} \right) \\
\underline{\mathcal{H}}_{met} &= \underline{\mathcal{H}}_{diel} = -b_p \frac{n_{met}}{\eta_0} \underline{u}_s
\end{aligned} \right\} .
\tag{2.29}
$$

2.2.6 Time-Averaged Poynting Vector

The instantaneous Poynting vector is defined as the cross-product $\underline{\tilde{E}}(\underline{r}, t) \times \underline{\tilde{H}}(\underline{r}, t)$. When the fields are time-harmonic, i.e.

$$
\left. \begin{aligned}
\underline{\tilde{E}}(\underline{r}, t) &= \mathrm{Re}\{\underline{E}(\underline{r}) \exp(-i\omega t)\} \\
\underline{\tilde{H}}(\underline{r}, t) &= \mathrm{Re}\{\underline{H}(\underline{r}) \exp(-i\omega t)\}
\end{aligned} \right\} ,
\tag{2.30}
$$

the time-averaged Poynting vector is given by

$$
\underline{P}(\underline{r}) = (1/2)\mathrm{Re}\{\underline{E}(\underline{r}) \times \underline{H}^*(\underline{r})\}.
\tag{2.31}
$$

The asterisk denotes the complex conjugate.

From the expressions obtained in Section 2.2.5, the time-averaged Poynting vector of a simple SPP wave turns out be as follows:

$$
\underline{P}(x, y, z) =
\begin{cases}
\dfrac{|b_p|^2}{2\omega\mu_0} \mathrm{Re}\left\{ \left(\dfrac{n_{met}^*}{n_{met}} \right) (q\underline{u}_{prop} - \alpha_{met}\underline{u}_z) \right\} \\
\quad \times \exp\left[-2\mathrm{Im}\{q\}\underline{u}_{prop} \boldsymbol{\cdot} \underline{r} \right] \exp[2\mathrm{Im}\{\alpha_{met}\}z], & z < 0, \\[2ex]
\dfrac{|b_p|^2}{2\omega\mu_0} |n_{met}|^2 \mathrm{Re}\left\{ \left(\dfrac{1}{n_{diel}} \right)^2 (q\underline{u}_{prop} + \alpha_{diel}\underline{u}_z) \right\} \\
\quad \times \exp[-2\mathrm{Im}\{q\}\underline{u}_{prop} \boldsymbol{\cdot} \underline{r}] \exp[-2\mathrm{Im}\{\alpha_{diel}\}z], & z > 0.
\end{cases}
\tag{2.32}
$$

2.2.7 Wavenumbers

Equations (2.16) and (2.26) yield the simple expressions

$$
\alpha_{met} = \frac{k_0 \epsilon_{met}}{\sqrt{\epsilon_0(\epsilon_{met} + \epsilon_{diel})}}
\tag{2.33}
$$

and

$$
\alpha_{diel} = \frac{k_0 \epsilon_{diel}}{\sqrt{\epsilon_0(\epsilon_{met} + \epsilon_{diel})}},
\tag{2.34}
$$

in which the signs of the square roots must be chosen to satisfy the restrictions (2.13) that are appropriate for a surface wave. Using either Eqs. (2.16)$_1$ and (2.33) or Eqs. (2.16)$_2$ and (2.34), we get

$$q^2 = k_0^2 \left[\frac{\epsilon_{\mathrm{met}}\epsilon_{\mathrm{diel}}}{\epsilon_0(\epsilon_{\mathrm{met}} + \epsilon_{\mathrm{diel}})} \right] \tag{2.35}$$

as the square of the SPP wavenumber. Both the real and the imaginary parts of q must be of the same sign. The positive sign indicates an SPP wave traveling parallel to $\underline{u}_{\mathrm{prop}}$; the negative sign indicates propagation parallel to $-\underline{u}_{\mathrm{prop}}$.

2.2.8 Phase Speed and Characteristic Lengths

The foregoing quantities may be used to calculate other quantities which give insight into the surface wave's characteristics. Two quantities characterizing the propagation of the surface wave are related to the wavenumber q of the surface wave. The first is the phase speed of the surface wave given by

$$v_p = \frac{\omega}{\mathrm{Re}\{q\}}. \tag{2.36}$$

Sometimes, this phase speed is normalized by the phase speed of a plane wave in the material occupying the half-space $z > 0$, and the following relative phase speed is defined:

$$v_{\mathrm{rel}} = k_0 \frac{\mathrm{Re}\{n_{\mathrm{diel}}\}}{\mathrm{Re}\{q\}}. \tag{2.37}$$

The second is the propagation length, i.e. the distance along the direction of propagation over which the field amplitudes of the surface wave decrease by a factor of $\exp(-1)$. This propagation length is given by

$$\Delta_{\mathrm{prop}} = \frac{1}{\mathrm{Im}\{q\}}, \tag{2.38}$$

by virtue of Eqs. (2.6), (2.7), and (2.10). According to Eq. (2.32), the time-averaged Poynting vector decreases by the same factor over half that distance.

The decay of the surface wave along the direction perpendicular to the interface $z = 0$ is described by two decay rates: $\mathrm{Im}\{\alpha_{\mathrm{met}}\}$ and $\mathrm{Im}\{\alpha_{\mathrm{diel}}\}$. A decay rate has the dimension of inverse length. Thus, two penetration depths are defined as

$$\Delta_{\mathrm{met}} = \frac{1}{\mathrm{Im}\{\alpha_{\mathrm{met}}\}} \tag{2.39}$$

and

$$\Delta_{\mathrm{diel}} = \frac{1}{\mathrm{Im}\{\alpha_{\mathrm{diel}}\}}, \tag{2.40}$$

which represent distances—along the direction perpendicular to the interface in the two partnering materials—over which the fields decrease by a factor of $\exp(-1)$. Again, Eq. (2.32) indicates that the time-averaged Poynting vector decreases twice as rapidly as the fields normal to the interface.

2.2.9 Characteristics of Simple SPP Waves

At this juncture, let the material in the half-space $z < 0$ be a metal, while the other half-space be occupied by a dielectric material—in other words, we specialize the remainder of the analysis to a simple SPP wave.

Typically, $\mathrm{Im}\{\epsilon_{diel}\}$ is usually small enough in comparison to $\mathrm{Re}\{\epsilon_{diel}\}$ that the former can be ignored. If we also ignore $\mathrm{Im}\{\epsilon_{met}\}$ for the moment, Eq. (2.26) implies that either (i) $\epsilon_{met} < 0$ and $\epsilon_{diel} > 0$ or (ii) $\epsilon_{met} > 0$ and $\epsilon_{diel} < 0$. The first choice is the one made conventionally, and then the subscripts $_{diel}$ and $_{met}$ do stand for *dielectric* and *metal*, respectively.

In real metals, $\mathrm{Im}\{\epsilon_{met}\}$ is substantial and cannot be ignored. The existence of this non-zero imaginary part has two particularly noteworthy consequences for the SPP wave: (i) the SPP wavenumber q is accordingly complex valued and the propagation length is finite; and (ii) α_{met} and α_{diel} accordingly have non-zero real parts which cause oscillations with distance from the interface.

With $k_0^2 \epsilon_{diel}/\epsilon_0$ purely real valued, the relation

$$\mathrm{Re}\{\alpha_{diel}\}\,\mathrm{Im}\{\alpha_{diel}\} = -\mathrm{Re}\{q\}\,\mathrm{Im}\{q\} \tag{2.41}$$

follows from Eq. (2.16)$_2$. For a surface wave, both $\mathrm{Re}\{q\}$ and $\mathrm{Im}\{q\}$ must be of the same sign, as noted in Section 2.2.7. Hence, $\mathrm{Re}\{\alpha_{diel}\}$ and $\mathrm{Im}\{\alpha_{diel}\}$ have to be of opposite signs; accordingly,

$$\mathrm{Re}\{\alpha_{diel}\} < 0 \tag{2.42}$$

by virtue of Eq. (2.13)$_2$.

As $\mathrm{Im}\{\epsilon_{met}\}$ cannot usually be ignored in favor of $\mathrm{Re}\{\epsilon_{met}\}$, Eq. (2.16)$_1$ yields

$$\mathrm{Re}\{\alpha_{met}\}\,\mathrm{Im}\{\alpha_{met}\} = (1/2)k_0^2\,\mathrm{Im}\{\epsilon_{met}/\epsilon_0\} - \mathrm{Re}\{q\}\,\mathrm{Im}\{q\}. \tag{2.43}$$

Furthermore, $\mathrm{Im}\{\alpha_{met}\} > 0$ because of Eq. (2.13)$_1$ and $\mathrm{Im}\{\epsilon_{met}\} > 0$ for a passive material. Therefore,

$$\mathrm{Re}\{\alpha_{met}\} \begin{cases} >0, & k_0^2\,\mathrm{Im}\{\epsilon_{met}/\epsilon_0\} > 2\mathrm{Re}\{q\}\,\mathrm{Im}\{q\}, \\ =0, & k_0^2\,\mathrm{Im}\{\epsilon_{met}/\epsilon_0\} = 2\mathrm{Re}\{q\}\,\mathrm{Im}\{q\}, \\ <0, & k_0^2\,\mathrm{Im}\{\epsilon_{met}/\epsilon_0\} < 2\mathrm{Re}\{q\}\,\mathrm{Im}\{q\}, \end{cases} \tag{2.44}$$

so that the signs of $\mathrm{Re}\{\alpha_{met}\}$ and $\mathrm{Im}\{\alpha_{met}\}$ may or may not be the same, and it is even possible that $\mathrm{Re}\{\alpha_{met}\} = 0$. All three possibilities are exemplified numerically in Table 2.1.

The dependencies of q, α_{diel}, and α_{met} on n_{diel} are illustrated in Figure 2.2 for fixed n_{met}. In the limit $n_{diel} \to \infty$, we get $q \to k_0 n_{met}$, $\alpha_{diel} \to -k_0 n_{diel}$, and $\alpha_{met} \to 0$, according to Eqs. (2.33)–(2.35). The graphs in Figure 2.2 are in consonance with these limits. Most importantly, the almost linear variation of $\mathrm{Re}\{q\}/k_0$ with respect to n_{diel} for small values of n_{diel} is the feature that is commonly exploited for optical biosensing.

A metal is often used for SPP-wave applications at an operating angular frequency ω below that metal's plasma frequency, because then $\mathrm{Re}\{\epsilon_{met}\} < 0$ and numerous

Table 2.1 Numerical illustrations of the three possibilities delineated in Eq. (2.44).

$\epsilon_{diel}/\epsilon_0$	$\epsilon_{met}/\epsilon_0$	q/k_0	α_{diel}/k_0	α_{met}/k_0
4	$-15 + 6i$	$2.267 + 0.135i$	$-0.279 + 1.095i$	$0.595 + 4.525i$
	$-7.0405 + 2.5991i$	$2.697 + 0.482i$	$-0.693 + 1.876i$	$3.752i$
	$-3 + i$	$1.572 + 2.544i$	$-3.108 + 1.287i$	$-2.009 + 1.742i$

dielectric materials can be used as that metal's partner. The metals used most commonly are aluminum, copper, gold, and silver. In the visible regime [77,219], silver performs better than the other metals, insofar as it has the least damping. Damping in the case of gold is less when the free-space wavelength $\lambda_0 \gtrsim 500$ nm in the visible regime and into the near-infrared regime [219,220]. Copper could be considered for $\lambda_0 \gtrsim 600$ nm [219], except for problems with rapid oxidation. At the other end of the spectrum, aluminum has suitably low damping in the blue and ultraviolet regimes for $\lambda_0 \gtrsim 200$ nm. Palladium [221], platinum [222], alkali metals [223], and metallic alloys [224] have been considered for special applications. Semiconductors [225] can also be used, but their use is restricted to the far-infrared regime. Finally, graphene has emerged recently as a plasmonic material [226] but is unlikely to outperform gold for SPP-wave applications [227].

Table 2.2 shows the phase speeds and characteristic lengths of SPP waves guided by interfaces formed with various isotropic materials. The table indicates that the penetration depth into the metal is not a strong function of the refractive index of the partnering dielectric material, with $\Delta_{met} \approx 25$ nm. However, v_{rel}, Δ_{prop}, and Δ_{diel}, all depend strongly on the refractive index of the partnering dielectric material.

Figure 2.3a and b shows the spatial variations of the magnitudes of the non-zero Cartesian components[1] of the electric and magnetic field phasors of an SPP wave along the z axis. The variations are smooth, following the same pattern as illustrated schematically in Figure 1.1. The field phasors decay more rapidly in the metal than in the partnering dielectric material. In both partnering materials, Figure 2.3c confirms that the time-averaged Poynting vector decays twice as fast as the field phasors do. The maximum magnitudes of the field phasors and the time-averaged Poynting vector are at the interface $z = 0$.

2.2.10   Fano Wave

As mentioned in Section 2.2.9, Im$\{\epsilon_{diel}\}$ is usually small enough in comparison to Re$\{\epsilon_{diel}\}$ that the former can be ignored. Suppose that Im$\{\epsilon_{met}\}$ is also set equal to zero. When both $\epsilon_{diel} > 0$ and $\epsilon_{met} < 0$ are purely real, Eqs. (2.33) and (2.34) require

$$\epsilon_{met} + \epsilon_{diel} < 0 \qquad\qquad\qquad (2.45)$$

---

[1] The Cartesian components of a vector $\underline{a}$ are denoted by $a_x = \underline{u}_x \cdot \underline{a}$, $a_y = \underline{u}_y \cdot \underline{a}$, and $a_z = \underline{u}_x \cdot \underline{a}$; thus, $\underline{a} = a_x \underline{u}_x + a_y \underline{u}_y + a_z \underline{u}_z$.

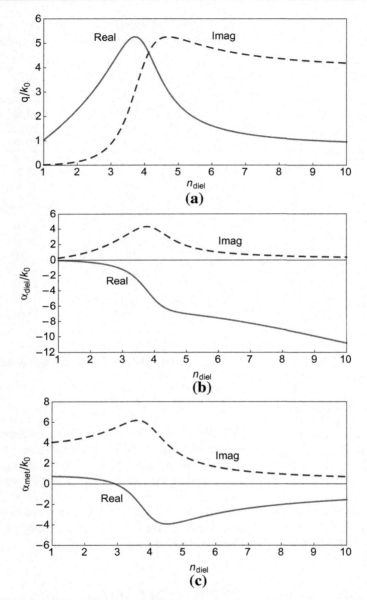

Figure 2.2 Normalized wavenumbers (a) q/k_0, (b) α_{diel}/k_0, and (c) α_{met}/k_0 as functions of real-valued n_{diel} when the metal is aluminum ($n_{\text{met}} = 0.75 + 3.9i$) and $\lambda_0 = 633$ nm.

in order to ensure that Eqs. (2.13) are satisfied. Furthermore, as q must be real because the permittivities of both partnering materials do not have imaginary parts, Eq. (2.35) then requires

$$\epsilon_{\text{met}} \epsilon_{\text{diel}} < 0. \qquad (2.46)$$

Table 2.2 Phase speeds and characteristic lengths of SPP waves at $\lambda_0 = 633$ nm calculated for a selection of partnering materials. The refractive indexes for metals come from measurements on thin films.

Metal	n_{met}	Dielectric material	n_{diel}	v_{rel}	Δ_{prop} (μm)	Δ_{met} (nm)	Δ_{diel} (nm)
Aluminum	$0.75 + 3.9i$	Water	1.33	0.95	3.09	24.36	219.16
		Quartz	1.54	0.93	1.88	23.85	160.60
		Diamond	2.41	0.82	0.34	20.88	58.97
Copper	$0.27 + 3.4i$	Water	1.33	0.92	4.95	27.29	179.89
		Quartz	1.54	0.89	2.91	26.45	130.18
		Diamond	2.41	0.72	0.39	21.16	42.90
Gold	$0.18 + 3.4i$	Water	1.33	0.92	7.35	27.28	178.96
		Quartz	1.54	0.89	4.31	26.43	129.40
		Diamond	2.41	0.71	0.57	21.02	42.18
Silver	$0.056 + 4.3i$	Water	1.33	0.95	52.31	22.28	232.94
		Quartz	1.54	0.93	31.89	21.88	170.59
		Diamond	2.41	0.83	5.81	19.41	61.80

Therefore, $-\epsilon_{met} > \epsilon_{diel}$ for surface-wave propagation to occur. The SPP wave then is called a Fano wave [19].

2.2.11   Zenneck Wave

The Zenneck wave is related to the SPP wave in the following manner. Suppose the metal of permittivity ϵ_{met} involved in the metal/dielectric interface of Figure 2.1 were to be replaced by a lossy dielectric material with permittivity $\epsilon_{\ell d}$. Whereas $\mathrm{Re}\{\epsilon_{met}\} < 0$, $\mathrm{Re}\{\epsilon_{\ell d}\} > 0$. Then, a surface wave guided by the interface is not an SPP wave but a Zenneck wave.

The existence of the Zenneck wave remains controversial at microwave frequencies [9,10]. At optical frequencies, the Zenneck wave has never been observed. Nevertheless, a brief description is provided in Appendix C.

2.3   Optical Excitation of Simple SPP Waves

The observation of loss of energy of electrons after impinging on a thin metal film resulted in the prediction of surface plasmons [16]. Although an electron beam directed at the interface of a metal and a partnering dielectric material may excite an SPP wave, this method of excitation is inconvenient for practical applications. The complex apparatus needed and the necessity of working at a very low pressure make excitation by electron bombardment almost always cumbersome—and next to impossible in

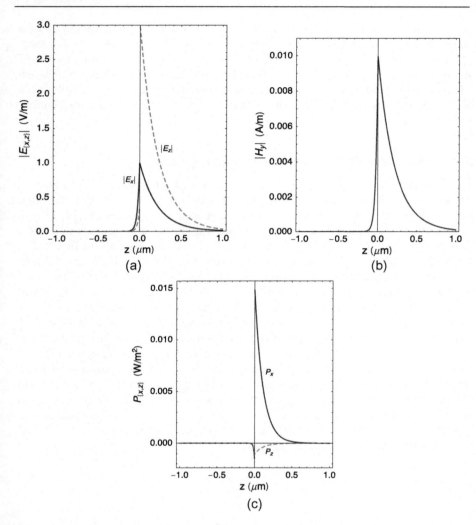

Figure 2.3 Spatial variations with z of the non-zero Cartesian components of (a) $\underline{E}(0, 0, z)$, (b) $\underline{H}(0, 0, z)$, and (c) $\underline{P}(0, 0, z)$ of an SPP wave guided by the planar interface $z = 0$ of a half-space filled with aluminum and a half-space filled with water. The free-space wavelength $\lambda_0 = 633$ nm, the angle $\psi = 0$ so that $\underline{u}_{\text{prop}} = \underline{u}_x$ and $\underline{u}_s = \underline{u}_y$, $n_{\text{met}} = 0.75 + 3.9i$, and $n_{\text{diel}} = 1.33$. Calculations were made with $b_p = k_0 n_{\text{met}}/\alpha_{\text{met}}$ so that $\underline{u}_x \cdot \underline{E}(0, 0, 0) = 1$ V m^{-1}.

some cases exemplified by biochemical sensors which use liquid partnering dielectric materials.

The modern sensor exploiting the SPP-wave phenomenon owes its ubiquity to the development of much simpler optical means of SPP-wave excitation. The fundamental problem of optical excitation is the matching of the component of a wave vector parallel to the interface of the exciting optical beam with the SPP wavenumber q.

From Eq. (2.35), it follows that $\mathrm{Re}\{q\}$ is greater than the wavenumber $k_0 n_{\mathrm{diel}}$ of a plane wave traveling in the bulk partnering dielectric material (assumed to be lossless). There are two general approaches to optical excitation of SPP waves: either the use of an evanescent wave or the use of a wave traveling in a waveguide, both of which have a wavenumber exceeding $k_0 n_{\mathrm{diel}}$. Regardless of the method of excitation, the SPP waves considered in the remainder of this chapter have simple field profiles, and are therefore simple SPP waves.

2.3.1 Turbadar-Kretschmann-Raether Configuration

The most popular method of SPP-wave excitation employing an evanescent wave is accomplished with the Turbadar-Kretschmann-Raether configuration [20, 21, 23], often called the Kretschmann configuration. This arrangement—consisting of a prism with a refractive index greater than that of the partnering dielectric material, a thin metal film, and the partnering dielectric material—is shown schematically in Figure 2.4. Figure 2.5 is a photograph of an automated arrangement. A collimated beam of almost monochromatic light is directed through one of the two slanted faces of the prism. This beam bends inside the prism to impinge on the metal film. If the metal film is sufficiently thin, the field amplitude at the other face of the metal film is significant. If the angle of incidence θ_{inc} inside the prism at the prism/metal interface is greater than the critical angle $\theta_{\mathrm{cr}} = \sin^{-1}(n_{\mathrm{diel}}/n_{\mathrm{prism}})$ for total internal reflection at an interface of the prism material and the partnering dielectric material, where n_{prism} is the refractive index of the prism material, light enters the partnering dielectric material as an evanescent wave. Since the Maxwell postulates require continuity of the components of the electric and magnetic fields parallel to an interface, the component of the wave vector parallel to the metal/dielectric interface must remain constant as light travels from the prism to the metal to the partnering dielectric material. The component of the wave vector parallel to the metal/dielectric interface is denoted by

$$\kappa = n_{\mathrm{prism}} k_0 \sin \theta_{\mathrm{inc}}. \tag{2.47}$$

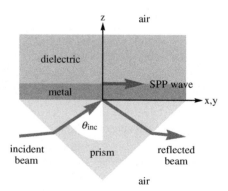

Figure 2.4 Schematic of the Turbadar-Kretschmann-Raether configuration.

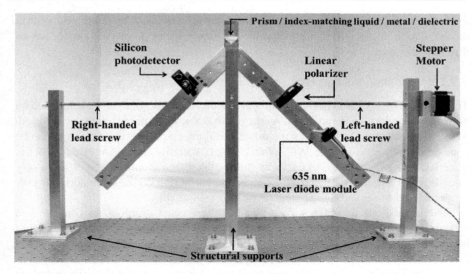

Figure 2.5 Automated implementation of Turbadar-Kretschmann-Raether configuration. A stepper motor rotates the two arms of an inverted V synchronously. Light from a laser diode and a linear polarizer mounted on the right arm is incident on one slanted face of a stationary prism affixed to the vertex of the inverted V. The intensity of the light exiting the other slanted face of the prism is measured by a silicon photodetector mounted on the left arm. The bottom face of a glass slide is affixed to the base of the prism by an index-matching liquid. The top face of the glass slide has a metal film deposited on it. A layer of the partnering dielectric material is deposited on top of the metal film.

With a sufficiently large value of n_{prism} and $\theta_{inc} \geqslant \theta_{cr}$, an SPP wave is excited when

$$\kappa \approx \mathrm{Re}\{q\}. \tag{2.48}$$

The excitation of the SPP wave by the collimated optical beam is known as surface plasmon resonance (SPR).

Resonance can be achieved by satisfying Eq. (2.48) in either of two ways: either the angle of incidence θ_{inc} or the angular frequency of light may be varied. The variation of angle of incidence to achieve resonance is known as *angular interrogation*, while the variation of frequency is known as *frequency interrogation*. At first glance, both κ and q depend linearly on k_0, as evidenced by Eqs. (2.35) and (2.47), and the two dependencies cancel. However, metals are highly dispersive in the optical regime—which means that ϵ_{met} depends strongly on the angular frequency ω. Both $\epsilon_{prism} = n_{prism}^2 \epsilon_0$ and ϵ_{diel} also depend on ω, if not strongly then weakly in the relevant spectral regime. The dependencies of the permittivities make frequency interrogation possible.

Generally, monitoring the light reflected from the prism/metal interface and emerging from the second slanted face of the prism allows for SPR-based sensing. In conventional prism-coupled sensing, the intensity of the light is monitored as a function of either the angle of incidence or the free-space wavelength of light. A prominent dip in the reflectance indicates SPR, as the energy needed to excite the SPP wave is absorbed from the incident light. It is also possible to monitor the phase of the light reflected from the prism/metal interface. As with mechanical systems, a sudden phase shift in the oscillation of the system relative to the exciting force occurs as the system passes through resonance.

2.3.1.1 Boundary-Value Problem

The response of the Turbadar-Kretschmann-Raether configuration can be calculated by solving a boundary-value problem. As the prism is thick compared to the decay length of the SPP wave perpendicular to the prism/metal interface, the prism material can be considered to be semi-infinite. With a thin layer of a material as the partnering dielectric material, the finite thickness of the layer must be taken into account [228]. When a very thick layer of a dielectric material serves as the partnering dielectric material, it may be considered of semi-infinite extent as well.

Several methods exist for solving the boundary-value problem. In Chapter 3, a straightforward method based on 4×4 matrices is developed, for use when the materials are not isotropic and/or homogeneous. Let us use in this chapter a recursive method similar to that developed by Parratt [229] to study the effect that layers of materials on the surface of a bulk material have on X-rays. Much of current research in surface waves involves multiple layers made of different materials. A recursive approach is particularly convenient for these configurations, since it is a simple matter to develop a computer program which uses the same formula as it steps through each layer.

Figure 2.6 schematically illustrates a stack of N planar layers of different materials. The ℓth layer is made of an isotropic dielectric material with refractive index n_ℓ and extends from $z_{\ell-1}$ to $z_\ell = z_{\ell-1} + d_\ell$, $\ell \in \{1, 2, 3, \ldots, N\}$, $z_0 = 0$. A plane wave is incident on the stack from the half-space $z < 0$ at an angle θ_{inc} to the z axis. The field

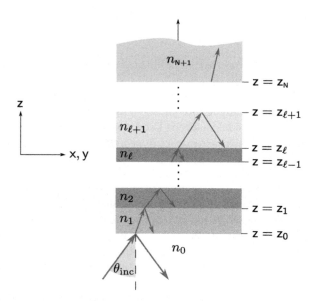

Figure 2.6 Schematic showing up-going and down-going waves in each layer of a stack of planar layers of different materials.

within the ℓth layer comprises, in general, an up-going wave originating from the $(\ell - 1)$th layer and a down-going wave originating from the $(\ell + 1)$th layer. The up-going wave propagates in the $+z$ direction, whereas the down-going wave propagates in the $-z$ direction.

For convenience, let us label the half-space $z < 0$ by $\ell = 0$ and denote the refractive index of the isotropic, homogeneous dielectric material occupying that half-space by n_0. For application to the Turbadar-Kretschmann-Raether configuration, $n_0 = n_{prism}$. Likewise, the label $\ell = N+1$ and the refractive index n_{N+1} belong to the half-space $z > z_N$.

2.3.1.2 p-Polarized Incident Plane Wave

Let us consider a p-polarized incident plane wave first. Its electric and magnetic field phasors are written as

$$\left.\begin{aligned}
\underline{E}_{inc}(\underline{r}) &= a_p(-\underline{u}_{prop}\cos\theta_{inc} + \underline{u}_z \sin\theta_{inc}) \\
&\quad \times \exp\{ik_0 n_0[\underline{u}_{prop}\cdot\underline{r}\sin\theta_{inc} + z\cos\theta_{inc}]\} \\
\underline{H}_{inc}(\underline{r}) &= -a_p\frac{n_0}{\eta_0}\underline{u}_s \exp\{ik_0 n_0[\underline{u}_{prop}\cdot\underline{r}\sin\theta_{inc} + z\cos\theta_{inc}]\} \\
&\quad z < 0,
\end{aligned}\right\}$$

(2.49)

where a_p is the incidence amplitude—which is supposed to be known.

The field phasors of the reflected plane wave are given by

$$\left.\begin{aligned}
\underline{E}_{ref}(\underline{r}) &= r_p(\underline{u}_{prop}\cos\theta_{inc} + \underline{u}_z \sin\theta_{inc}) \\
&\quad \times \exp\{ik_0 n_0[\underline{u}_{prop}\cdot\underline{r}\sin\theta_{inc} - z\cos\theta_{inc}]\} \\
\underline{H}_{ref}(\underline{r}) &= -r_p\frac{n_0}{\eta_0}\underline{u}_s \exp\{ik_0 n_0[\underline{u}_{prop}\cdot\underline{r}\sin\theta_{inc} - z\cos\theta_{inc}]\} \\
&\quad z < 0,
\end{aligned}\right\}$$

(2.50)

where r_p is the unknown reflection amplitude. The field phasors of the transmitted plane wave are given by

$$\left.\begin{aligned}
\underline{E}_{tr}(\underline{r}) &= t_p(-\underline{u}_{prop}\cos\theta_{tr} + \underline{u}_z \sin\theta_{tr}) \\
&\quad \times \exp\{ik_0 n_{N+1}[\underline{u}_{prop}\cdot\underline{r}\sin\theta_{tr} + (z - z_N)\cos\theta_{tr}]\} \\
\underline{H}_{tr}(\underline{r}) &= -t_p\frac{n_{N+1}}{\eta_0}\underline{u}_s \exp\{ik_0 n_{N+1}[\underline{u}_{prop}\cdot\underline{r}\sin\theta_{tr} + (z - z_N)\cos\theta_{tr}]\} \\
&\quad z > z_N,
\end{aligned}\right\}$$

(2.51)

where t_p is the unknown transmission amplitude and

$$\sin\theta_{tr} = \frac{n_0}{n_{N+1}}\sin\theta_{inc},$$

(2.52)

by virtue of the satisfaction of phase-matching conditions commonly known as Snel's laws.[2] It is possible that θ_{tr} is a complex angle, depending on the value of the ratio n_0/n_{N+1}.

[2]The name of the Dutch scientist Willebrord van Snel van Royen is often wrongly spelled as *Snell* in the English-language optics literature. Snel found the law of refraction in 1621, after Thomas Harriot (1601) but before René Descartes (1637). However, the law of refraction had been discovered six centuries earlier, by Ibn Sahl [230, p. 478].

The reflection and the transmission coefficients of the stack of N layers for p-polarized incidence are defined as

$$\left.\begin{array}{l} r_{pp} = r_p/a_p \\ t_{pp} = t_p/a_p \end{array}\right\}. \tag{2.53}$$

The corresponding linear reflectance and transmittance are

$$\left.\begin{array}{l} R_{pp} = |r_{pp}|^2 \\ T_{pp} = \dfrac{n_{N+1}}{n_0} \dfrac{\mathrm{Re}\,\{\cos\theta_{tr}\}}{\cos\theta_{inc}} |t_{pp}|^2 \end{array}\right\}, \tag{2.54}$$

while the linear absorptance of the stack for p-polarized incidence is

$$A_p = 1 - \left(R_{pp} + T_{pp}\right). \tag{2.55}$$

The principle of conservation of energy requires that $A_p \in [0, 1]$, if all $N+1$ materials are assumed to be passive.

The electric and magnetic field phasors in the ℓth layer may be decomposed as

$$\left.\begin{array}{l} \underline{E}(r) = \underline{E}_\ell^{up}(r) + \underline{E}_\ell^{dn}(r) \\ \underline{H}(r) = \underline{H}_\ell^{up}(r) + \underline{H}_\ell^{dn}(r) \end{array}\right\}, \quad z_{\ell-1} < z < z_\ell, \quad \ell \in [1, N], \tag{2.56}$$

where

$$\left.\begin{array}{l} \underline{E}_\ell^{up}(r) = a_p^{(\ell)}\left(\dfrac{-\alpha_\ell \underline{u}_{prop} + \kappa\underline{u}_z}{k_0 n_\ell}\right)\exp\{i[\kappa\underline{u}_{prop}\bullet\underline{r} + \alpha_\ell(z - z_{\ell-1})]\} \\[2ex] \underline{E}_\ell^{dn}(r) = b_p^{(\ell)}\left(\dfrac{\alpha_\ell \underline{u}_{prop} + \kappa\underline{u}_z}{k_0 n_\ell}\right)\exp\{i[\kappa\underline{u}_{prop}\bullet\underline{r} - \alpha_\ell(z - z_{\ell-1})]\} \end{array}\right\} \tag{2.57}$$

and

$$\left.\begin{array}{l} \underline{H}_\ell^{up}(r) = -a_p^{(\ell)}\dfrac{n_\ell}{\eta_0}\underline{u}_s \exp\{i[\kappa\underline{u}_{prop}\bullet\underline{r} + \alpha_\ell(z - z_{\ell-1})]\} \\[2ex] \underline{H}_\ell^{dn}(r) = -b_p^{(\ell)}\dfrac{n_\ell}{\eta_0}\underline{u}_s \exp\{i[\kappa\underline{u}_{prop}\bullet\underline{r} - \alpha_\ell(z - z_{\ell-1})]\} \end{array}\right\}. \tag{2.58}$$

In these equations, the coefficients $a_p^{(\ell)}$ and $b_p^{(\ell)}$ have yet to be determined,

$$\kappa = n_0 k_0 \sin\theta_{inc}, \tag{2.59}$$

and the imaginary part of

$$\alpha_\ell = \sqrt{(k_0 n_\ell)^2 - \kappa^2} \tag{2.60}$$

is restricted to be non-negative:

$$\mathrm{Im}\,\{\alpha_\ell\} \geqslant 0. \tag{2.61}$$

In order to develop a compact algorithm, let us extend Eqs. (2.56)–(2.61) to the two half-spaces with the stipulations

$$
\left.
\begin{aligned}
a_p^{(0)} &= a_p, & b_p^{(0)} &= r_p \\
a_p^{(N+1)} &= t_p, & b_p^{(N+1)} &= 0
\end{aligned}
\right\}.
\tag{2.62}
$$

Let us also expunge the exponential factors $\exp{(\pm i\alpha_0 z_{-1})}$ from Eqs. (2.57) and (2.58) for $\ell = 0$.

Across the interface $z = z_{\ell-1}$, the following boundary conditions must hold:

$$
\left.
\begin{aligned}
\underline{u}_{\text{prop}} \cdot \left[\underline{E}_{\ell-1}^{\text{up}}(x, y, z_{\ell-1} -) + \underline{E}_{\ell-1}^{\text{dn}}(x, y, z_{\ell-1} -) \right] \\
= \underline{u}_{\text{prop}} \cdot \left[\underline{E}_{\ell}^{\text{up}}(x, y, z_{\ell-1} +) + \underline{E}_{\ell}^{\text{dn}}(x, y, z_{\ell-1} +) \right] \\
\underline{u}_s \cdot \left[\underline{H}_{\ell-1}^{\text{up}}(x, y, z_{\ell-1} -) + \underline{H}_{\ell-1}^{\text{dn}}(x, y, z_{\ell-1} -) \right] \\
= \underline{u}_s \cdot \left[\underline{H}_{\ell}^{\text{up}}(x, y, z_{\ell-1} +) + \underline{H}_{\ell}^{\text{dn}}(x, y, z_{\ell-1} +) \right]
\end{aligned}
\right\}, \quad \ell \in [1, N+1].
\tag{2.63}
$$

With the shorthand notations

$$
r_p^{(\ell)} = \frac{b_p^{(\ell)}}{a_p^{(\ell)}}, \quad \upsilon_\ell = \frac{\alpha_\ell}{k_0 n_\ell}, \quad \beta_\ell = \begin{cases} 1, & \ell = 0 \\ \exp{(i\alpha_\ell d_\ell)}, & \ell \in [1, N] \end{cases},
\tag{2.64}
$$

Eqs. (2.63) yield

$$
\left.
\begin{aligned}
\frac{a_p^{(\ell-1)}}{a_p^{(\ell)}} &= \beta_{\ell-1} \frac{\upsilon_\ell}{\upsilon_{\ell-1}} \frac{r_p^{(\ell)} - 1}{r_p^{(\ell-1)} - \beta_{\ell-1}^2} \\
\frac{a_p^{(\ell-1)}}{a_p^{(\ell)}} &= \beta_{\ell-1} \frac{n_\ell}{n_{\ell-1}} \frac{r_p^{(\ell)} + 1}{r_p^{(\ell-1)} + \beta_{\ell-1}^2}
\end{aligned}
\right\}, \quad \ell \in [1, N+1].
\tag{2.65}
$$

Equating the right sides of Eqs. $(2.65)_1$ and $(2.65)_2$, we obtain the recursion relation

$$
r_p^{(\ell-1)} = \beta_{\ell-1}^2 \frac{r_p^{(\ell)} \left(1 + \gamma_p^{(\ell)}\right) + \left(1 - \gamma_p^{(\ell)}\right)}{r_p^{(\ell)} \left(1 - \gamma_p^{(\ell)}\right) + \left(1 + \gamma_p^{(\ell)}\right)}, \quad \ell \in [1, N+1],
\tag{2.66}
$$

where

$$
\gamma_p^{(\ell)} = \frac{\upsilon_\ell}{\upsilon_{\ell-1}} \frac{n_{\ell-1}}{n_\ell}.
\tag{2.67}
$$

In order to exploit the recursion relation (2.66), let us recall from Eqs. (2.62) that

$$
r_p^{(N+1)} \equiv 0
\tag{2.68}
$$

which leads to

$$
r_p^{(N)} = \beta_N^2 \frac{1 - \gamma_p^{(N+1)}}{1 + \gamma_p^{(N+1)}}.
\tag{2.69}
$$

Thereafter, $r_p^{(N-1)}$, $r_p^{(N-2)}$, etc. can be calculated in succession, the series ending with

$$r_p = r_p^{(0)} = \frac{r_p^{(1)}\left(1+\gamma_p^{(1)}\right) + \left(1-\gamma_p^{(1)}\right)}{r_p^{(1)}\left(1-\gamma_p^{(1)}\right) + \left(1+\gamma_p^{(1)}\right)}. \tag{2.70}$$

Finally, the transmission coefficient t_p can be found by using either Eq. (2.65)$_1$ or (2.65)$_2$. For instance, the latter can be applied as

$$a_p^{(\ell)} = \frac{a_p^{(\ell-1)}}{\beta_{\ell-1}}\frac{n_{\ell-1}}{n_\ell}\frac{r_p^{(\ell-1)} + \beta_{\ell-1}^2}{r_p^{(\ell)}+1} \tag{2.71}$$

to calculate $a_p^{(1)}$ from $a_p^{(0)} = a_p$, $a_p^{(2)}$ from $a_p^{(1)}$, and so on, until $t_p = a_p^{(N+1)}$ is calculated from $a_p^{(N)}$.

2.3.1.3   s-Polarized Incident Plane Wave

The electric and magnetic field phasors of an s-polarized incident plane wave are written as

$$\left.\begin{aligned}
\underline{E}_{\text{inc}}(\underline{r}) &= a_s\underline{u}_s \exp\{ik_0n_0[\underline{u}_{\text{prop}}\cdot\underline{r}\sin\theta_{\text{inc}} + z\cos\theta_{\text{inc}}]\} \\
\underline{H}_{\text{inc}}(\underline{r}) &= a_s\frac{n_0}{\eta_0}(-\underline{u}_{\text{prop}}\cos\theta_{\text{inc}} + \underline{u}_z\sin\theta_{\text{inc}}) \\
&\quad \times \exp\{ik_0n_0[\underline{u}_{\text{prop}}\cdot\underline{r}\sin\theta_{\text{inc}} + z\cos\theta_{\text{inc}}]\} \\
&\qquad z < 0,
\end{aligned}\right\} \tag{2.72}$$

while those of the reflected and transmitted plane waves, respectively, are

$$\left.\begin{aligned}
\underline{E}_{\text{ref}}(\underline{r}) &= r_s\underline{u}_s \exp\{ik_0n_0[\underline{u}_{\text{prop}}\cdot\underline{r}\sin\theta_{\text{inc}} - z\cos\theta_{\text{inc}}]\} \\
\underline{H}_{\text{ref}}(\underline{r}) &= r_s\frac{n_0}{\eta_0}(\underline{u}_{\text{prop}}\cos\theta_{\text{inc}} + \underline{u}_z\sin\theta_{\text{inc}}) \\
&\quad \times \exp\{ik_0n_0[\underline{u}_{\text{prop}}\cdot\underline{r}\sin\theta_{\text{inc}} - z\cos\theta_{\text{inc}}]\} \\
&\qquad z < 0,
\end{aligned}\right\} \tag{2.73}$$

and

$$\left.\begin{aligned}
\underline{E}_{\text{tr}}(\underline{r}) &= t_s\underline{u}_s \exp\{ik_0n_{N+1}[\underline{u}_{\text{prop}}\cdot\underline{r}\sin\theta_{\text{tr}} + (z-z_N)\cos\theta_{\text{tr}}]\} \\
\underline{H}_{\text{tr}}(\underline{r}) &= t_s\frac{n_{N+1}}{\eta_0}(-\underline{u}_{\text{prop}}\cos\theta_{\text{tr}} + \underline{u}_z\sin\theta_{\text{tr}}) \\
&\quad \times \exp\{ik_0n_{N+1}[\underline{u}_{\text{prop}}\cdot\underline{r}\sin\theta_{\text{tr}} + (z-z_N)\cos\theta_{\text{tr}}]\} \\
&\qquad z > z_N.
\end{aligned}\right\} \tag{2.74}$$

Here, r_s and t_s are unknown reflection and transmission amplitudes to be calculated in terms of the incidence amplitude a_s.

The reflection and the transmission coefficients of the stack for s-polarized incidence are defined as

$$\left.\begin{aligned}
r_{\text{ss}} &= r_s/a_s \\
t_{\text{ss}} &= t_s/a_s
\end{aligned}\right\}, \tag{2.75}$$

respectively. The corresponding linear reflectance and transmittance are

$$\left.\begin{array}{l} R_{ss} = |r_{ss}|^2 \\ T_{ss} = \dfrac{n_{N+1}}{n_0} \dfrac{\text{Re}\{\cos\theta_{tr}\}}{\cos\theta_{inc}} |t_{ss}|^2 \end{array}\right\}. \tag{2.76}$$

The linear absorptance of the stack for s-polarized incidence is given by

$$A_s = 1 - (R_{ss} + T_{ss}). \tag{2.77}$$

If all $N+1$ materials are taken to be passive, then $A_s \in [0, 1]$ by virtue of the principle of conservation of energy.

Equations (2.56) apply for s-polarized incidence, but with

$$\left.\begin{array}{l} \underline{E}_\ell^{up}(\underline{r}) = a_s^{(\ell)} \underline{u}_s \exp\{i[\kappa\underline{u}_{prop} \cdot \underline{r} + \alpha_\ell(z - z_{\ell-1})]\} \\ \underline{E}_\ell^{dn}(\underline{r}) = a_s^{(\ell)} r_s^{(\ell)} \underline{u}_s \exp\{i[\kappa\underline{u}_{prop} \cdot \underline{r} - \alpha_\ell(z - z_{\ell-1})]\} \end{array}\right\} \tag{2.78}$$

and

$$\left.\begin{array}{l} \underline{H}_\ell^{up}(\underline{r}) = a_s^{(\ell)} \dfrac{n_\ell}{\eta_0} \left(\dfrac{-\alpha_\ell\underline{u}_{prop} + \kappa\underline{u}_z}{k_0 n_\ell}\right) \exp\{i[\kappa\underline{u}_{prop} \cdot \underline{r} + \alpha_\ell(z - z_{\ell-1})]\} \\ \underline{H}_\ell^{dn}(\underline{r}) = a_s^{(\ell)} r_s^{(\ell)} \dfrac{n_\ell}{\eta_0} \left(\dfrac{\alpha_\ell\underline{u}_{prop} + \kappa\underline{u}_z}{k_0 n_\ell}\right) \exp\{i[\kappa\underline{u}_{prop} \cdot \underline{r} - \alpha_\ell(z - z_{\ell-1})]\} \end{array}\right\}. \tag{2.79}$$

As in Section 2.3.1.2, Eqs. (2.56), (2.78), and (2.79) are extended to $\ell \in \{0, N+1\}$ with the stipulations

$$\left.\begin{array}{l} a_s^{(0)} = a_s, \quad r_s^{(0)} = r_s \\ a_s^{(N+1)} = t_s, \quad r_s^{(N+1)} = 0 \end{array}\right\}, \tag{2.80}$$

and the exponential factors $\exp(\pm i\alpha_0 z_{-1})$ are expunged for $\ell = 0$.

Application of the boundary conditions

$$\left.\begin{array}{l} \underline{u}_s \cdot [\underline{E}_{\ell-1}^{up}(x, y, z_{\ell-1} -) + \underline{E}_{\ell-1}^{dn}(x, y, z_{\ell-1} -)] \\ = \underline{u}_s \cdot [\underline{E}_\ell^{up}(x, y, z_{\ell-1} +) + \underline{E}_\ell^{dn}(x, y, z_{\ell-1} +)] \\ \underline{u}_{prop} \cdot [\underline{H}_{\ell-1}^{up}(x, y, z_{\ell-1} -) + \underline{H}_{\ell-1}^{dn}(x, y, z_{\ell-1} -)] \\ = \underline{u}_{prop} \cdot [\underline{H}_\ell^{up}(x, y, z_{\ell-1} +) + \underline{H}_\ell^{dn}(x, y, z_{\ell-1} +)] \end{array}\right\}, \quad \ell \in [1, N+1], \tag{2.81}$$

yields the twin relations

$$
\left.
\begin{aligned}
\frac{a_s^{(\ell-1)}}{a_s^{(\ell)}} &= \beta_{\ell-1} \frac{r_s^{(\ell)} + 1}{r_s^{(\ell-1)} + \beta_{\ell-1}^2} \\
\frac{a_s^{(\ell-1)}}{a_s^{(\ell)}} &= \beta_{\ell-1} \gamma_s^{(\ell)} \frac{r_s^{(\ell)} - 1}{r_s^{(\ell-1)} - \beta_{\ell-1}^2}
\end{aligned}
\right\},
\tag{2.82}
$$

with

$$
\gamma_s^{(\ell)} = \frac{v_\ell}{v_{\ell-1}} \frac{n_\ell}{n_{\ell-1}}.
\tag{2.83}
$$

The recursion relation

$$
r_s^{(\ell-1)} = \beta_{\ell-1}^2 \frac{r_s^{(\ell)} \left(1 + \gamma_s^{(\ell)}\right) + \left(1 - \gamma_s^{(\ell)}\right)}{r_s^{(\ell)} \left(1 - \gamma_s^{(\ell)}\right) + \left(1 + \gamma_s^{(\ell)}\right)}, \quad \ell \in [1, N+1],
\tag{2.84}
$$

follows from Eqs. (2.82). Starting from

$$
r_s^{(N+1)} \equiv 0, \quad r_s^{(N)} = \beta_N^2 \frac{1 - \gamma_s^{(N+1)}}{1 + \gamma_s^{(N+1)}},
\tag{2.85}
$$

we can calculate $r_s^{(N-1)}, r_s^{(N-2)}$, etc., in succession to finally get $r_s = r_s^{(0)}$. Thereafter, the recursion relation

$$
a_s^{(\ell)} = \frac{a_s^{(\ell-1)}}{\beta_{\ell-1}} \frac{r_s^{(\ell-1)} + \beta_{\ell-1}^2}{r_s^{(\ell)} + 1}
\tag{2.86}
$$

can be used to calculate $a_s^{(1)}$ from $a_s^{(0)} = a_s$, $a_s^{(2)}$ from $a_s^{(1)}$, and so on, until $t_s = a_s^{(N+1)}$ is calculated from $a_s^{(N)}$.

2.3.1.4   Illustrative Results

The boundary-value problem for the Turbadar-Kretschmann-Raether configuration requires $N = 1$ when the partnering dielectric material is very thick. Suppose that the region $z > d_{\mathrm{met}}$ is filled with water ($n_2 = n_{\mathrm{diel}} = 1.33$); the region $z < 0$ is filled with rutile ($n_0 = n_{\mathrm{prism}} = 2.6$), a material often used to make prisms; the metal film $0 < z < d_{\mathrm{met}}$ is made by physical vapor deposition of aluminum ($n_1 = n_{\mathrm{met}} = 0.75 + 3.9i$). The free-space wavelength $\lambda_0 = 633$ nm.

For the specific choice of materials, $\theta_{\mathrm{cr}} = \sin^{-1}(n_{\mathrm{diel}}/n_{\mathrm{prism}}) = 30.766°$. According to Eq. (2.35), $q/k_0 = 1.4041 + 0.0326i$. Then, Eqs. (2.47) and (2.48) allow us to predict the occurrence of SPR at $\theta_{\mathrm{inc}} \approx \sin^{-1}(\mathrm{Re}\{q\}/k_0 n_{\mathrm{prism}}) = 32.685°$, which indeed exceeds θ_{cr}.

When $\theta_{\mathrm{inc}} > \theta_{\mathrm{cr}}$, the magnitude of $\sin\theta_{\mathrm{tr}}$ exceeds unity and $\cos\theta_{\mathrm{tr}}$ is purely imaginary; hence,

$$
\left.
\begin{aligned}
T_{\mathrm{pp}} &= 0, \quad A_p = 1 - R_{\mathrm{pp}} \\
T_{\mathrm{ss}} &= 0, \quad A_s = 1 - R_{\mathrm{ss}}
\end{aligned}
\right\}
\tag{2.87}
$$

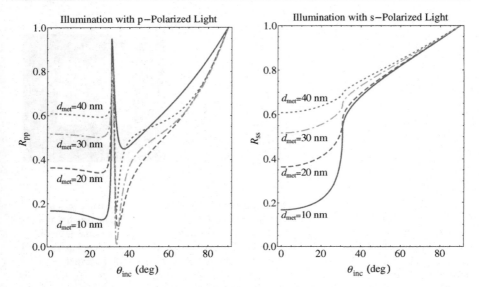

Figure 2.7 Calculated linear reflectances R_{pp} and R_{ss} as functions of the angle of incidence θ_{inc} at $\lambda_0 = 633$ nm for the Turbadar-Kretschmann-Raether configuration wherein rutile and water occupy half-spaces separated by a thin film of aluminum of thickness d_{met}.

for $\theta_{inc} > \theta_{cr}$. Plots of R_{pp} and R_{ss} as functions of $\theta_{inc} > \theta_{cr}$, whether obtained experimentally or theoretically, therefore are valuable in determining the excitation of SPP waves.

The linear reflectances R_{pp} and R_{ss} are plotted in Figure 2.7 as functions of the angle of incidence θ_{inc}, for $d_{met} \in \{10, 20, 30, 40\}$ nm. Each plot of R_{pp} as a function of θ_{inc} contains a sharp rise as θ_{inc} approaches θ_{cr}, the rise being followed by a strong dip that indicates the excitation of the SPP wave. The minimum of the dip occurs at an angle of incidence θ_{inc}^{SPR}. The strong dip in R_{pp} indicates a very high value of A_p, according to Eq. (2.87), the peak absorptance occurring at θ_{inc}^{SPR}, the resonance angle at which the most efficient excitation of the SPP wave occurs. Clearly, θ_{inc}^{SPR} is very close to the predicted value $32.685°$.

The presence of the prism/metal interface has a small effect, and θ_{inc}^{SPR} varies slightly with the thickness d_{met} of the metal film. The width and the depth of the dip, which affect the resolution of SPR-based sensing, are also influenced by d_{met}. The choice of this thickness is important when optimizing SPR sensors.

A similar dip is not present in the plots of R_{ss} as a function of θ_{inc} in Figure 2.7. This illustrates that an s-polarized incident plane wave cannot excite an SPP wave guided by the planar interface of a metal and a homogeneous, isotropic, dielectric material.

2.3.1.5   SPR-Based Prism-Coupled Sensing

According to Figure 2.2a, Re $\{q\}/k_0$ varies almost linearly with respect to n_{diel} for small values of $n_{diel} > 1$ and fixed n_{met}. Making use of Eqs. (2.47) and (2.48), one can

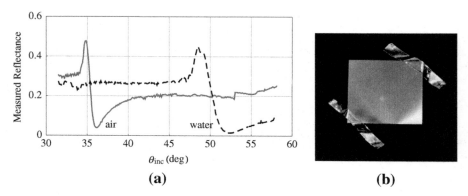

(a)   **(b)**

Figure 2.8 (a) Linear reflectance measured as a function of the angle of incidence θ_{inc} at $\lambda_0 = 635$ nm in the Turbadar–Kretschmann–Raether configuration of Figure 2.5. The prism was made of dense flint glass ($n_{prism} = 1.78471$), the ~ 30-nm-thick metal film of aluminum was deposited on a substrate made of dense flint glass, the dielectric partnering material was either air or water, and the incident light was p polarized. The two plots should not be compared for magnitude, as different gain settings on the silicon photodetector were used for air and water. (b) Aluminum thin film deposited on the substrate used for the experiment.

predict that the product $n_{prism} \sin \theta_{inc}^{SPR}$ must also vary almost linearly with respect to n_{diel} for small values of $n_{diel} > 1$ and fixed n_{met}. With n_{prism} also fixed, the variation of θ_{inc}^{SPR} with small changes in n_{diel} would have to be almost linear as well. This feature is commonly exploited for optical biosensing in the Turbadar-Kretschmann-Raether configuration.

Figure 2.8a presents two plots of the reflectance at $\lambda_0 = 635$ nm with respect to θ_{inc}, measured using p-polarized incident light and a prism made of dense flint glass ($n_{prism} = 1.78471$) in the apparatus depicted in Figure 2.5. The measured reflectance is the intensity of light leaving the second slanted face of the prism divided by the intensity of light incident on the first slanted face of the prism. With a ~ 30 nm-thick aluminum film as the metallic partnering material and air as the dielectric partnering material, the measured reflectance has a dip at $\theta_{inc} = 36.01°$. With water as the dielectric partner, the reflectance dip shifts to $\theta_{inc} = 52.07°$. The ratio $(\sin 52.07°)/(\sin 36.01°)$ equals 1.3416, which is very close to the ratio of the refractive index of water to that of air.

2.3.1.6   Fiber-Optic Coupling

Although the Turbadar-Kretschmann-Raether configuration can be used to make chemical sensors which are extremely sensitive to changes in refractive index of the partnering dielectric material [4], it suffers from some disadvantages. Generally, the configuration is bulky and complicated, qualities not very amenable for use in the field. Also, a sensor based on this configuration requires a significant volume of the analyte to be present. Many researchers have pursued the use of optical fibers as a solution to these problems. Geometries and sensing modalities differ, but the various schemes involve evaporating a thin metal film on an exposed portion of the core of an optical fiber.

The core, with a high refractive index, then acts like the high-refractive-index prism of the Turbadar-Kretschmann-Raether configuration.

The metal film can be applied to the fiber core in a variety of ways. One method [231,232] is to simply remove the fiber cladding and evaporate a uniform metallic film around the cylindrical core. Other researchers [233,234] taper a section of the fiber before depositing the metal film which extends the spectral regime of operation. In some cases, the metal film is only deposited on one side. Another method is to polish a flat on the core and deposit the metal film on the flat. This is known as the side-polish method. Cylindrically asymmetric schemes, such as of the side-polished fiber, have the disadvantage of being polarization sensitive, whereas cylindrically symmetric schemes are polarization insensitive. Prism coupling not only couples the exciting light beam to the SPP wave but also couples the SPP wave to an outgoing re-radiated wave. The use of an optical fiber cleaved at a steep angle reduces re-radiation (or loading), while capable of exciting an SPP wave in a small microstructure [235].

Fiber-optic coupling also offers some versatility in sensing based on the excitation of SPP waves. With light propagating in a narrow range of angles relative to the axis of the optical fiber, white light may be used. After insertion at one end of the optical fiber, light that passes the portion of the optical fiber that interfaces with the deposited metal and has a wavelength that satisfies the coupling condition to excite the SPP wave will be absorbed, thereby producing a dip in the output spectrum. The other end of the optical fiber can therefore be coupled to a spectrograph for analysis. For remote sensing, some researchers [231] have deposited a mirror surface at one end of the optical fiber. Thus, with splitters, the insertion of light and the sensing after interaction with the SPP wave can be done at one end of the optical fiber. In addition, the light passes the interface twice and increases the absorption of light. Monochromatic light may also be used to excite the SPP wave. The frequency can be scanned to look for the spectral location of the resonance [236]. A sensor may use just a single wavelength, with the intensity of output light indicating a shift toward or away from resonance [232].

Although there are many advantages to fiber-optic coupling, there are some disadvantages as well. Fiber-optic coupling is noisier than the Turbadar-Kretschmann-Raether configuration [237]. Sensors with a fixed wavelength source can only detect SPP waves over a limited range of refractive index of the partnering dielectric material, but this limited range can be shifted by the application of a thin layer of dielectric material between the fiber core and the metal film [232,236].

2.3.2   Turbadar-Otto Configuration

A schematic diagram of the Turbadar-Otto configuration [22] is shown in Figure 2.9. This is a less frequently used configuration than the Turbadar-Kretschmann-Raether configuration. As in that configuration, a collimated beam of almost monochromatic light is directed through one of the two slanted faces of a prism in the Turbadar-Otto configuration, but the partnering materials are reversed. A layer of the partnering dielectric material is interposed between the prism and a thick metal film.

If the angle of incidence θ_{inc} inside the prism at the prism/dielectric interface is greater than the critical angle $\theta_{cr} = \sin^{-1}(n_{diel}/n_{prism})$ for total internal reflection,

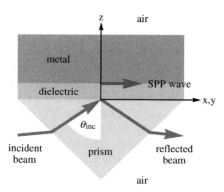

Figure 2.9 Schematic of the Turbadar-Otto configuration.

an evanescent wave is created in the partnering dielectric material. If this material is present as a sufficiently thin layer, the evanescent wave reaches the metal/dielectric interface with a significant amplitude. When

$$n_{\text{prism}}k_0 \sin \theta_{\text{inc}} \approx \text{Re}\,\{q\}, \tag{2.88}$$

the resonance condition (2.48) is satisfied and the SPP wave is excited—if the incident light is p polarized.

The thickness of the layer of the partnering dielectric material must be quite small in order for the evanescent wave to arrive at the metal with sufficient amplitude. Typically, in the optical regime, this thickness is at most a few hundred nanometers. Whereas fabrication of this layer is straightforward using a variety of techniques [111], maintaining such a small spacing and keeping it free of contamination for liquid dielectric materials—which is the type of dielectric material most common in SPP sensors—is quite difficult. Accordingly, the Turbadar-Otto configuration is not very popular for optical sensing applications.

Reflection from the Turbadar-Otto configuration may be theoretically treated in the same way as the Turbadar-Kretschmann-Raether configuration—i.e. by using the algorithms provided in Sections 2.3.1.2 and 2.3.1.3. However, $N = 2$ is needed with both partnering materials of finite thickness.

As an example, suppose that the region $z < 0$ is filled with rutile ($n_0 = n_{\text{prism}} = 2.6$); the region $0 < z < d_{\text{diel}}$ is occupied by a material of refractive index $n_1 = n_{\text{diel}} = 1.33$, which was chosen to afford comparison with Figure 2.7; the region $d_{\text{diel}} < z < d_{\text{diel}} + d_{\text{met}}$ is a thin film of aluminum ($n_2 = n_{\text{met}} = 0.75 + 3.9i$); and the region $z > d_{\text{diel}} + d_{\text{met}}$ is occupied by rutile ($n_3 = 2.6$) also. If the metal film is sufficiently thick, the region labeled $N + 1 = 3$ can be occupied by any material because a thick layer of a metal is a very effective electromagnetic isolator. The free-space wavelength $\lambda_0 = 633$ nm.

As in Section 2.3.1.4, $\theta_{\text{cr}} = 30.766°$, $q/k_0 = 1.4041 + 0.0326i$, and SPR can be predicted to occur at $\theta_{\text{inc}} \approx 32.685°$. The linear reflectances R_{pp} and R_{ss} are plotted in Figure 2.10 as functions of θ_{inc} for $d_{\text{met}} = 70$ and $d_{\text{diel}} \in \{100, 200, 300, 400\}$ nm. Each plot of R_{pp} as a function of θ_{inc} contains a strong dip that indicates the excitation of a p-polarized SPP wave, the minimum of the dip occuring at an angle of incidence

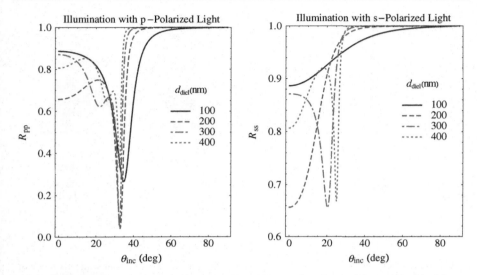

Figure 2.10 Calculated linear reflectances R_{pp} and R_{ss} as functions of the angle of incidence θ_{inc} at $\lambda_0 = 633$ nm of a rutile/dielectric-material/aluminum/rutile system in the Turbadar-Otto configuration, for $d_{diel} \in \{100, 200, 300, 400\}$ and $d_{met} = 70$ nm. The dielectric material is of refractive index $n_1 = n_{diel} = 1.33$.

θ_{inc}^{SPR} that is very close to the predicted value $32.685°$. Let us note that after the thickness of the partnering dielectric material increases beyond some threshold, θ_{inc}^{SPR} does not alter. Moreover, the dip narrows as d_{diel} increases.

No dip is present in the plots of R_{ss} against θ_{inc} in Figure 2.10 that indicates the excitation of an s-polarized SPP wave. The non-SPP dips in these plots can be attributed to interference effects across the layer of the partnering dielectric material.

2.3.3 Sarid Configuration

As a metal is one of the partnering materials forming the interface, the propagation length of an SPP wave is short—typically on the order of a few micrometers, according to Table 2.2. This short range limits the usefulness of SPP waves for many applications. However, if one is willing to consider guiding the SPP wave with two parallel metal/dielectric interfaces in close proximity, the propagation length can be radically enhanced [238].

The Sarid configuration also requires the use of a prism as do the two configurations originated by Turbadar [20,21]. The prism abuts a layer of a dielectric material which, in turn, is in contact with a metal film. The other face of the metal film is in contact with another layer of a dielectric material. Often, the first dielectric layer is called the *cover* and the the second dielectric layer is called the *substrate*. The substrate is sufficiently thick to be considered as occupying a half-space. Figure 2.11 presents a schematic of the Sarid configuration.

When the metal film is very thick, an SPP wave may be guided by the cover/metal interface. For moderate thickness of the metal film, two SPP waves may be guided

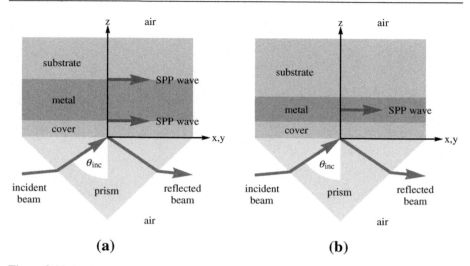

Figure 2.11 Schematics of the Sarid configuration when the cover and the substrate are made of identical materials. (a) Two distinct SPP waves can be excited if the metal film is sufficiently thick. (b) As the metal film becomes thinner, the two SPP waves hybridize to form an LRSPP wave.

separately, one by the cover/metal interface and the other by the metal/substrate interface [239]. As the metal film is made even thinner, the fields of these two SPP waves begin to overlap, and eventually it is no longer possible to consider the two SPP waves as separate entities. The two waves then hybridize into different waves that also propagate parallel to $\underline{u}_{\mathrm{prop}}$.

Often, both the cover and the substrate are made of the same dielectric material. When the metal film is sufficiently thin, the profile of the electric field of one hybrid wave is symmetric about the center of the metal film, whereas the same profile of the other hybrid wave is anti-symmetric. The symmetric SPP wave penetrates very far into the dielectric materials, and only a smaller fraction of its energy resides in the metal film. Since a metal is highly dissipative while dissipation is negligibly small in both the cover and the substrate, the propagation length of the symmetric SPP wave is large. In fact, the propagation length of the symmetric SPP wave is not only large compared to that of the anti-symmetric SPP wave, but is large compared to an SPP wave guided by a single interface of the same metal and dielectric material. This type of SPP wave is now known as a long-range SPP (LRSPP) wave, while the anti-symmetric wave is known as a short-range SPP (SRSPP) wave. The propagation lengths of LRSPP waves can be as large as a few hundred micrometers [239].

Since the initial investigation into LRSPP waves by Sarid [238], many researchers have developed schemes to increase the propagation length to as much as a few centimeters [216]. One method [240, 241] is to choose materials with different refractive indexes for the cover and the substrate. This structure, theoretically, can guide LRSPP waves with propagation lengths up to three orders of magnitude larger than waves guided by the Sarid configuration with identical materials for the cover and the substrate.

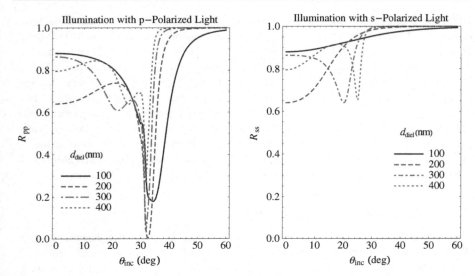

Figure 2.12 Calculated linear reflectances R_{pp} and R_{ss} as functions of the angle of incidence θ_{inc} at $\lambda_0 = 633$ nm of a rutile/cover/aluminum/substrate system in the Sarid configuration, for $d_1 = d_{diel} \in \{100, 200, 300, 400\}$ and $d_2 = d_{met} = 60$ nm. The cover ($n_1 = n_{diel} = 1.33$) and the substrate ($n_3 = n_{diel} = 1.33$) have the same refractive index.

The algorithms provided in Sections 2.3.1.2 and 2.3.1.3 can be used to compute R_{pp} and R_{ss} as functions of θ_{inc}. Since the substrate is very thick, it can be assumed to be semi-infinitely thick for calculations, and $N = 2$ then suffices.

As an example, suppose that the region $z < 0$ is filled with rutile ($n_0 = n_{prism} = 2.6$), and the region $d_1 < z < d_1 + d_2$ is a thin film of aluminum ($n_2 = n_{met} = 0.75 + 3.9i$) of thickness $d_{met} = d_2$. The cover $0 < z < d_1 = d_{diel}$ and the substrate $z > d_1 + d_2$ are occupied by a material of refractive index $n_1 = n_3 = n_{diel} = 1.33$, which was chosen to afford comparison with Figures 2.7 and 2.10. The free-space wavelength $\lambda_0 = 633$ nm. Plots of R_{pp} and R_{ss} as functions of θ_{inc} in Figure 2.12 clearly show the excitation of a p-polarized SPP wave at an angle of incidence slightly greater than $\theta_{cr} = 30.766°$.

Calculations have shown that placing an intermediate thin dielectric layer between the metal film and the cover and another intermediate thin dielectric layer between the metal films and the substrate may also increase the propagation length by several orders of magnitude [242,243]. In addition, by using intermediate layers, large propagation lengths may be obtained with thicker metal films. This may be significant since reducing the thickness of the metal film to obtain larger propagation lengths in the Sarid configuration is met with the technological limit of economically fabricating uniform films thinner than about 15 nm. More recently, researchers have considered sandwiching the metal thin film between periodically layered dielectric materials [155,244]. Both s-polarized and p-polarized SPP waves are possible with this approach.

Although the LRSPP wave has attracted much attention, the SRSPP mode also may be used to advantage [245]. One problem with biochemical sensing is that changes in the refractive index of the bulk solution containing the analyte, due to changes in the temperature and the concentrations of non-analyte components, can interfere with the signal from the analyte which binds to the recognition molecule on the surface of the sensor. Since the SRSPP wave is more closely bound to the guiding interface, it is less affected by changes in the bulk solution. Monitoring both the SRSPP and LRSPP waves allows separation of artifacts from the sensing of the analyte, thereby reducing sensing errors.

Finally, it should be mentioned that another method of LRSPP propagation has received some attention, namely, the hybrid SPP waveguide [246, 247]. As originally proposed [246], this configuration consisted of a dielectric cylinder above a metal. However, more practical planar geometries [247] are now being considered along with efficient means of coupling to dielectric waveguides [248].

2.3.4   Grating-Coupled Configuration

Total internal reflection, exploited in both the Turbadar-Kretschmann-Raether and the Turbadar-Otto configurations, provides one method of producing an evanescent wave with a component κ of the wave vector parallel to the interface that is greater in magnitude than the wavenumber $k_0 n_{\text{diel}}$ of a plane wave traveling in the bulk dielectric material. Surface-relief gratings also can produce waves that satisfy that requirement, and researchers have explored, experimentally [249] and theoretically [250–252], the use of these gratings to excite SPP waves. The use of surface-relief gratings for the excitation of SPP waves is particularly attractive for harvesting solar energy using thin-film silicon solar cells whose metallic back-surface reflectors are periodically corrugated [207–209].

The prediction of SPP-wave excitation in the grating-coupled configuration requires the solution of a boundary-value problem that is more complicated than those encountered for the Turbadar-Kretschmann-Raether, Turbadar-Otto, and Sarid configurations. The solution can be attempted by a variety of numerical techniques, such as the extinction boundary condition method [81, 253], the method of covariant spatial transformation [254, 255], and the rigorous coupled-wave approach (RCWA) [250, 256–258, 153, 259]. Although all three numerical techniques work well when the partnering dielectric material is isotropic and homogeneous, the RCWA is especially well suited for extension to periodically nonhomogeneous and/or bianisotropic materials, as presented in Section 3.8.5.

Figure 2.13 shows a schematic of the grating-coupled configuration. For simplicity, the half-spaces $z < 0$ and $z > d_1 + d_2$ are occupied by air which is approximated by free space (i.e. vacuum). For every $x \in (-\infty, \infty)$, the region $0 < z < d = d_1 + g(x)$ is occupied by an isotropic, homogeneous, dielectric material of permittivity ϵ_{diel}, while the partnering metal occupies the region $d_1 + g(x) < z < d_1 + d_2$. The grating function

$$g(x) = g(x \pm L_x) \tag{2.89}$$

has a period L_x along the x axis; the magnitude of its minimum value $g_{\text{min}} < 0$ is less than d_1, and its maximum value $g_{\text{max}} > 0$ is less than d_2. Thus, the xz plane is the grating plane because it completely captures the cross-sectional shape of the grating, and the surface-relief grating is singly periodic, with a trough-to-crest height

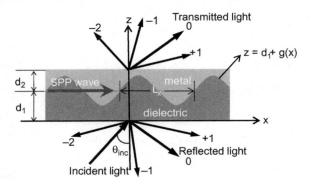

Figure 2.13 Schematic of the boundary-value problem for the grating-coupled configuration to excite an SPP wave guided by the periodically corrugated interface $z = d_1 + g(x)$ of a metal and an isotropic, homogeneous, dielectric material. The propagation plane and the grating plane are the xz plane. The reflected field phasors comprise specular components (identified by 0) and nonspecular components (identified by ± 1, etc.), and the transmitted field phasors may also have specular and nonspecular components. The trough-to-crest height of the periodically corrugated interface is typically a small fraction of the free-space wavelength, whereas the period is on the same order as the free-space wavelength.

$L_g = g_{max} - g_{min}$ that is typically a small fraction of λ_0. Typically also, the period L_x is on the order of λ_0 but does not exceed it [207,208].

The surface-relief grating may be either simple or compound. A simple surface-relief grating [260] is exemplified by a sinusoidal grating: $g(x) = (L_g/2) \sin(2\pi x/L_x)$. A sawtooth grating furnishes another example. Each period of a compound grating comprises two or more simple gratings, each several periods long. A compound grating can help excite SPP waves over a range of free-space wavelengths and for several different incidence conditions [85,86].

If the grating is doubly periodic, i.e. it is periodically corrugated along both the x and the y axes, the procedure provided in Section 3.8.5 should be followed for analysis. The same procedure is useful when the half-spaces $z < 0$ and $z > d_1 + d_2$ are taken to be occupied by different isotropic, homogeneous, and dielectric materials.

2.3.4.1 Incident Plane Wave

Let the incident plane wave be assumed to propagate in the xz plane at an angle θ_{inc} relative to the z axis in the vacuous half-space $z < 0$. Thus the plane of propagation of the incident plane wave and the grating plane are the same.

The incident field phasors are written in terms of a Fourier series with respect to x as

$$
\left.
\begin{aligned}
\underline{E}_{inc}(\underline{r}) &= \sum_{n=-\infty}^{\infty} \left(a_s^{(n)} \underline{s}^{(n)} + a_p^{(n)} \underline{p}_{inc}^{(n)} \right) \exp\left[i\left(\kappa^{(n)} x + \alpha^{(n)} z \right) \right] \\
\underline{H}_{inc}(\underline{r}) &= \eta_0^{-1} \sum_{n=-\infty}^{\infty} \left(a_s^{(n)} \underline{p}_{inc}^{(n)} - a_p^{(n)} \underline{s}^{(n)} \right) \exp\left[i\left(\kappa^{(n)} x + \alpha^{(n)} z \right) \right]
\end{aligned}
\right\}, \quad z < 0,
$$

$$(2.90)$$

where the amplitudes

$$\left.\begin{array}{l} a_s^{(n)} = a_s^{(0)} \delta_{n0} \\ a_p^{(n)} = a_p^{(0)} \delta_{n0} \end{array}\right\} \tag{2.91}$$

involve the Kronecker delta

$$\delta_{nn'} = \begin{cases} 1, & n = n', \\ 0, & n \neq n'. \end{cases} \tag{2.92}$$

The other quantities appearing in Eqs. (2.90) are as follows:

$$\left.\begin{array}{l} \kappa^{(n)} = k_0 \sin\theta_{\mathrm{inc}} + n\left(2\pi/L_x\right) \\ \alpha^{(n)} = +\sqrt{k_0^2 - \left(\kappa^{(n)}\right)^2} \\ \underline{s}^{(n)} = \underline{u}_y \\ \underline{p}_{\mathrm{inc}}^{(n)} = \dfrac{-\alpha^{(n)}\underline{u}_x + \kappa^{(n)}\underline{u}_z}{k_0} \end{array}\right\}. \tag{2.93}$$

Each term in this representation is a linear Floquet harmonic, the qualifier *linear* indicating that the fields are expressed in terms of s- and p-polarization states. If the incident plane wave is s polarized, then $a_p^{(0)} = 0$; if it is p polarized, then $a_s^{(0)} = 0$. Alternatively, one can employ circular Floquet harmonics and circular polarization states, as shown in Section 3.7.4, but these entities are not very useful because all materials involved here are isotropic and achiral [17].

2.3.4.2   Reflected and Transmitted Field Phasors

The reflected field phasors are written as

$$\left.\begin{array}{l} \underline{E}_{\mathrm{ref}}(\underline{r}) = \displaystyle\sum_{n=-\infty}^{\infty} \left(r_s^{(n)}\underline{s}^{(n)} + r_p^{(n)}\underline{p}_{\mathrm{ref}}^{(n)}\right) \exp\left[i\left(\kappa^{(n)}x - \alpha^{(n)}z\right)\right] \\[2mm] \underline{H}_{\mathrm{ref}}(\underline{r}) = \eta_0^{-1} \displaystyle\sum_{n=-\infty}^{\infty} \left(r_s^{(n)}\underline{p}_{\mathrm{ref}}^{(n)} - r_p^{(n)}\underline{s}^{(n)}\right) \exp\left[i\left(\kappa^{(n)}x - \alpha^{(n)}z\right)\right] \end{array}\right\}, \quad z < 0,$$

$$\tag{2.94}$$

where

$$\underline{p}_{\mathrm{ref}}^{(n)} = \frac{\alpha^{(n)}\underline{u}_x + \kappa^{(n)}\underline{u}_z}{k_0}, \tag{2.95}$$

while $r_s^{(n)}$ and $r_p^{(n)}$ are unknown reflection amplitudes. Similarly to the incident field phasors, the transmitted field phasors are written as

$$\left.\begin{array}{l} \underline{E}_{\mathrm{tr}}(\underline{r}) = \displaystyle\sum_{n=-\infty}^{\infty} \left(t_s^{(n)}\underline{s}^{(n)} + t_p^{(n)}\underline{p}_{\mathrm{inc}}^{(n)}\right) \exp\left\{i\left[\kappa^{(n)}x + \alpha^{(n)}(z - d_1 - d_2)\right]\right\} \\[2mm] \underline{H}_{\mathrm{tr}}(\underline{r}) = \eta_0^{-1} \displaystyle\sum_{n=-\infty}^{\infty} \left(t_s^{(n)}\underline{p}_{\mathrm{inc}}^{(n)} - t_p^{(n)}\underline{s}^{(n)}\right) \exp\left\{i\left[\kappa^{(n)}x + \alpha^{(n)}(z - d_1 - d_2)\right]\right\} \end{array}\right\},$$

$$z > d_1 + d_2, \quad (2.96)$$

where $t_s^{(n)}$ and $t_p^{(n)}$ are unknown transmission amplitudes.

The specular terms in the sums on the right sides of Eqs. (2.94) and (2.96) are the ones for which $n = 0$. The remaining terms in these sums are nonspecular terms which arise due to the periodic nature of the function $g(x)$. If this function does not vary with x, then $g(x) = 0$ and all nonspecular terms can be discarded.

2.3.4.3   Linear Reflectances and Transmittances

As the plane of propagation of the incident plane wave is the same as the grating plane, no depolarization can occur on reflection and transmission, which means that

$$
\left.
\begin{aligned}
r_s^{(n)} &= r_{ss}^{(n)} a_s^{(0)}, \quad t_s^{(n)} = t_{ss}^{(n)} a_s^{(0)} \\
r_p^{(n)} &= r_{pp}^{(n)} a_p^{(0)}, \quad t_p^{(n)} = t_{pp}^{(n)} a_p^{(0)}
\end{aligned}
\right\},
\tag{2.97}
$$

where $r_{ss}^{(n)}$ and $r_{pp}^{(n)}$ are reflection coefficients, and $t_{ss}^{(n)}$ and $t_{pp}^{(n)}$ are transmission coefficients. Linear reflectances and transmittances of order $n \in (-\infty, \infty)$ are defined as

$$
\left.
\begin{aligned}
R_{ss}^{(n)} &= |r_{ss}^{(n)}|^2, \quad T_{ss}^{(n)} = \frac{\mathrm{Re}\{\alpha^{(n)}\}}{\alpha^{(0)}}|t_{ss}^{(n)}|^2 \\
R_{pp}^{(n)} &= |r_{pp}^{(n)}|^2, \quad T_{pp}^{(n)} = \frac{\mathrm{Re}\{\alpha^{(n)}\}}{\alpha^{(0)}}|t_{pp}^{(n)}|^2
\end{aligned}
\right\}.
\tag{2.98}
$$

Absorptances for s- and p-polarized incidences are calculated as

$$
\left.
\begin{aligned}
A_s &= 1 - \sum_{n=-\infty}^{\infty} \left(R_{ss}^{(n)} + T_{ss}^{(n)} \right) \\
A_p &= 1 - \sum_{n=-\infty}^{\infty} \left(R_{pp}^{(n)} + T_{pp}^{(n)} \right)
\end{aligned}
\right\}.
\tag{2.99}
$$

Neither of the two linear absorptances can be negative nor can they exceed unity, by virtue of the principle of conservation of energy, provided that both partnering materials are passive.

2.3.4.4   Rigorous Coupled-Wave Approach

The RCWA requires that all field phasors as well as the permittivity in the region $0 \leqslant z \leqslant d_1 + d_2$ be expressed as Fourier series with respect to x.

The constitutive relations of the two partnering materials are jointly written as

$$
\left.
\begin{aligned}
\underline{D}(\underline{r}) &= \epsilon(x, z)\underline{E}(\underline{r}) \\
\underline{B}(\underline{r}) &= \mu_0 \underline{H}(\underline{r})
\end{aligned}
\right\}, \quad z \in (0, d_1 + d_2),
\tag{2.100}
$$

where

$$
\epsilon(x, z) =
\begin{cases}
\epsilon_{\mathrm{diel}}, & z \in (0, d_1 + g_{min}], \\
\epsilon_{\mathrm{diel}} + (\epsilon_{\mathrm{met}} - \epsilon_{\mathrm{diel}})\, \mathcal{U}\left[z - g(x)\right], & z \in (d_1 + g_{min}, d_1 + g_{max}), \\
\epsilon_{\mathrm{met}}, & z \in [d_1 + g_{max}, d_1 + d_2),
\end{cases}
\tag{2.101}
$$

and

$$\mathcal{U}(\sigma) = \begin{cases} 1, & \sigma \geqslant 0 \\ 0, & \sigma < 0 \end{cases} \tag{2.102}$$

is the unit step function. Accordingly, we get the Fourier series

$$\epsilon(x, z) = \sum_{n=-\infty}^{\infty} \epsilon^{(n)}(z) \exp{(i2\pi nx/L_x)}, \quad z \in (0, d_1 + d_2), \tag{2.103}$$

where

$$\epsilon^{(n)}(z) = \begin{cases} (\epsilon_{met} - \epsilon_{diel}) \Upsilon^{(n)}(z), & z \in (d_1 + g_{min}, d_1 + g_{max}), \\ 0, & z \notin (d_1 + g_{min}, d_1 + g_{max}), \end{cases} \tag{2.104}$$

except that

$$\epsilon^{(0)}(z) = \begin{cases} \epsilon_{diel}, & z \in (0, d_1 + g_{min}], \\ \epsilon_{diel} + (\epsilon_{met} - \epsilon_{diel}) \Upsilon^{(0)}(z), & z \in (d_1 + g_{min}, d_1 + g_{max}), \\ \epsilon_{met}, & z \in [d_1 + g_{max}, d_1 + d_2), \end{cases} \tag{2.105}$$

and

$$\Upsilon^{(n)}(z) = \frac{1}{L_x} \int_0^{L_x} \mathcal{U}[z - g(x)] \exp{(-i2\pi nx/L_x)} dx. \tag{2.106}$$

The field phasors in the region occupied by either partnering material are expressed as Fourier series with respect to x too, just like the reflected and transmitted field phasors; thus,

$$\left. \begin{aligned} \underline{E}(\underline{r}) &= \sum_{n=-\infty}^{\infty} [\underline{e}^{(n)}(z) \exp{(i\kappa^{(n)}x)}] \\ \underline{H}(\underline{r}) &= \sum_{n=-\infty}^{\infty} [\underline{h}^{(n)}(z) \exp{(i\kappa^{(n)}x)}] \end{aligned} \right\}, \quad z \in (0, d_1 + d_2), \tag{2.107}$$

where the vector functions $\underline{e}^{(n)}(z) = e_x^{(n)} \underline{u}_x + e_y^{(n)} \underline{u}_y + e_z^{(n)} \underline{u}_z$, and $\underline{h}^{(n)}(z) = h_x^{(n)} \underline{u}_x + h_y^{(n)} \underline{u}_y + h_z^{(n)} \underline{u}_z$ remain to be determined for all n.

Substitution of Eqs. (2.100), (2.103), and (2.107) into the source-free frequency-domain Maxwell curl postulates

$$\left. \begin{aligned} \nabla \times \underline{E}(\underline{r}) &= i\omega\mu_0 \underline{H}(\underline{r}) \\ \nabla \times \underline{H}(\underline{r}) &= -i\omega\epsilon(x, z)\underline{E}(\underline{r}) \end{aligned} \right\} \tag{2.108}$$

results in a system of four ordinary differential equations

$$
\left.
\begin{aligned}
&\frac{d}{dz}e_x^{(n)}(z) - i\kappa^{(n)}e_z^{(n)} = i\omega\mu_0 h_y^{(n)}(z) \\
&\frac{d}{dz}e_y^{(n)}(z) = -i\omega\mu_0 h_x^{(n)}(z) \\
&\frac{d}{dz}h_x^{(n)}(z) - i\kappa^{(n)}h_z^{(n)} = -i\omega \sum_{m=-\infty}^{\infty} \epsilon^{(n-m)}(z)e_y^{(m)}(z) \\
&\frac{d}{dz}h_y^{(n)}(z) = i\omega \sum_{m=-\infty}^{\infty} \epsilon^{(n-m)}(z)e_x^{(m)}(z)
\end{aligned}
\right\}, \quad n \in (-\infty, \infty),
$$

(2.109)

and two algebraic equations

$$
\left.
\begin{aligned}
&\kappa^{(n)}e_y^{(n)} = \omega\mu_0 h_z^{(n)}(z) \\
&\kappa^{(n)}h_y^{(n)}(z) = -\omega \sum_{m=-\infty}^{\infty} \epsilon^{(n-m)}(z)e_z^{(m)}(z)
\end{aligned}
\right\}, \quad n \in (-\infty, \infty),
$$

(2.110)

for all $z \in (0, d_1 + d_2)$.

The sums in Eqs. (2.109) and (2.110) contain an infinite number of terms. Moreover, the equations themselves are infinite in number. This infinite set of equations cannot be implemented in practice. As with any Fourier series, an approximation is obtained by truncating the series. Therefore, we restrict the index $n \in [-N_t, N_t]$ in Eq. (2.107), and define $(2N_t + 1)$-column vectors[3]

$$
\left[\breve{\underline{e}}_\sigma(z)\right] = \left[e_\sigma^{(-N_t)}(z), e_\sigma^{(-N_t+1)}(z), \ldots, e_\sigma^{(N_t-1)}(z), e_\sigma^{(N_t)}(z)\right]^T,
$$

(2.111)

and

$$
\left[\breve{\underline{h}}_\sigma(z)\right] = \left[h_\sigma^{(-N_t)}(z), h_\sigma^{(-N_t+1)}(z), \ldots, h_\sigma^{(N_t-1)}(z), h_\sigma^{(N_t)}(z)\right]^T,
$$

(2.112)

where $\sigma \in \{x, y, z\}$. Furthermore, after the introduction of $(2N_t + 1) \times (2N_t + 1)$ matrixes

$$
\left[\breve{\underline{\underline{K}}}\right] = \mathrm{diag}\left[\kappa^{(-N_t)}, \kappa^{(-N_t+1)}, \ldots, \kappa^{(N_t-1)}, \kappa^{(N_t)}\right],
$$

(2.113)

and

$$
\left[\breve{\underline{\underline{\epsilon}}}(z)\right] =
\begin{bmatrix}
\epsilon^{(0)}(z) & \epsilon^{(-1)}(z) & \epsilon^{(-2)}(z) & \cdots & \epsilon^{(-2N_t+1)}(z) & \epsilon^{(-2N_t)}(z) \\
\epsilon^{(1)}(z) & \epsilon^{(0)}(z) & \epsilon^{(-1)}(z) & \cdots & \epsilon^{(-2N_t+2)}(z) & \epsilon^{(-2N_t+1)}(z) \\
\cdots & \cdots & \cdots & \cdots & \cdots & \cdots \\
\epsilon^{(2N_t-1)}(z) & \epsilon^{(2N_t-2)}(z) & \epsilon^{(2N_t-3)}(z) & \cdots & \epsilon^{(0)}(z) & \epsilon^{(-1)}(z) \\
\epsilon^{(2N_t)}(z) & \epsilon^{(2N_t-1)}(z) & \epsilon^{(2N_t-2)}(z) & \cdots & \epsilon^{(1)}(z) & \epsilon^{(0)}(z)
\end{bmatrix}.
$$

(2.114)

---

[3] The symbol ˘ identifies quantities associated with the RCWA.

Equations (2.110) yield

$$\left[\check{\underline{h}}_z(z)\right] = (\omega\mu_0)^{-1}\left[\underline{\underline{\check{K}}}\right] \cdot \left[\check{\underline{e}}_y(z)\right] \tag{2.115}$$

and

$$\left[\check{\underline{e}}_z(z)\right] = -\omega^{-1}[\underline{\underline{\check{\varepsilon}}}(z)]^{-1} \cdot \left[\underline{\underline{\check{K}}}\right] \cdot \left[\check{\underline{h}}_y(z)\right]. \tag{2.116}$$

With Eqs. (2.115) and (2.116), $e_z^{(n)}(z)$ and $h_z^{(n)}(z)$ can be eliminated from Eqs. (2.109) $\forall\, n \in (-\infty, \infty)$, thereby obtaining the matrix ordinary differential equation

$$\frac{d}{dz}\left[\check{\underline{f}}(z)\right] = i\left[\underline{\underline{\check{P}}}(z)\right] \cdot \left[\check{\underline{f}}(z)\right], \quad z \in (0, d_1 + d_2), \tag{2.117}$$

where the $4(2N_t + 1)$-column vector

$$\left[\check{\underline{f}}(z)\right] = \begin{bmatrix} \left[\check{\underline{e}}_x(z)\right] \\ \left[\check{\underline{e}}_y(z)\right] \\ \left[\check{\underline{h}}_x(z)\right] \\ \left[\check{\underline{h}}_y(z)\right] \end{bmatrix} \tag{2.118}$$

and the $4(2N_t + 1) \times 4(2N_t + 1)$ matrix

$$\left[\underline{\underline{\check{P}}}(z)\right] = \omega \begin{bmatrix} \left[\underline{\underline{\check{0}}}\right] & \left[\underline{\underline{\check{0}}}\right] & \left[\underline{\underline{\check{0}}}\right] & \mu_0\left[\underline{\underline{\check{I}}}\right] \\ \left[\underline{\underline{\check{0}}}\right] & \left[\underline{\underline{\check{0}}}\right] & -\mu_0\left[\underline{\underline{\check{I}}}\right] & \left[\underline{\underline{\check{0}}}\right] \\ \left[\underline{\underline{\check{0}}}\right] & -\left[\underline{\underline{\check{\varepsilon}}}(z)\right] & \left[\underline{\underline{\check{0}}}\right] & \left[\underline{\underline{\check{0}}}\right] \\ \left[\underline{\underline{\check{\varepsilon}}}(z)\right] & \left[\underline{\underline{\check{0}}}\right] & \left[\underline{\underline{\check{0}}}\right] & \left[\underline{\underline{\check{0}}}\right] \end{bmatrix}$$

$$+ \frac{1}{\omega} \begin{bmatrix} \left[\underline{\underline{\check{0}}}\right] & \left[\underline{\underline{\check{0}}}\right] & \left[\underline{\underline{\check{0}}}\right] & -\left[\underline{\underline{\check{K}}}\right] \cdot \left[\underline{\underline{\check{\varepsilon}}}(z)\right]^{-1} \cdot \left[\underline{\underline{\check{K}}}\right] \\ \left[\underline{\underline{\check{0}}}\right] & \left[\underline{\underline{\check{0}}}\right] & \left[\underline{\underline{\check{0}}}\right] & \left[\underline{\underline{\check{0}}}\right] \\ \left[\underline{\underline{\check{0}}}\right] & \mu_0^{-1}\left[\underline{\underline{\check{K}}}\right] \cdot \left[\underline{\underline{\check{K}}}\right] & \left[\underline{\underline{\check{0}}}\right] & \left[\underline{\underline{\check{0}}}\right] \\ \left[\underline{\underline{\check{0}}}\right] & \left[\underline{\underline{\check{0}}}\right] & \left[\underline{\underline{\check{0}}}\right] & \left[\underline{\underline{\check{0}}}\right] \end{bmatrix}. \tag{2.119}$$

All submatrixes in the definition of $\left[\underline{\underline{\check{P}}}(z)\right]$ are $(2N_t + 1) \times (2N_t + 1)$ in dimension, with $\left[\underline{\underline{\check{0}}}\right]$ as the null matrix and $\left[\underline{\underline{\check{I}}}\right]$ the identity matrix.

The boundary values $\left[\check{\underline{f}}(0-)\right]$ and $\left[\check{\underline{f}}(d_1 + d_2 +)\right]$ of $\left[\check{\underline{f}}(z)\right]$, which are needed to implement the boundary conditions in solving Eq. (2.117), can be assembled using Eqs. (2.90), (2.94), and (2.96). If the phasor amplitudes for the s-polarized and p-polarized components of the incident, reflected, and transmitted fields are collected into $2(2N_t + 1)$-column vectors as

$$\left[\breve{\underline{A}}\right] = \left[a_s^{(-N_t)}, a_s^{(-N_t+1)}, \ldots, a_s^{(N_t-1)}, a_s^{(N_t)}, a_p^{(-N_t)}, a_p^{(-N_t+1)}, \ldots, \right.$$
$$\left. a_p^{(N_t-1)}, a_p^{(N_t)}\right]^T,$$
(2.120)

$$\left[\breve{\underline{R}}\right] = \left[r_s^{(-N_t)}, r_s^{(-N_t+1)}, \ldots, r_s^{(N_t-1)}, r_s^{(N_t)}, r_p^{(-N_t)}, r_p^{(-N_t+1)}, \ldots, \right.$$
$$\left. r_p^{(N_t-1)}, r_p^{(N_t)}\right]^T,$$
(2.121)

and

$$\left[\breve{\underline{T}}\right] = \left[t_s^{(-N_t)}, t_s^{(-N_t+1)}, \ldots, t_s^{(N_t-1)}, t_s^{(N_t)}, t_p^{(-N_t)}, t_p^{(-N_t+1)}, \right.$$
$$\left. \ldots, t_p^{(N_t-1)}, t_p^{(N_t)}\right]^T,$$
(2.122)

then,

$$\left[\breve{\underline{f}}(0-)\right] = \begin{bmatrix} [\breve{\underline{0}}] & -k_0^{-1}[\breve{\underline{\alpha}}] & [\breve{\underline{0}}] & k_0^{-1}[\breve{\underline{\alpha}}] \\ [\breve{\underline{I}}] & [\breve{\underline{0}}] & [\breve{\underline{I}}] & [\breve{\underline{0}}] \\ -\eta_0^{-1}k_0^{-1}[\breve{\underline{\alpha}}] & [\breve{\underline{0}}] & \eta_0^{-1}k_0^{-1}[\breve{\underline{\alpha}}] & [\breve{\underline{0}}] \\ [\breve{\underline{0}}] & -\eta_0^{-1}[\breve{\underline{I}}] & [\breve{\underline{0}}] & -\eta_0^{-1}[\breve{\underline{I}}] \end{bmatrix} \cdot \begin{bmatrix} [\breve{\underline{A}}] \\ [\breve{\underline{R}}] \end{bmatrix},$$
(2.123)

and

$$\left[\breve{\underline{f}}(d_1 + d_2 +)\right] = \begin{bmatrix} [\breve{\underline{0}}] & -k_0^{-1}[\breve{\underline{\alpha}}] \\ [\breve{\underline{I}}] & [\breve{\underline{0}}] \\ -\eta_0^{-1}k_0^{-1}[\breve{\underline{\alpha}}] & [\breve{\underline{0}}] \\ [\breve{\underline{0}}] & -\eta_0^{-1}[\breve{\underline{I}}] \end{bmatrix} \cdot [\breve{\underline{T}}],$$
(2.124)

where the $(2N_t + 1) \times (2N_t + 1)$ matrix

$$\left[\breve{\underline{\alpha}}\right] = \text{diag}\left[\alpha^{(-N_t)}, \alpha^{(-N_t+1)}, \ldots, \alpha^{(N_t-1)}, \alpha^{(N_t)}\right].$$
(2.125)

For convenience, the boundary values $\left[\breve{\underline{f}}(0-)\right]$ and $\left[\breve{\underline{f}}(d_1 + d_2 +)\right]$ are recast compactly as

$$\left[\breve{\underline{f}}(0-)\right] = \begin{bmatrix} [\breve{\underline{Y}}_e^{\text{inc}}] & [\breve{\underline{Y}}_e^{\text{ref}}] \\ [\breve{\underline{Y}}_h^{\text{inc}}] & [\breve{\underline{Y}}_h^{\text{ref}}] \end{bmatrix} \cdot \begin{bmatrix} [\breve{\underline{A}}] \\ [\breve{\underline{R}}] \end{bmatrix},$$
(2.126)

and

$$\left[\breve{\underline{f}}(d_1 + d_2 +)\right] = \begin{bmatrix} [\breve{\underline{Y}}_e^{\text{tr}}] \\ [\breve{\underline{Y}}_h^{\text{tr}}] \end{bmatrix} \cdot [\breve{\underline{T}}].$$
(2.127)

The $2(2N_t + 1) \times 2(2N_t + 1)$ matrixes appearing in Eq. (2.126) can be synthesized from Eq. (2.123) by inspection; thus,

$$\left[\underline{\underline{\check{Y}}}_e^{\text{inc}}\right] = \begin{bmatrix} [\underline{\underline{\check{0}}}] & -k_0^{-1}[\underline{\underline{\check{\alpha}}}] \\ [\underline{\underline{I}}] & [\underline{\underline{\check{0}}}] \end{bmatrix}, \quad \left[\underline{\underline{\check{Y}}}_h^{\text{inc}}\right] = -\eta_0^{-1}\begin{bmatrix} k_0^{-1}[\underline{\underline{\check{\alpha}}}] & [\underline{\underline{\check{0}}}] \\ [\underline{\underline{\check{0}}}] & [\underline{\underline{I}}] \end{bmatrix} \qquad (2.128)$$

and

$$\left[\underline{\underline{\check{Y}}}_e^{\text{ref}}\right] = \begin{bmatrix} [\underline{\underline{\check{0}}}] & k_0^{-1}[\underline{\underline{\check{\alpha}}}] \\ [\underline{\underline{I}}] & [\underline{\underline{\check{0}}}] \end{bmatrix}, \quad \left[\underline{\underline{\check{Y}}}_h^{\text{ref}}\right] = \eta_0^{-1}\begin{bmatrix} k_0^{-1}[\underline{\underline{\check{\alpha}}}] & [\underline{\underline{\check{0}}}] \\ [\underline{\underline{\check{0}}}] & -[\underline{\underline{I}}] \end{bmatrix}. \qquad (2.129)$$

Since the half-space $z > d_1 + d_2$ is occupied by the same medium—i.e. air—as the half-space $z < 0$ in the present case,

$$\left[\underline{\underline{\check{Y}}}_e^{\text{tr}}\right] = \left[\underline{\underline{\check{Y}}}_e^{\text{inc}}\right], \quad \left[\underline{\underline{\check{Y}}}_h^{\text{tr}}\right] = \left[\underline{\underline{\check{Y}}}_h^{\text{inc}}\right]. \qquad (2.130)$$

As a solution of Eq. (2.117) in closed form is infeasible, we proceed from here with an approximate numerical method. The region $0 \leqslant z \leqslant d_1 + g_{\min}$ is divided into N_{diel} subregions, the region $d_1 + g_{\min} \leqslant z \leqslant d_1 + g_{\max}$ into N_g subregions, and the region $d_1 + g_{\max} \leqslant z \leqslant d_2$ into N_{met} subregions. Thus, the total number of subregions is $N_s = N_{\text{diel}} + N_g + N_{\text{met}}$, the ℓth subregion being bounded by the planes $z = z_{\ell-1}$ and $z = z_\ell, \ell \in [1, N_s]$, where $z_0 = 0$ and $z_{N_s} = d_1 + d_2$. Often, but not always, $N_{\text{diel}} = 1$ and/or $N_{\text{met}} = 1$ suffice. An electrically thick region requires a larger number of subregions for stable computations of the exponentials of matrices with complex-valued elements.

In the ℓth subregion, the approximation

$$\left[\underline{\underline{\check{P}}}(z)\right] \approx \left[\underline{\underline{\check{P}}}\right]^{(\ell)} \equiv \left[\underline{\underline{\check{P}}}\left(\frac{z_\ell + z_{\ell-1}}{2}\right)\right], \quad z \in (z_\ell, z_{\ell-1}) \qquad (2.131)$$

is made. This approximation is called the piecewise-uniform approximation, and it permits the solution

$$\left[\underline{\check{f}}(z_\ell)\right] = \left[\underline{\underline{\check{W}}}\right]^{(\ell)} \cdot \left[\underline{\check{f}}(z_{\ell-1})\right], \quad \ell \in [1, N_s], \qquad (2.132)$$

where

$$\left[\underline{\underline{\check{W}}}\right]^{(\ell)} = \exp\left\{i\left(z_\ell - z_{\ell-1}\right)\left[\underline{\underline{\check{P}}}\right]^{(\ell)}\right\}, \quad \ell \in [1, N_s]. \qquad (2.133)$$

Repeated use of Eq. (2.132) yields

$$\left[\underline{\check{f}}(d_1 + d_2 -)\right] = \left[\underline{\underline{\check{W}}}\right]^{(N_s)} \cdot \left[\underline{\underline{\check{W}}}\right]^{(N_s-1)} \cdot \cdots \cdot \left[\underline{\underline{\check{W}}}\right]^{(2)} \cdot \left[\underline{\underline{\check{W}}}\right]^{(1)} \cdot \left[\underline{\check{f}}(0 +)\right]. \qquad (2.134)$$

As the boundary conditions

$$\left[\underline{\check{f}}(0 -)\right] = \left[\underline{\check{f}}(0 +)\right], \quad \left[\underline{\check{f}}(d_1 + d_2 -)\right] = \left[\underline{\check{f}}(d_1 + d_2 +)\right], \qquad (2.135)$$

must be satisfied, Eq. (2.134) can be rewritten as

$$\left[\check{\underline{f}}(d_1 + d_2 +)\right] = \left[\check{\underline{\underline{W}}}\right]^{(N_s)} \cdot \left[\check{\underline{\underline{W}}}\right]^{(N_s-1)} \cdot \cdots \cdot \left[\check{\underline{\underline{W}}}\right]^{(2)} \cdot \left[\check{\underline{\underline{W}}}\right]^{(1)} \cdot \left[\check{\underline{f}}(0 -)\right].$$

(2.136)

Equations (2.126), (2.127), and (2.136) lead to the algebraic equation

$$\begin{bmatrix} \left[\check{\underline{\underline{Y}}}_e^{\mathrm{tr}}\right] \\ \left[\check{\underline{\underline{Y}}}_h^{\mathrm{tr}}\right] \end{bmatrix} \cdot \left[\check{\underline{T}}\right] = \left[\check{\underline{\underline{W}}}\right]^{(N_s)} \cdot \left[\check{\underline{\underline{W}}}\right]^{(N_s-1)} \cdot \cdots$$

$$\cdot \left[\check{\underline{\underline{W}}}\right]^{(2)} \cdot \left[\check{\underline{\underline{W}}}\right]^{(1)} \cdot \begin{bmatrix} \left[\check{\underline{\underline{Y}}}_e^{\mathrm{inc}}\right] & \left[\check{\underline{\underline{Y}}}_e^{\mathrm{ref}}\right] \\ \left[\check{\underline{\underline{Y}}}_h^{\mathrm{inc}}\right] & \left[\check{\underline{\underline{Y}}}_h^{\mathrm{ref}}\right] \end{bmatrix} \cdot \begin{bmatrix} \left[\check{\underline{A}}\right] \\ \left[\check{\underline{R}}\right] \end{bmatrix},$$

(2.137)

which may be solved for $\left[\check{\underline{R}}\right]$ and $\left[\check{\underline{T}}\right]$ using standard matrix techniques [261].

2.3.4.5   Stable RCWA Algorithm

Numerical problems bedevil straightforward attempts to solve Eq. (2.137) on a digital computer, particularly when either the metal is very thick and/or the incidence is highly oblique. A stable algorithm has been devised to overcome these problems [251,252, 256–258], based on the assumption that $\left[\check{\underline{\underline{P}}}\right]^{(\ell)}$ is diagonalizable [262] for all $\ell \in [1, N_s]$. The matrix $\left[\check{\underline{\underline{P}}}\right]^{(\ell)}$ is indeed diagonalizable if the ℓth subregion lies either in the region $0 \leqslant z \leqslant d_1 + g_{\min}$ or in the region $d_1 + g_{\max} \leqslant z \leqslant d_2$. Let us also assume that it is also diagonalizable if the ℓth subregion lies in the region $d_1 + g_{\min} \leqslant z \leqslant d_1 + g_{\max}$. Then,

$$\left[\check{\underline{\underline{P}}}\right]^{(\ell)} = \left[\check{\underline{\underline{V}}}\right]^{(\ell)} \cdot \left[\check{\underline{\underline{G}}}\right]^{(\ell)} \cdot \left(\left[\check{\underline{\underline{V}}}\right]^{(\ell)}\right)^{-1}, \quad \ell \in [1, N_s],$$

(2.138)

where the diagonal matrix $\left[\check{\underline{\underline{G}}}\right]^{(\ell)}$ contains the eigenvalues of $\left[\check{\underline{\underline{P}}}\right]^{(\ell)}$ either in decreasing order or in increasing order of the magnitude of the imaginary part, and $\left[\check{\underline{\underline{V}}}\right]^{(\ell)}$ is a square matrix comprising the eigenvectors of $\left[\check{\underline{\underline{P}}}\right]^{(\ell)}$ as its columns, arranged so that each eigenvector is in the same position as the corresponding eigenvalue in $\left[\check{\underline{\underline{G}}}\right]^{(\ell)}$. Accordingly, Eq. (2.133) is recast as

$$\left[\check{\underline{\underline{W}}}\right]^{(\ell)} = \left[\check{\underline{\underline{V}}}\right]^{(\ell)} \cdot \exp\left\{i(z_\ell - z_{\ell-1})\left[\check{\underline{\underline{G}}}\right]^{(\ell)}\right\} \cdot \left(\left[\check{\underline{\underline{V}}}\right]^{(\ell)}\right)^{-1}.$$

(2.139)

We now define the column vector $\left[\check{\underline{T}}\right]^{(\ell)}$ and the auxiliary transmission matrix $\left[\check{\underline{\underline{Z}}}\right]^{(\ell)}$ for the ℓth subregion through the relationship

$$\left[\check{\underline{f}}(z_\ell)\right] = \left[\check{\underline{\underline{Z}}}\right]^{(\ell)} \cdot \left[\check{\underline{T}}\right]^{(\ell)}, \quad \ell \in [0, N_s].$$

(2.140)

Although both $\left[\breve{\underline{\underline{T}}}\right]^{(\ell)}$ and $\left[\breve{\underline{\underline{Z}}}\right]^{(\ell)}$ are not known in general at this stage, we choose to set

$$
\left[\breve{\underline{\underline{T}}}\right]^{(N_s)} = \left[\underline{\underline{T}}\right], \quad \left[\breve{\underline{\underline{Z}}}\right]^{(N_s)} = \left[\begin{array}{c} \left[\breve{\underline{\underline{Y}}}_e^{\mathrm{tr}}\right] \\ \left[\breve{\underline{\underline{Y}}}_h^{\mathrm{tr}}\right] \end{array} \right],
\tag{2.141}
$$

consistently with Eq. (2.127). Use of Eqs. (2.139) and (2.140) in Eq. (2.132) yields

$$
\begin{aligned}
\left[\breve{\underline{\underline{Z}}}\right]^{(\ell-1)} \cdot \left[\breve{\underline{\underline{T}}}\right]^{(\ell-1)} &= \left[\underline{\underline{\breve{V}}}\right]^{(\ell)} \\
&\cdot \left[\begin{array}{cc} \exp\left\{-i(z_\ell - z_{\ell-1}) \left[\breve{\underline{\underline{G}}}_{\mathrm{upper}}\right]^{(\ell)}\right\} & \left[\breve{\underline{\underline{o}}}\right] \\ \left[\breve{\underline{\underline{o}}}\right] & \exp\left\{-i(z_\ell - z_{\ell-1}) \left[\breve{\underline{\underline{G}}}_{\mathrm{lower}}\right]^{(\ell)}\right\} \end{array} \right] \\
&\cdot \left(\left[\underline{\underline{\breve{V}}}\right]^{(\ell)}\right)^{-1} \cdot \left[\breve{\underline{\underline{Z}}}\right]^{(\ell)} \cdot \left[\breve{\underline{\underline{T}}}\right]^{(\ell)}, \quad \ell \in [1, N_s],
\end{aligned}
\tag{2.142}
$$

wherein the $4(2N_t + 1) \times 4(2N_t + 1)$ matrix $\left[\breve{\underline{\underline{G}}}\right]^{(\ell)}$ has been partitioned into the following four $2(2N_t + 1) \times 2(2N_t + 1)$ matrixes: two null matrixes $\left[\breve{\underline{\underline{o}}}\right]$, an upper diagonal matrix $\left[\breve{\underline{\underline{G}}}_{\mathrm{upper}}\right]^{(\ell)}$, and a lower diagonal matrix $\left[\breve{\underline{\underline{G}}}_{\mathrm{lower}}\right]^{(\ell)}$.

With the extra degree of freedom afforded by the two sets of created unknowns, $\left[\breve{\underline{\underline{Z}}}\right]^{(\ell)}$ and $\left[\breve{\underline{\underline{T}}}\right]^{(\ell)}$, we impose the relationship

$$
\left[\breve{\underline{\underline{T}}}\right]^{(\ell-1)} = \exp\left\{-i(z_\ell - z_{\ell-1}) \left[\breve{\underline{\underline{G}}}_{\mathrm{upper}}\right]^{(\ell)}\right\} \cdot \left[\breve{\underline{\underline{X}}}_{\mathrm{upper}}\right]^{(\ell)} \cdot \left[\breve{\underline{\underline{T}}}\right]^{(\ell)},
\tag{2.143}
$$

where the square matrix $\left[\breve{\underline{\underline{X}}}_{\mathrm{upper}}\right]^{(\ell)}$ and its partner $\left[\breve{\underline{\underline{X}}}_{\mathrm{lower}}\right]^{(\ell)}$ are defined through

$$
\left[\begin{array}{c} \left[\breve{\underline{\underline{X}}}_{\mathrm{upper}}\right]^{(\ell)} \\ \left[\breve{\underline{\underline{X}}}_{\mathrm{lower}}\right]^{(\ell)} \end{array} \right] = \left(\left[\underline{\underline{\breve{V}}}\right]^{(\ell)}\right)^{-1} \cdot \left[\breve{\underline{\underline{Z}}}\right]^{(\ell)}.
\tag{2.144}
$$

Now, substitution of Eq. (2.143) in (2.142) leads to the relation

$$
\left[\breve{\underline{\underline{Z}}}\right]^{(\ell-1)} = \left[\underline{\underline{\breve{V}}}\right]^{(\ell)} \cdot \left[\begin{array}{c} \left[\underline{\underline{\breve{1}}}\right] \\ \left[\breve{\underline{\underline{U}}}\right]^{(\ell)} \end{array} \right], \quad \ell \in [1, N_s],
\tag{2.145}
$$

where $\left[\underline{\underline{\breve{1}}}\right]$ is the $2(2N_t + 1) \times 2(2N_t + 1)$ identity matrix and

$$
\begin{aligned}
\left[\breve{\underline{\underline{U}}}\right]^{(\ell)} &= \exp\left\{-i(z_\ell - z_{\ell-1}) \left[\breve{\underline{\underline{G}}}_{\mathrm{lower}}\right]^{(\ell)}\right\} \cdot \left[\breve{\underline{\underline{X}}}_{\mathrm{lower}}\right]^{(\ell)} \cdot \left(\left[\breve{\underline{\underline{X}}}_{\mathrm{upper}}\right]^{(\ell)}\right)^{-1} \\
&\cdot \exp\left\{i(z_\ell - z_{\ell-1}) \left[\breve{\underline{\underline{G}}}_{\mathrm{upper}}\right]^{(\ell)}\right\}, \quad \ell \in [1, N_s].
\end{aligned}
\tag{2.146}
$$

Iteration of Eq. (2.145), with the use of Eq. (2.144), yields $\left[\breve{\underline{\underline{Z}}}\right]^{(0)}$, which, after partitioning, may be written as

$$\left[\breve{\underline{\underline{Z}}}\right]^{(0)} = \left[\begin{array}{c} \left[\breve{\underline{\underline{Z}}}_{\text{upper}}\right]^{(0)} \\ \left[\breve{\underline{\underline{Z}}}_{\text{lower}}\right]^{(0)} \end{array}\right]. \tag{2.147}$$

Setting the right side of Eq. (2.140) with $\ell = 0$ equal to the right side of Eq. (2.126) gives

$$\left[\begin{array}{c} \left[\breve{\underline{\underline{Z}}}_{\text{upper}}\right]^{(0)} \\ \left[\breve{\underline{\underline{Z}}}_{\text{lower}}\right]^{(0)} \end{array}\right] \cdot \left[\breve{\underline{T}}\right]^{(0)} = \left[\begin{array}{cc} \left[\breve{\underline{\underline{Y}}}_e^{\text{inc}}\right] & \left[\breve{\underline{\underline{Y}}}_e^{\text{ref}}\right] \\ \left[\breve{\underline{\underline{Y}}}_h^{\text{inc}}\right] & \left[\breve{\underline{\underline{Y}}}_h^{\text{ref}}\right] \end{array}\right] \cdot \left[\begin{array}{c} \left[\underline{A}\right] \\ \left[\underline{R}\right] \end{array}\right], \tag{2.148}$$

which may be rearranged to yield

$$\left[\begin{array}{c} \left[\breve{\underline{T}}\right]^{(0)} \\ \left[\underline{R}\right] \end{array}\right] = \left[\begin{array}{c} \left[\breve{\underline{\underline{Z}}}_{\text{upper}}\right]^{(0)} - \left[\breve{\underline{\underline{Y}}}_e^{\text{ref}}\right] \\ \left[\breve{\underline{\underline{Z}}}_{\text{lower}}\right]^{(0)} - \left[\breve{\underline{\underline{Y}}}_h^{\text{ref}}\right] \end{array}\right] \cdot \left[\begin{array}{c} \left[\breve{\underline{\underline{Y}}}_e^{\text{inc}}\right] \\ \left[\breve{\underline{\underline{Y}}}_h^{\text{inc}}\right] \end{array}\right] \cdot \left[\underline{A}\right]. \tag{2.149}$$

Equation (2.149) delivers $\left[\breve{\underline{R}}\right]$. After $\left[\breve{\underline{T}}\right]^{(0)}$ is known, $\left[\breve{\underline{T}}\right] = \left[\breve{\underline{T}}\right]^{(N_s)}$ is found by reversing the sense of iterations in Eq. (2.143).

Moreover, with $\left[\breve{\underline{f}}(z_0)\right] = \left[\breve{\underline{f}}(0-)\right]$ now known by virtue of Eq. (2.126), $\left[\breve{\underline{f}}\left(z_\ell\right)\right]$ can be calculated for all $\ell \in [1, N_s]$ by repeated use of Eq. (2.132). With $\left[\breve{\underline{f}}\left(z_\ell\right)\right]$ known, $\left[\breve{\underline{h}}_z\left(z_\ell\right)\right]$ can be obtained from Eq. (2.115) and $\left[\breve{\underline{e}}_z\left(z_\ell\right)\right]$ likewise from Eq. (2.116). After the further use of Eqs. (2.107), $\underline{E}(\underline{r})$ and $\underline{H}(\underline{r})$ can be mapped in the region $0 \leqslant z \leqslant d_1 + d_2$.

The sufficiency of N_t requires convergence tests. As N_t increases, one must test that the reflectances $R_{\text{sp}}^{(n)}$, etc., and the transmittances $T_{\text{sp}}^{(n)}$, etc., converge within satisfactory tolerance limits (say, 0.1%). The principle of conservation of energy also must be satisfied simultaneously.

2.3.4.6 Excitation of an SPP Wave

In order to investigate the excitation of an SPP wave in the grating-coupled configuration at a specified value of the free-space wavelength, the linear absorptances A_p and A_s must be plotted as functions of the angle of incidence θ_{inc}. Such plots will contain many peaks. Each peak has to survive two tests before it can be considered indicative of the excitation of an SPP wave.

First, the angular location of a certain absorptance peak must remain about the same regardless of the average thickness d_1 of the partnering dielectric material above a threshold, when the average thickness d_2 of the partnering metal is similar in magnitude to an adequate value of d_{met} in the Turbadar-Kretschmann-Raether configuration. Then, if at that angular location, a value of $n \in \{0, \pm 1, \pm 2, \ldots\}$ can be found such that

$$\kappa^{(n)} \approx \pm \text{Re}\,\{q\}, \tag{2.150}$$

where q^2 is given by Eq. (2.35), an SPP wave is excited as a Floquet harmonic of order n. Furthermore, if $\kappa^{(n)} \approx \text{Re}\{q\}$, the SPP wave propagates along the $+x$ axis; if $\kappa^{(n)} \approx -\text{Re}\{q\}$, the SPP wave propagates along the $-x$ axis. Such an SPP wave will always be p polarized, as no A_s-peaks will pass the twin tests—the partnering dielectric material being homogeneous. At least one value of n will enable the satisfaction of Eq. (2.150). But, that equation can possibly be satisfied by more than one value of n; thus, two or more A_p-peaks can indicate the excitation of the same SPP wave.

2.3.4.7   Illustrative Results

The left panel in Figure 2.14 shows the computed values of A_p plotted against the angle of incidence θ_{inc} of an air/dielectric/aluminum/air system in the grating-coupled configuration. The dielectric material has a refractive index the same as water and the metal is aluminum in the thin-film form, in order to facilitate comparison with the numerical results for other configurations in this chapter. The dielectric material's average thickness $d_1 \in \{1015, 1215\}$ nm, whereas the thin-film aluminum's average thickness $d_2 = 45$ nm. The free-space wavelength $\lambda_0 = 633$ nm, and the grating function is sinusoidal with a period $L_x = 633$ nm and trough-to-crest height of 30 nm.

As noted in Section 2.3.1.4, the solution of the canonical boundary-value problem yields $q/k_0 = 1.4041 + 0.0326i$ as the normalized SPP wavenumber for the chosen partnering materials. The left panel in Figure 2.14 contains an A_p-peak at $\theta_{\text{inc}} = 25.9°$ for both values of d_1. When $\theta_{\text{inc}} = 25.9°$, Eq. (2.93)$_1$ yields $\kappa^{(1)}/k_0 = 1.4368$, which is close enough to $\text{Re}\{q\}/k_0 = 1.4041$ to let us conclude that a p-polarized SPP wave is excited as a Floquet harmonic of order $n = 1$, when the incident plane wave is p polarized.

If the grating were to be removed by setting $g(x) \equiv 0 \ \forall x$, the plots of A_p against θ_{inc} would not contain any peak that would survive the twin tests mentioned in Section

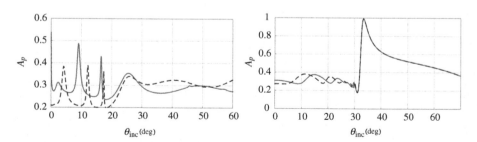

Figure 2.14 Calculated linear absorptance A_p as a function of the angle of incidence θ_{inc} at $\lambda_0 = 633$ nm. Left: air/dielectric/aluminum/air system in the grating-coupled configuration, for $d_1 = 1015$ nm (red solid line) or 1215 nm (blue dashed line), and $d_2 = 45$ nm. The grating function $g(x) = L_g \sin(2\pi x/L_x)$ with $L_g = 30$ nm and $L_x = \lambda_0$. Right: same as left, except that $d_1 = 1000$ nm (red solid line) or 1200 nm (blue dashed line), $d_2 = 30$ nm, $g(x) \equiv 0 \ \forall x$, and the air above and below the partnering materials has been replaced by rutile. (For interpretation of the references to color in this figure legend, the reader is referred to the web version of this book.)

2.3.4.6, and no SPP wave would be excited. In that case, the half-space $z < 0$ would have to be filled with a dielectric material of refractive index greater than that of the partnering dielectric material, in order to have the right conditions to excite an SPP wave. The resulting system would be in a modified Turbadar-Kretschmann-Raether configuration [228], wherein the thickness of the partnering dielectric material is large but finite.

For the right panel of Figure 2.14, the metal/dielectric interface is planar and both half-spaces $z < 0$ and $z > d_1 + d_2$ are occupied by rutile ($n_{\text{rutile}} = 2.6$). This panel contains an A_p-peak at $\theta_{\text{inc}} \simeq 33.4°$ for two large values of d_1 and a sufficiently large value of d_2, the metal/dielectric interface being flat. The angle of incidence predicted for SPR by the solution of the canonical problem is $\sin^{-1}\left[\text{Re}\left\{q\right\}/n_{\text{rutile}}k_0\right] = 32.685°$, which is very close to the angular location of the A_p-peak in the right panel of Figure 2.14.

2.3.5 Waveguide-Coupled Configuration

The confinement of SPP waves to subwavelength dimensions offers hope for miniaturized optical circuitry with much higher densities of components and devices than currently available with electronic and optical circuitry. Integration of SPP waveguides with current planar electronic and optical technology has attracted the attention of many researchers. Waveguide-coupled configurations allow for miniaturization, and many schemes are possible to implement waveguide coupling.

The end-fire configuration depicted in Figure 1.6a was devised [87] to couple photonic waveguides to plasmonic waveguides. In this configuration, a parallel plate or a rectangular waveguide feeds a metal/dielectric structure. Analytical treatment of the involved boundary-value problem is difficult unless drastic assumptions are made [263], and numerical methods such as the finite-difference time-domain (FDTD) method [264] must be used for analysis.

A popular approach is to place rectangular dielectric and plasmonic waveguides parallel to each other on an integrated circuit, as shown in Figure 1.6b. Often, the two waveguides are separated by a small gap. This type of coupling, known as directional coupling, can be implemented in either a horizontal [265–268] or vertical [269,270] format. Much of the effort has been directed at exciting LRSPP waves [269,266–268], with reported efficiencies as high as 99% [268].

2.4 Nonlinear Dielectric Materials

New characteristics and potential practical applications of SPP waves emerge when an optically nonlinear material [271] is chosen as the partnering dielectric material. Theoretical research began in the 1970s and this area became vibrant in the following decade [272–276]. Much of the current work is aimed more at issues related to practical applications of the phenomenons.

Although many investigations have been and are focused on nonlinear anisotropic materials, nonlinear isotropic materials have also been considered as partnering materials. Such materials, exemplified by carbon disulfide [277] and polydiacetylene [278], exhibit an intensity-dependent refractive index: the refractive index n_{ID} varies from a base value n_o proportionally to the square of the magnitude of the electric phasor [279], i.e.

$$n_{\text{ID}} = n_o + \delta_n |\underline{E}(\underline{r})|^2, \tag{2.151}$$

where δ_n is the proportionality constant. Such a material can be classified as positive nonlinear ($\delta_n > 0$) or negative nonlinear ($\delta_n < 0$).

Some properties of the SPP wave do not change upon introducing a partnering dielectric material with an intensity-dependent refractive index. A single interface of a metal and a nonlinear, isotropic, dielectric material can guide SPP waves that can only be p polarized [274], whether $\delta_n = 0$ or not. However—unlike the case for a metal/dielectric interface formed with a linear, isotropic dielectric partnering material—theoretical investigations [272] show that two branches of SPP waves may be observed when scanning in the Turbadar-Kretschmann-Raether configuration using angular interrogation when a nonlinear dielectric material is employed. SPR may occur at a higher angle of incidence with $\delta_n > 0$, when the angle of incidence is increased from low to high values rather than high to low. The reverse is true when $\delta_n < 0$; then scanning the angle of incidence from low to high values may result in a lower value of Re $\{q\}$ than scanning from high to low. The occurrence of this optical hysteresis [280] is dependent on initial conditions [272]. Bistability and switching have also been extensively studied [281,282].

In the Sarid configuration, the use of nonlinear dielectric materials [273,274] results in two interesting phenomenons. First, with positive nonlinear dielectric materials on both sides of a thin metal film, s-polarized SPP waves are theoretically possible [274]. However, achievement of the required change in the refractive index of the nonlinear material induced by light intensity is challenging from a practical perspective. Perhaps ongoing or future developments with nanoengineered materials may change this assessment. Second, when both dielectric materials bounding the metal film are nonlinear, additional propagation modes appear above minimum power thresholds.

3 General Theory of Surface-Wave Propagation

3.1 Introduction

Chapter 2 was chiefly devoted to the propagation of the surface-plasmon-polariton wave guided by the planar interface of a metal and a homogeneous, isotropic, and dielectric material. The important features of this SPP wave emerge from solving a canonical boundary-value problem, although practical implementation requires the use of a prism-coupled configuration, a grating-coupled configuration, or a waveguide-coupled configuration.

If the partnering dielectric material were to be anisotropic and periodically non-homogeneous normal to the metal/dielectric interface, multiple SPP-wave modes may be guided by that interface along a specified direction, all of the same frequency. Isotropy of the partnering dielectric material eliminates the dependence on the direction of propagation but not the multiplicity of SPP-wave modes. If both partnering materials were to be homogeneous dielectric materials, their planar interface may guide Dyakonov waves, provided that at least one of the two materials is anisotropic. Periodic nonhomogeneity normal to the interface in one or both partnering materials introduces the possibilities of Tamm waves and Dyakonov-Tamm waves. All of these surface waves are classified in Table 1.1.

Underlying all of these diverse surface-wave phenomenons is the same canonical boundary-value problem. The formulation of this problem admits not only metals and dielectric materials but also the more general bianisotropic materials.

3.2 Bianisotropic Materials

An isotropic medium has electromagnetic properties which are the same in all directions. However, the notion of isotropy, as encountered in elementary treatments of electromagnetics, is an abstraction which requires qualification when applied to real materials. For examples, liquids and random composite materials may be isotropic on a statistical basis, while cubic crystals are isotropic when viewed at macroscopic length-scales. Electromagnetically isotropic materials are characterized simply by scalar constitutive parameters which relate the field phasors $\underline{D}$ and $\underline{B}$ to the field phasors $\underline{E}$ and $\underline{H}$.

Electromagnetic Surface Waves. http://dx.doi.org/10.1016/B978-0-12-397024-4.00003-7

Often, naturally occurring materials and artificially constructed materials are more accurately described as anisotropic rather than isotropic. Anisotropic mediums exhibit directionally dependent electromagnetic properties, such that $\underline{D}$ and $\underline{E}$ are not aligned or $\underline{B}$ and $\underline{H}$ are not aligned. Instead of scalar constitutive parameters, constitutive dyadics are needed to relate $\underline{D}$ to $\underline{E}$ and $\underline{B}$ to $\underline{H}$ in anisotropic materials.

Bianisotropy [17] is the natural generalization of anisotropy. In the electromagnetic description of a bianisotropic material, *both* $\underline{D}$ and $\underline{H}$ are anisotropically coupled to *both* $\underline{E}$ and $\underline{B}$. Hence, in general, a linear bianisotropic material is characterized by four constitutive dyadics. Though seldom described in standard textbooks, bianisotropy is commonplace. Suppose a certain material is characterized as isotropic dielectric by an observer in an inertial reference frame Σ. The same material generally exhibits bianisotropic properties when viewed by an observer in another reference frame that translates at uniform velocity with respect to Σ. Aside from relativistic scenarios, bianisotropic effects are observed at low frequencies and temperatures in a host of naturally occurring minerals [283,284]. Furthermore, the phenomenon of bianisotropy looks set to play an increasingly important role in the rapidly burgeoning fields relating to complex composite materials. In particular, bianisotropic materials readily emerge from the process of homogenization of a composite of two or more constituent materials [17].

Moreover, the concepts of anisotropy and bianisotropy emerge when an external agency affects the optical response characteristics of an isotropic dielectric material. For instance, doped semiconductors are isotropic dielectric materials, but become gyrotropic uniaxial dielectric materials under the action of a quasistatic magnetic field [285]. Strain induced either piezoelectrically [286] or by electrostriction [287] can be used to significantly alter the optical response characteristics of a material [288]. The electro-optic effect [271] can be applied to turn an isotropic dielectric material into an anisotropic one [289]. Such effects are important for modulating surface-wave propagation in practical applications [172,290,291].

3.2.1 Maxwell Postulates

The basic description of electromagnetic anisotropy and bianisotropy is constructed in terms of the four macroscopic electromagnetic fields $\tilde{\underline{E}}(\underline{r}, t)$, $\tilde{\underline{D}}(\underline{r}, t)$, $\tilde{\underline{B}}(\underline{r}, t)$, and $\tilde{\underline{H}}(\underline{r}, t)$. These are piecewise differentiable vector functions of position $\underline{r}$ and time t which arise as spatial averages of microscopic fields and bound sources. The fields $\tilde{\underline{E}}(\underline{r}, t)$ and $\tilde{\underline{B}}(\underline{r}, t)$ are directly measurable quantities which produce the Lorentz force [292]. Accordingly, $\tilde{\underline{E}}(\underline{r}, t)$ and $\tilde{\underline{B}}(\underline{r}, t)$ are viewed as the *primitive* fields. The fields $\tilde{\underline{D}}(\underline{r}, t)$ and $\tilde{\underline{H}}(\underline{r}, t)$ develop within a material in response to the primitive fields; hence, they are considered as *induction* fields. Conventionally, $\tilde{\underline{E}}(\underline{r}, t)$ and $\tilde{\underline{D}}(\underline{r}, t)$ are called the electric field and the dielectric displacement, respectively. The conventional terms for $\tilde{\underline{B}}(\underline{r}, t)$ and $\tilde{\underline{H}}(\underline{r}, t)$, namely the magnetic induction and magnetic field, respectively, are confusing and are avoided in this book.

In order to distinguish matter from vacuum or free space, two macroscopic fields are defined. One is called polarization

$$\tilde{\underline{P}}(\underline{r}, t) = \tilde{\underline{D}}(\underline{r}, t) - \epsilon_0 \tilde{\underline{E}}(\underline{r}, t), \tag{3.1}$$

while the other is called magnetization

$$\tilde{\underline{M}}(\underline{r}, t) = \mu_0^{-1} \tilde{\underline{B}}(\underline{r}, t) - \tilde{\underline{H}}(\underline{r}, t). \tag{3.2}$$

Polarization must not be confused with the term *polarization state*, which is an essential attribute of an electromagnetic plane wave [17].

The physical principles governing the behavior of $\tilde{\underline{E}}(\underline{r}, t)$, $\tilde{\underline{D}}(\underline{r}, t)$, $\tilde{\underline{B}}(\underline{r}, t)$, and $\tilde{\underline{H}}(\underline{r}, t)$ are encapsulated by the Maxwell curl postulates

$$\left. \begin{array}{l} \nabla \times \tilde{\underline{H}}(\underline{r}, t) - \dfrac{\partial}{\partial t} \tilde{\underline{D}}(\underline{r}, t) = \tilde{\underline{J}}_e(\underline{r}, t) \\[2mm] \nabla \times \tilde{\underline{E}}(\underline{r}, t) + \dfrac{\partial}{\partial t} \tilde{\underline{B}}(\underline{r}, t) = \underline{0} \end{array} \right\} \tag{3.3}$$

and divergence postulates

$$\left. \begin{array}{l} \nabla \cdot \tilde{\underline{D}}(\underline{r}, t) = \tilde{\rho}_e(\underline{r}, t) \\[2mm] \nabla \cdot \tilde{\underline{B}}(\underline{r}, t) = 0 \end{array} \right\}. \tag{3.4}$$

The terms on the right sides of Eqs. (3.3) and (3.4) represent sources of fields: $\tilde{\underline{J}}_e(\underline{r}, t)$ and $\tilde{\rho}_e(\underline{r}, t)$ are the externally impressed electric current and electric charge densities, respectively. In consonance with the macroscopic viewpoint, the source terms are also piecewise differentiable and satisfy the continuity relation

$$\nabla \cdot \tilde{\underline{J}}_e(\underline{r}, t) + \frac{\partial}{\partial t} \tilde{\rho}_e(\underline{r}, t) = 0. \tag{3.5}$$

Under the presumption of source continuity, there is no need to explicitly consider the Maxwell divergence postulates (3.4).

Time-domain problems are notoriously difficult to solve when a physical material is involved. A widely used technique to circumvent various analytical difficulties (often without loss of essential physics) is to introduce the temporal Fourier transforms via

$$\mathcal{Z}(\underline{r}, \omega) = \int_{-\infty}^{\infty} \tilde{\mathcal{Z}}(\underline{r}, t) \exp(i\omega t) dt, \tag{3.6}$$

where ω is the angular frequency and $i = \sqrt{-1}$. Whereas $\tilde{\mathcal{Z}}(\underline{r}, t)$ is a field, $\mathcal{Z}(\underline{r}, \omega)$ is the corresponding field phasor. The frequency-domain Maxwell curl postulates emerge from Eqs. (3.3) accordingly as

$$\left. \begin{array}{l} \nabla \times \underline{H}(\underline{r}, \omega) + i\omega \underline{D}(\underline{r}, \omega) = \underline{J}_e(\underline{r}, \omega) \\[2mm] \nabla \times \underline{E}(\underline{r}, \omega) - i\omega \underline{B}(\underline{r}, \omega) = \underline{0} \end{array} \right\}. \tag{3.7}$$

3.2.2 Linear Constitutive Relations

The constitutive relations of a linear, nonhomogeneous, and bianisotropic material are as follows:

$$\left. \begin{array}{l} \underline{D}(\underline{r}, \omega) = \underline{\underline{\epsilon}}(\underline{r}, \omega) \cdot \underline{E}(\underline{r}, \omega) + \underline{\underline{\xi}}(\underline{r}, \omega) \cdot \underline{H}(\underline{r}, \omega) \\[2mm] \underline{B}(\underline{r}, \omega) = \underline{\underline{\zeta}}(\underline{r}, \omega) \cdot \underline{E}(\underline{r}, \omega) + \underline{\underline{\mu}}(\underline{r}, \omega) \cdot \underline{H}(\underline{r}, \omega) \end{array} \right\}. \tag{3.8}$$

The dielectric properties of the material are captured by the permittivity dyadic $\underline{\underline{\epsilon}}(\underline{r}, \omega)$, the magnetic properties by the permeability dyadic $\underline{\underline{\mu}}(\underline{r}, \omega)$, and the magnetoelectric properties jointly by the dyadics $\underline{\underline{\xi}}(\underline{r}, \omega)$ and $\underline{\underline{\zeta}}(\underline{r}, \omega)$. If the material were to be homogeneous, the constitutive dyadics would be independent of the position $\underline{r}$.

Each of these four dyadics can be represented as a 3×3 matrix in a Cartesian coordinate system; thus,

$$\underline{\underline{\epsilon}}(\underline{r}, \omega) \equiv \begin{bmatrix} \epsilon_{xx}(\underline{r}, \omega) & \epsilon_{xy}(\underline{r}, \omega) & \epsilon_{xz}(\underline{r}, \omega) \\ \epsilon_{yx}(\underline{r}, \omega) & \epsilon_{yy}(\underline{r}, \omega) & \epsilon_{yz}(\underline{r}, \omega) \\ \epsilon_{zx}(\underline{r}, \omega) & \epsilon_{zy}(\underline{r}, \omega) & \epsilon_{zz}(\underline{r}, \omega) \end{bmatrix}, \tag{3.9}$$

etc. The physical properties of a linear material are reflected in the structure of the constitutive dyadics as well as in their mutual relationships. Detailed descriptions of all constitutive dyadics are available elsewhere [17,293]. Most importantly in the present context, a material is non-dissipative (i.e. lossless) at a certain angular frequency, if all three conditions

$$\left. \begin{array}{l} \underline{\underline{\epsilon}}(\underline{r}, \omega) = \underline{\underline{\epsilon}}^{\dagger}(\underline{r}, \omega) \\ \underline{\underline{\zeta}}(\underline{r}, \omega) = \underline{\underline{\xi}}^{\dagger}(\underline{r}, \omega) \\ \underline{\underline{\mu}}(\underline{r}, \omega) = \underline{\underline{\mu}}^{\dagger}(\underline{r}, \omega) \end{array} \right\} \tag{3.10}$$

hold true, where $\underline{\underline{\xi}}^{\dagger}$ is the conjugate transpose of $\underline{\underline{\xi}}$, etc.

Explicit mention of the dependence on the angular frequency ω is not made in the rest of the book for notational compactness.

3.2.3   Periodic Nonhomogeneity

In the context of surface-wave propagation in this book, bianisotropic materials with periodic nonhomogeneity along the z axis are of particular interest. Their frequency-domain constitutive relations are written as

$$\left. \begin{array}{l} \underline{D}(\underline{r}) = \underline{\underline{\epsilon}}(z) \cdot \underline{E}(\underline{r}) + \underline{\underline{\xi}}(z) \cdot \underline{H}(\underline{r}) \\ \underline{B}(\underline{r}) = \underline{\underline{\zeta}}(z) \cdot \underline{E}(\underline{r}) + \underline{\underline{\mu}}(z) \cdot \underline{H}(\underline{r}) \end{array} \right\}, \tag{3.11}$$

where

$$\underline{\underline{\epsilon}}(z \pm 2\Omega) = \underline{\underline{\epsilon}}(z), \quad \underline{\underline{\xi}}(z \pm 2\Omega) = \underline{\underline{\xi}}(z), \quad \underline{\underline{\zeta}}(z \pm 2\Omega) = \underline{\underline{\zeta}}(z),$$
$$\underline{\underline{\mu}}(z \pm 2\Omega) = \underline{\underline{\mu}}(z), \tag{3.12}$$

and 2Ω is the period.

3.2.4   Homogeneous Bianisotropic Materials

The constitutive dyadics of a homogeneous bianisotropic material are independent of the position $\underline{r}$, and so the constitutive relations may be stated simply as

$$\left. \begin{array}{l} \underline{D}(\underline{r}) = \underline{\underline{\epsilon}} \cdot \underline{E}(\underline{r}) + \underline{\underline{\xi}} \cdot \underline{H}(\underline{r}) \\ \underline{B}(\underline{r}) = \underline{\underline{\zeta}} \cdot \underline{E}(\underline{r}) + \underline{\underline{\mu}} \cdot \underline{H}(\underline{r}) \end{array} \right\}. \tag{3.13}$$

3.3 Propagation in a Homogeneous Bianisotropic Material

Suppose a wave propagates in a homogeneous bianisotropic material described by Eqs. (3.13). The spatial variation in the xy plane of the wave is described by

$$\exp\left[iq(x\cos\psi + y\sin\psi)\right], \tag{3.14}$$

where q is a complex-valued wavenumber and the angle $\psi \in [0°, 360°)$. But, the spatial variation along the z axis is yet to be determined. Equations (3.7) are applicable with $\underline{J}_e(\underline{r}) = \underline{0}$ everywhere.

For all $z \in (-\infty, \infty)$, then the field phasors may be written as

$$\left.\begin{aligned} \underline{E}(\underline{r}) &= \underline{e}(z)\exp\left[iq(x\cos\psi + y\sin\psi)\right] \\ \underline{H}(\underline{r}) &= \underline{h}(z)\exp\left[iq(x\cos\psi + y\sin\psi)\right] \end{aligned}\right\}, \tag{3.15}$$

where

$$\left.\begin{aligned} \underline{e}(z) &= e_x(z)\underline{u}_x + e_y(z)\underline{u}_y + e_z(z)\underline{u}_z \\ \underline{h}(z) &= h_x(z)\underline{u}_x + h_y(z)\underline{u}_y + h_z(z)\underline{u}_z \end{aligned}\right\} \tag{3.16}$$

are auxiliary phasors.

3.3.1 Matrix Ordinary Differential Equation

Substitution of Eqs. (3.13) and (3.15) in (3.7) with $\underline{J}_e(\underline{r}) = \underline{0}$ everywhere yields the four ordinary differential equations

$$\left.\begin{aligned} \frac{d}{dz}e_x(z) &= i\omega\left[\zeta_{yx}e_x(z) + \zeta_{yy}e_y(z) + \left(\zeta_{yz} + \frac{q}{\omega}\cos\psi\right)e_z(z)\right.\\ &\quad \left. + \mu_{yx}h_x(z) + \mu_{yy}h_y(z) + \mu_{yz}h_z(z)\right] \\ \frac{d}{dz}e_y(z) &= -i\omega\left[\zeta_{xx}e_x(z) + \zeta_{xy}e_y(z) + \left(\zeta_{xz} - \frac{q}{\omega}\sin\psi\right)e_z(z)\right.\\ &\quad \left. + \mu_{xx}h_x(z) + \mu_{xy}h_y(z) + \mu_{xz}h_z(z)\right] \\ \frac{d}{dz}h_x(z) &= -i\omega\left[\epsilon_{yx}e_x(z) + \epsilon_{yy}e_y(z) + \epsilon_{yz}e_z(z)\right.\\ &\quad \left. + \xi_{yx}h_x(z) + \xi_{yy}h_y(z) + \left(\xi_{yz} - \frac{q}{\omega}\cos\psi\right)h_z(z)\right] \\ \frac{d}{dz}h_y(z) &= i\omega\left[\epsilon_{xx}e_x(z) + \epsilon_{xy}e_y(z) + \epsilon_{xz}e_z(z)\right.\\ &\quad \left. + \xi_{xx}h_x(z) + \xi_{xy}h_y(z) + \left(\xi_{xz} + \frac{q}{\omega}\sin\psi\right)h_z(z)\right] \end{aligned}\right\} \tag{3.17}$$

and the two algebraic equations

$$\left.\begin{aligned} \epsilon_{zz}e_z(z) + \xi_{zz}h_z(z) &= -\epsilon_{zx}e_x(z) - \epsilon_{zy}e_y(z) \\ &\quad - \left(\xi_{zx} - \frac{q}{\omega}\sin\psi\right)h_x(z) - \left(\xi_{zy} + \frac{q}{\omega}\cos\psi\right)h_y(z) \\ \zeta_{zz}e_z(z) + \mu_{zz}h_z(z) &= -\left(\zeta_{zx} + \frac{q}{\omega}\sin\psi\right)e_x(z) - \left(\zeta_{zy} - \frac{q}{\omega}\cos\psi\right)e_y(z) \\ &\quad - \mu_{zx}h_x(z) - \mu_{zy}h_y(z) \end{aligned}\right\}. \tag{3.18}$$

With the assumption that

$$\epsilon_{zz}\mu_{zz} \neq \xi_{zz}\zeta_{zz}, \tag{3.19}$$

the solution of the two algebraic equations may be stated as

$$\left.\begin{array}{l} e_z(z) = v_{zx}^{ee}e_x(z) + v_{zy}^{ee}e_y(z) + v_{zx}^{eh}h_x(z) + v_{zy}^{eh}h_y(z) \\ h_z(z) = v_{zx}^{he}e_x(z) + v_{zy}^{he}e_y(z) + v_{zx}^{hh}h_x(z) + v_{zy}^{hh}h_y(z) \end{array}\right\}, \tag{3.20}$$

where

$$\left.\begin{array}{l} v_{zx}^{ee} = -\dfrac{\mu_{zz}\epsilon_{zx} - \xi_{zz}\left[\zeta_{zx} + (q/\omega)\sin\psi\right]}{\epsilon_{zz}\mu_{zz} - \xi_{zz}\zeta_{zz}} \\[2mm] v_{zy}^{ee} = -\dfrac{\mu_{zz}\epsilon_{zy} - \xi_{zz}\left[\zeta_{zy} - (q/\omega)\cos\psi\right]}{\epsilon_{zz}\mu_{zz} - \xi_{zz}\zeta_{zz}} \\[2mm] v_{zx}^{eh} = \dfrac{\xi_{zz}\mu_{zx} - \mu_{zz}\left[\xi_{zx} - (q/\omega)\sin\psi\right]}{\epsilon_{zz}\mu_{zz} - \xi_{zz}\zeta_{zz}} \\[2mm] v_{zy}^{eh} = \dfrac{\xi_{zz}\mu_{zy} - \mu_{zz}\left[\xi_{zy} + (q/\omega)\cos\psi\right]}{\epsilon_{zz}\mu_{zz} - \xi_{zz}\zeta_{zz}} \\[2mm] v_{zx}^{he} = \dfrac{\zeta_{zz}\epsilon_{zx} - \epsilon_{zz}\left[\zeta_{zx} + (q/\omega)\sin\psi\right]}{\epsilon_{zz}\mu_{zz} - \xi_{zz}\zeta_{zz}} \\[2mm] v_{zy}^{he} = \dfrac{\zeta_{zz}\epsilon_{zy} - \epsilon_{zz}\left[\zeta_{zy} - (q/\omega)\cos\psi\right]}{\epsilon_{zz}\mu_{zz} - \xi_{zz}\zeta_{zz}} \\[2mm] v_{zx}^{hh} = -\dfrac{\epsilon_{zz}\mu_{zx} - \zeta_{zz}\left[\xi_{zx} - (q/\omega)\sin\psi\right]}{\epsilon_{zz}\mu_{zz} - \xi_{zz}\zeta_{zz}} \\[2mm] v_{zy}^{hh} = -\dfrac{\epsilon_{zz}\mu_{zy} - \zeta_{zz}\left[\xi_{zy} + (q/\omega)\cos\psi\right]}{\epsilon_{zz}\mu_{zz} - \xi_{zz}\zeta_{zz}} \end{array}\right\}. \tag{3.21}$$

Substitution of Eqs. (3.20) in (3.17) yields the 4×4 matrix ordinary differential equation

$$\frac{d}{dz}\left[\underline{f}(z)\right] = i\left[\underline{\underline{P}}\right] \cdot \left[\underline{f}(z)\right], \tag{3.22}$$

where the column 4-vector

$$\left[\underline{f}(z)\right] = \begin{bmatrix} e_x(z) \\ e_y(z) \\ h_x(z) \\ h_y(z) \end{bmatrix} \tag{3.23}$$

and the matrix

$$[\underline{\underline{P}}] = \omega \begin{bmatrix} \zeta_{yx} & \zeta_{yy} & \mu_{yx} & \mu_{yy} \\ -\zeta_{xx} & -\zeta_{xy} & -\mu_{xx} & -\mu_{xy} \\ -\epsilon_{yx} & -\epsilon_{yy} & -\xi_{yx} & -\xi_{yy} \\ \epsilon_{xx} & \epsilon_{xy} & \xi_{xx} & \xi_{xy} \end{bmatrix}$$

$$+ \omega \begin{bmatrix} \zeta_{yz} + \dfrac{q}{\omega}\cos\psi & 0 & 0 & 0 \\ 0 & \zeta_{xz} - \dfrac{q}{\omega}\sin\psi & 0 & 0 \\ 0 & 0 & \epsilon_{yz} & 0 \\ 0 & 0 & 0 & \epsilon_{xz} \end{bmatrix} \cdot [\underline{\underline{J}}] \cdot \begin{bmatrix} v_{zx}^{ee} & 0 & 0 & 0 \\ 0 & v_{zy}^{ee} & 0 & 0 \\ 0 & 0 & v_{zx}^{eh} & 0 \\ 0 & 0 & 0 & v_{zy}^{eh} \end{bmatrix}$$

$$+ \omega \begin{bmatrix} \mu_{yz} & 0 & 0 & 0 \\ 0 & \mu_{xz} & 0 & 0 \\ 0 & 0 & \xi_{yz} - \dfrac{q}{\omega}\cos\psi & 0 \\ 0 & 0 & 0 & \xi_{xz} + \dfrac{q}{\omega}\sin\psi \end{bmatrix} \cdot [\underline{\underline{J}}] \cdot \begin{bmatrix} v_{zx}^{he} & 0 & 0 & 0 \\ 0 & v_{zy}^{he} & 0 & 0 \\ 0 & 0 & v_{zx}^{hh} & 0 \\ 0 & 0 & 0 & v_{zy}^{hh} \end{bmatrix} \quad (3.24)$$

employs the shorthand notation

$$[\underline{\underline{J}}] = \begin{bmatrix} 1 & 1 & 1 & 1 \\ 1 & 1 & 1 & 1 \\ 1 & 1 & 1 & 1 \\ 1 & 1 & 1 & 1 \end{bmatrix}. \tag{3.25}$$

As the matrix $[\underline{\underline{P}}]$ on the right side of Eq. (3.22) is independent of z, the solution of this equation is [262, Section 2.5]

$$[\underline{f}(z)] = \exp\left\{ i\,[\underline{\underline{P}}]z \right\} \cdot [\underline{f}(0)], \tag{3.26}$$

where

$$\exp\left\{ i\,[\underline{\underline{P}}]z \right\} = \sum_{m=0}^{\infty} \frac{i^m}{m!} [\underline{\underline{P}}]^m \, z^m. \tag{3.27}$$

Instead of computing it as a sum, the exponential of a matrix can be computed in Mathematica™ using the command MatrixExp, and in Matlab™ with expm. More generally than Eq. (3.26), the solution of Eq. (3.22) can be stated as

$$[\underline{f}(z)] = \exp\left\{ i\,[\underline{\underline{P}}]\,(z - u) \right\} \cdot [\underline{f}(u)]. \tag{3.28}$$

Yet another way to handle the right side of Eq. (3.26) conveniently is to invoke the very slightly restrictive assumption of the 4×4 matrix $[\underline{\underline{P}}]$ having four linearly independent eigenvectors. This assumption allows $[\underline{\underline{P}}]$ to be diagonalized [294, Section 7.15]. Suppose the mth, $m \in [1, 4]$, eigenvalue of $[\underline{\underline{P}}]$ is denoted by the scalar g_m, and the corresponding eigenvector is the column 4-vector

$$[\underline{v}]^{(m)} = \begin{bmatrix} v_{1m} \\ v_{2m} \\ v_{3m} \\ v_{4m} \end{bmatrix}. \tag{3.29}$$

Then,

$$[\underline{\underline{P}}] = [\underline{\underline{V}}] \cdot [\underline{\underline{G}}] \cdot [\underline{\underline{V}}]^{-1}, \tag{3.30}$$

where

$$[\underline{\underline{V}}] = \begin{bmatrix} v_{11} & v_{12} & v_{13} & v_{14} \\ v_{21} & v_{22} & v_{23} & v_{24} \\ v_{31} & v_{32} & v_{33} & v_{34} \\ v_{41} & v_{42} & v_{43} & v_{44} \end{bmatrix}, \quad [\underline{\underline{G}}] = \begin{bmatrix} g_1 & 0 & 0 & 0 \\ 0 & g_2 & 0 & 0 \\ 0 & 0 & g_3 & 0 \\ 0 & 0 & 0 & g_4 \end{bmatrix}. \tag{3.31}$$

The normalization

$$|v_{1m}|^2 + |v_{2m}|^2 = 1, \quad m \in [1,4], \tag{3.32}$$

is assumed to have been undertaken in order to force the magnitude of the electric field phasor associated with every eigenvector to be close to unity, thereby facilitating comparison of the spatial profiles of the fields associated with different eigenvectors.

By virtue of the definition of the exponential of a matrix, and in view of Eq. (3.30), it follows that

$$\exp\{i\,[\underline{\underline{P}}]z\} = [\underline{\underline{V}}] \cdot \exp\{i\,[\underline{\underline{G}}]z\} \cdot [\underline{\underline{V}}]^{-1}; \tag{3.33}$$

hence,

$$[\underline{f}(z)] = [\underline{\underline{V}}] \cdot \exp\{i\,[\underline{\underline{G}}]z\} \cdot [\underline{\underline{V}}]^{-1} \cdot [\underline{f}(0)]. \tag{3.34}$$

Furthermore, as the exponential of a diagonal matrix is itself a diagonal matrix, i.e.

$$\exp\{i\,[\underline{\underline{G}}]z\} = \begin{bmatrix} \exp(ig_1 z) & 0 & 0 & 0 \\ 0 & \exp(ig_2 z) & 0 & 0 \\ 0 & 0 & \exp(ig_3 z) & 0 \\ 0 & 0 & 0 & \exp(ig_4 z) \end{bmatrix}, \tag{3.35}$$

the right side of Eq. (3.34) is very tractable for analysis as well as computation.

3.3.2   Eigenmodes

Knowledge of $[\underline{\underline{V}}]$ and $[\underline{\underline{G}}]$ allows the formulation of an eigenmodal representation of field phasors. Thus, the mth eigenmode may be expressed as

$$\left. \begin{aligned} \underline{E}^{(m)}(\underline{r}) &= \left[v_{1m}\,\underline{u}_x + v_{2m}\,\underline{u}_y + \left(v_{zx}^{ee}\,v_{1m} + v_{zy}^{ee}\,v_{2m} + v_{zx}^{eh}\,v_{3m} + v_{zy}^{eh}\,v_{4m} \right)\underline{u}_z \right] \\ &\quad \times \exp\left[iq(x\cos\psi + y\sin\psi)\right]\exp(ig_m z) \\[1ex] \underline{H}^{(m)}(\underline{r}) &= \left[v_{3m}\,\underline{u}_x + v_{4m}\,\underline{u}_y + \left(v_{zx}^{he}\,v_{1m} + v_{zy}^{he}\,v_{2m} + v_{zx}^{hh}\,v_{3m} + v_{zy}^{hh}\,v_{4m} \right)\underline{u}_z \right] \\ &\quad \times \exp\left[iq(x\cos\psi + y\sin\psi)\right]\exp(ig_m z) \end{aligned} \right\}, \tag{3.36}$$

so that the complete representation of the phasors defined via Eqs. (3.15) is as follows:

$$
\left.\begin{aligned}
\underline{E}(\underline{r}) &= \sum_{m=1}^{4} C_m \underline{E}^{(m)}(\underline{r}) \\
\underline{H}(\underline{r}) &= \sum_{m=1}^{4} C_m \underline{H}^{(m)}(\underline{r})
\end{aligned}\right\}. \tag{3.37}
$$

Comparison of this representation with Eq. (3.34) yields the identity

$$
\begin{bmatrix} C_1 \\ C_2 \\ C_3 \\ C_4 \end{bmatrix} = [\underline{\underline{V}}]^{-1} \cdot [\underline{f}(0)]. \tag{3.38}
$$

Associated with each eigenmode is the time-averaged Poynting vector

$$
\underline{P}^{(m)}(\underline{r}) = (1/2)\mathrm{Re}\left\{ \underline{E}^{(m)}(\underline{r}) \times \left[\underline{H}^{(m)}(\underline{r})\right]^{*} \right\}, \quad m \in [1,4]. \tag{3.39}
$$

This expression follows from Eq. (2.31). The z-directed component of the time-averaged Poynting vector

$$
\begin{aligned}
P_z^{(m)}(\underline{r}) &= \underline{u}_z \cdot \underline{P}^{(m)}(\underline{r}) \tag{3.40} \\
&= (1/2)\mathrm{Re}\left\{ v_{1m} v_{4m}^{*} - v_{2m} v_{3m}^{*} \right\} \\
&\quad \times \exp\left\{ -2\left[\mathrm{Im}\{q\}(x\cos\psi + y\sin\psi) + \mathrm{Im}\{g_m\} z \right] \right\}, \\
&\qquad\qquad m \in [1,4], \tag{3.41}
\end{aligned}
$$

is needed in order to classify the eigenmodes into

i. those which transport energy along the $+z$ axis and
ii. those which transport energy along the $-z$ axis.

Eigenmodes which transport energy along the $+z$ axis are accorded the labels $m = 1$ and $m = 2$; thus,

$$
P_z^{(m)}(\underline{r}) > 0, \quad m \in [1,2]. \tag{3.42}
$$

Eigenmodes labeled $m \in [3,4]$ obey the constraints

$$
P_z^{(m)}(\underline{r}) < 0, \quad m \in [3,4], \tag{3.43}
$$

and transport energy along the $-z$ axis. This classification of four eigenmodes into two eigenmodes of each type represents an assumption rather than a proved fact.
Eigenmodes labeled $m \in [1,2]$ should also satisfy the constraint

$$
\mathrm{Im}\{g_m\} \geqslant 0, \quad m \in [1,2], \tag{3.44}
$$

the inequality indicating the decay of fields in a passive material as $z \to \infty$; likewise,

$$\text{Im}\,\{g_m\} \leqslant 0, \quad m \in [3, 4],\tag{3.45}$$

the inequality indicating the decay of fields in a passive material as $z \to -\infty$. The conditions (3.44) and (3.45) are consistent with the conditions (3.42) and (3.43).

Consistently with these labels, $\left[\,\underline{f}(z)\right]$ can be partitioned as

$$\left[\,\underline{f}(z)\right] = \left[\,\underline{f}^+(z)\right] + \left[\,\underline{f}^-(z)\right],\tag{3.46}$$

where

$$\left[\,\underline{f}^+(z)\right] = \begin{bmatrix} v_{11} & v_{12} \\ v_{21} & v_{22} \\ v_{31} & v_{32} \\ v_{41} & v_{42} \end{bmatrix} \cdot \begin{bmatrix} \exp(ig_1 z) & 0 \\ 0 & \exp(ig_2 z) \end{bmatrix} \cdot \begin{bmatrix} C_1 \\ C_2 \end{bmatrix}\tag{3.47}$$

represents fields transporting energy along the $+z$ axis and decaying as $z \to \infty$, and

$$\left[\,\underline{f}^-(z)\right] = \begin{bmatrix} v_{13} & v_{14} \\ v_{23} & v_{24} \\ v_{33} & v_{34} \\ v_{43} & v_{44} \end{bmatrix} \cdot \begin{bmatrix} \exp(ig_3 z) & 0 \\ 0 & \exp(ig_4 z) \end{bmatrix} \cdot \begin{bmatrix} C_3 \\ C_4 \end{bmatrix}\tag{3.48}$$

represents fields transporting energy along the $-z$ axis and decaying as $z \to -\infty$.

3.4 Propagation in a Periodically Nonhomogeneous Bianisotropic Material

Suppose a wave propagates in a periodically nonhomogeneous bianisotropic material described in Section 3.2.3 with its spatial variation in the xy plane described by $\exp\left[iq(x\cos\psi + y\sin\psi)\right]$. Most of the developments in Section 3.3 can be adapted to delineate the characteristics of this wave.

3.4.1 Matrix Ordinary Differential Equation

Equations (3.15)–(3.21) still apply but the elements of the constitutive dyadics are now functions of z. Accordingly, the matrix ordinary differential equation

$$\frac{d}{dz}\left[\,\underline{f}(z)\right] = i\left[\,\underline{\underline{P}}(z)\right] \cdot \left[\,\underline{f}(z)\right]\tag{3.49}$$

emerges. The matrix $\left[\,\underline{\underline{P}}(z)\right]$ is defined by Eq. (3.24), but with ϵ_{xy} replaced by $\epsilon_{xy}(z)$, etc.

The solution of Eq. (3.49) is not as straightforward as of Eq. (3.22), because $\left[\,\underline{\underline{P}}(z)\right]$ is a periodic function of z:

$$\left[\,\underline{\underline{P}}(z \pm 2\Omega)\right] = \left[\,\underline{\underline{P}}(z)\right].\tag{3.50}$$

As shown in Appendix D, application of the Floquet theory indicates that the solution of Eq. (3.49) must be of the form

$$[\underline{f}(z)] = [\underline{\underline{F}}(z)] \cdot \exp\{i[\underline{\underline{A}}]z\} \cdot [\underline{f}(0)], \tag{3.51}$$

where

$$[\underline{\underline{F}}(z \pm 2\Omega)] = [\underline{\underline{F}}(z)] \tag{3.52}$$

is a periodic matrix with $[\underline{\underline{F}}(0)] = [\underline{\underline{I}}]$, the identity matrix, and the matrix $[\underline{\underline{A}}]$ is independent of z. Unfortunately, this theory does not deliver actual expressions for $[\underline{\underline{F}}(z)]$ and $[\underline{\underline{A}}]$.

Even without resorting to Floquet theory, the solution of Eq. (3.49) has to lead to the relation

$$[\underline{f}(2\Omega)] = [\underline{\underline{Q}}] \cdot [\underline{f}(0)], \tag{3.53}$$

where the matrix $[\underline{\underline{Q}}]$ can be computed by the piecewise-uniform-approximation method, introduced in Section 2.3.4.4. Thereby, the continuous variation of the matrix $[\underline{\underline{P}}(z)]$ in a certain region is replaced by a set of stepwise variations, in each step of finite length the matrix being assumed to be constant. After choosing an integer $N > 1$, the following quantities are defined for all integers $n \in (-\infty, -1] \cup [1, \infty)$:

$$[\underline{\underline{W}}]^{(n)} = \begin{cases} \exp\left\{i(z_n - z_{n-1})\left[\underline{\underline{P}}\left(\dfrac{z_n + z_{n-1}}{2}\right)\right]\right\}, & n > 0, \\[2ex] \exp\left\{i(z_n - z_{n+1})\left[\underline{\underline{P}}\left(\dfrac{z_n + z_{n+1}}{2}\right)\right]\right\}, & n < 0, \end{cases} \tag{3.54}$$

where

$$z_n = 2\Omega\frac{n}{N}, \quad n \in (-\infty, \infty). \tag{3.55}$$

Thus, each period is divided into N subregions, each of thickness $2\Omega/N$. The matrix $[\underline{\underline{W}}]^{(n)}$ approximately describes propagation from the plane $z = z_{n-1}$ to the plane $z = z_n$ for $n > 0$, but from the plane $z = z_{n+1}$ to the plane $z = z_n$ for $n < 0$. The piecewise-uniform-approximation method therefore delivers

$$[\underline{\underline{Q}}] \cong [\underline{\underline{W}}]^{(N)} \cdot [\underline{\underline{W}}]^{(N-1)} \cdot \ldots \cdot [\underline{\underline{W}}]^{(2)} \cdot [\underline{\underline{W}}]^{(1)}. \tag{3.56}$$

The integer N should be neither too large nor too small. If it is too small, the continuous variations of $[\underline{\underline{P}}(z)]$ would not be captured well by the set of stepwise variations. If N is too large, the computation of $[\underline{\underline{Q}}]$ would take too much time. Therefore, the convergence of $[\underline{\underline{Q}}]$ with increasing N must be examined for a satisfactory choice.

Comparison of Eqs. (3.51) and (3.53) yields the identity

$$\left[\underline{\underline{Q}}\right] = \exp\left\{i2\Omega\left[\underline{\underline{A}}\right]\right\}. \tag{3.57}$$

Both $\left[\underline{\underline{Q}}\right]$ and $\left[\underline{\underline{A}}\right]$ share the same eigenvectors, and their eigenvalues are also related as follows. Let

$$\left[\underline{v}\right]^{(m)} = \begin{bmatrix} v_{1m} \\ v_{2m} \\ v_{3m} \\ v_{4m} \end{bmatrix} \tag{3.58}$$

be the eigenvector corresponding to the mth eigenvalue σ_m, $m \in [1, 4]$, of $\left[\underline{\underline{Q}}\right]$; then, the corresponding eigenvalue α_m of $\left[\underline{\underline{A}}\right]$ is given by

$$\alpha_m = -i\frac{\ln \sigma_m}{2\Omega}, \quad m \in [1, 4]. \tag{3.59}$$

Energy analysis is facilitated if the eigenvectors defined by Eq. (3.58) are normalized to obey Eq. (3.32).

In contrast to Section 3.3.1, the eigenvectors $\left[\underline{v}\right]^{(m)}$, $m \in [1, 4]$, defined in Eq. (3.58) are the eigenvectors of $\left[\underline{\underline{Q}}\right]$ and $\left[\underline{\underline{A}}\right]$ but not of $\left[\underline{\underline{P}}(z)\right]$. Therefore, although a matrix $\left[\underline{\underline{V}}\right]$ comprising these eigenvectors can be defined just as in Eqs. (3.31), it is not useful.

If $\left[\underline{f}(0)\right]$ is expressed in terms of the eigenvectors $\left[\underline{v}\right]^{(m)}$ of Eq. (3.58) as

$$\left[\underline{f}(0)\right] = \sum_{m=1}^{4} C_m \left[\underline{v}\right]^{(m)}, \tag{3.60}$$

where C_m, $m \in [1, 4]$, are unknown coefficients, then [295]

$$\left[\underline{f}(z)\right] = \sum_{m=1}^{4} C_m \left[\underline{t}\right]^{(m)}(z) \tag{3.61}$$

with

$$\left[\underline{t}\right]^{(m)}(z) \cong \exp\left\{i(z - z_n)\left[\underline{\underline{P}}\left(\frac{z_{n+1} + z_n}{2}\right)\right]\right\} \cdot \left[\underline{\underline{W}}\right]^{(n)} \cdot \left[\underline{\underline{W}}\right]^{(n-1)} \cdots$$
$$\cdot \left[\underline{\underline{W}}\right]^{(2)} \cdot \left[\underline{\underline{W}}\right]^{(1)} \cdot \left[\underline{v}\right]^{(m)}, \quad z \in [z_n, z_{n+1}], \quad n \in [1, \infty), \tag{3.62}$$

$$\left[\underline{t}\right]^{(m)}(z) \cong \exp\left\{iz\left[\underline{\underline{P}}\left(\frac{z_1}{2}\right)\right]\right\} \cdot \left[\underline{v}\right]^{(m)}, \quad z \in [0, z_1], \tag{3.63}$$

$$\left[\underline{t}\right]^{(m)}(z) \cong \exp\left\{iz\left[\underline{\underline{P}}\left(\frac{z_{-1}}{2}\right)\right]\right\} \cdot \left[\underline{v}\right]^{(m)}, \quad z \in [z_{-1}, 0], \tag{3.64}$$

and

$$\left[\underline{t}\right]^{(m)}(z) \cong \exp\left\{i(z - z_n)\left[\underline{\underline{P}}\left(\frac{z_{n-1} + z_n}{2}\right)\right]\right\} \cdot \left[\underline{\underline{W}}\right]^{(n)} \cdot \left[\underline{\underline{W}}\right]^{(n+1)} \cdots$$
$$\cdot \left[\underline{\underline{W}}\right]^{(-2)} \cdot \left[\underline{\underline{W}}\right]^{(-1)} \cdot \left[\underline{v}\right]^{(m)}, \quad z \in [z_{n-1}, z_n], \quad n \in (-\infty, -1]. \tag{3.65}$$

3.4.2   Eigenmodes

An eigenmodal representation of field phasors now follows as

$$
\left.
\begin{aligned}
\underline{E}(\underline{r}) &= \sum_{m=1}^{4} C_m \underline{E}^{(m)}(\underline{r}) \\
\underline{H}(\underline{r}) &= \sum_{m=1}^{4} C_m \underline{H}^{(m)}(\underline{r})
\end{aligned}
\right\}.
\tag{3.66}
$$

In this representation, the eigenmodes are given by

$$
\left.
\begin{aligned}
\underline{E}^{(m)}(\underline{r}) &= \big\{ t_{1m}(z)\underline{u}_x + t_{2m}(z)\underline{u}_y + \big[v_{zx}^{ee}(z)t_{1m}(z) \\
&\quad + v_{zy}^{ee}(z)t_{2m}(z) + v_{zx}^{eh}(z)t_{3m}(z) + v_{zy}^{eh}(z)t_{4m}(z) \big] \underline{u}_z \big\} \\
&\quad \times \exp\big[iq(x\cos\psi + y\sin\psi) \big] \\
\underline{H}^{(m)}(\underline{r}) &= \big\{ t_{3m}(z)\underline{u}_x + t_{4m}(z)\underline{u}_y + \big[v_{zx}^{he}(z)t_{1m}(z) \\
&\quad + v_{zy}^{he}(z)t_{2m}(z) + v_{zx}^{hh}(z)t_{3m}(z) + v_{zy}^{hh}(z)t_{4m}(z) \big] \underline{u}_z \big\} \\
&\quad \times \exp\big[iq(x\cos\psi + y\sin\psi) \big]
\end{aligned}
\right\},
\tag{3.67}
$$

wherein

$$
\left[\underline{t}\right]^{(m)}(z) =
\begin{bmatrix}
t_{1m}(z) \\
t_{2m}(z) \\
t_{3m}(z) \\
t_{4m}(z)
\end{bmatrix}
\tag{3.68}
$$

has been used.

Equation (3.39) remains valid to obtain the time-averaged Poynting vector associated with each eigenmode, but its z-directed component is not simply analogous to the right side of Eq. (3.41) because the eigenvectors $\left[\underline{v}\right]^{(m)}$ defined in Section 3.3.1 are different in character from those defined in Section 3.4.1.

Classification of the eigenmodes may require the assumption of dissipation in the periodically nonhomogeneous bianisotropic material. The assumption is not onerous, because, by virtue of the principle of causality, all passive materials must be dissipative. Occasionally it can be expedient to neglect dissipation and exploit Eqs. (3.10) for simplified analyses, especially if attention is confined to a narrow range of angular frequencies wherein dissipation is very small over the length-scales of interest. Nevertheless, that tiny dissipation must not be ignored for the classification of eigenmodes [295,296].

With the supposition that all four eigenvalues of $\left[\underline{\underline{A}}\right]$ have imaginary parts, eigenmodes that decay as $z \to \infty$ are accorded the labels $m = 1$ and $m = 2$; thus,

$$
\mathrm{Im}\{\alpha_m\} > 0, \quad m \in [1, 2].
\tag{3.69}
$$

The labels $m = 3$ and $m = 4$ are reserved for eigenmodes that decay as $z \to -\infty$ so that

$$
\mathrm{Im}\{\alpha_m\} < 0, \quad m \in [3, 4].
\tag{3.70}
$$

Consistently with these labels, $\left[\underline{f}(z)\right]$ can be partitioned into

$$\left[\underline{f}^{+}(z)\right] = \begin{bmatrix} t_{11}(z) & t_{12}(z) \\ t_{21}(z) & t_{22}(z) \\ t_{31}(z) & t_{32}(z) \\ t_{41}(z) & t_{42}(z) \end{bmatrix} \cdot \begin{bmatrix} C_1 \\ C_2 \end{bmatrix} \tag{3.71}$$

and

$$\left[\underline{f}^{-}(z)\right] = \begin{bmatrix} t_{13}(z) & t_{14}(z) \\ t_{23}(z) & t_{24}(z) \\ t_{33}(z) & t_{34}(z) \\ t_{43}(z) & t_{44}(z) \end{bmatrix} \cdot \begin{bmatrix} C_3 \\ C_4 \end{bmatrix}. \tag{3.72}$$

3.5 Canonical Boundary-Value Problem

The canonical boundary-value problem of surface-wave propagation can now be formulated quite simply to determine the relevant dispersion equation. The formulation presented in this section applies to SPP waves, Fano waves, Zenneck waves, Dyakonov waves, Tamm waves, and Dyakonov-Tamm waves.

3.5.1 Dispersion Equation

Let a bianisotropic material of type $\mathcal{A}$ occupy the half-space $z > 0$ and a different bianisotropic material, of type $\mathcal{B}$, occupy the half-space $z < 0$, as shown in Figure 3.1. The constitutive relations are stated as

$$\left. \begin{aligned} \underline{D}(\underline{r}) &= \begin{cases} \underline{\underline{\epsilon}}^{\mathcal{A}}(z) \cdot \underline{E}(\underline{r}) + \underline{\underline{\xi}}^{\mathcal{A}}(z) \cdot \underline{H}(\underline{r}), & z > 0 \\ \underline{\underline{\epsilon}}^{\mathcal{B}}(z) \cdot \underline{E}(\underline{r}) + \underline{\underline{\xi}}^{\mathcal{B}}(z) \cdot \underline{H}(\underline{r}), & z < 0 \end{cases} \\ \underline{B}(\underline{r}) &= \begin{cases} \underline{\underline{\zeta}}^{\mathcal{A}}(z) \cdot \underline{E}(\underline{r}) + \underline{\underline{\mu}}^{\mathcal{A}}(z) \cdot \underline{H}(\underline{r}), & z > 0 \\ \underline{\underline{\zeta}}^{\mathcal{B}}(z) \cdot \underline{E}(\underline{r}) + \underline{\underline{\mu}}^{\mathcal{B}}(z) \cdot \underline{H}(\underline{r}), & z < 0 \end{cases} \end{aligned} \right\}, \tag{3.73}$$

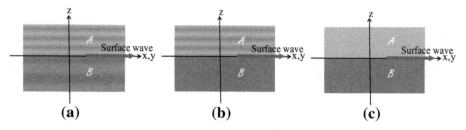

Figure 3.1 Schematics of the canonical boundary-value problem for surface-wave propagation guided by the planar interface of two dissimilar materials $\mathcal{A}$ and $\mathcal{B}$. (a) Both materials are periodically nonhomogeneous in the direction normal to the interface. (b) Only one is periodically nonhomogeneous, and the other is homogeneous. (c) Both materials are homogeneous.

where

$$\underline{\underline{\epsilon}}^A(z \pm 2\Omega^A) = \underline{\underline{\epsilon}}^A(z), \quad \underline{\underline{\xi}}^A(z \pm 2\Omega^A) = \underline{\underline{\xi}}^A(z)$$
$$\underline{\underline{\zeta}}^A(z \pm 2\Omega^A) = \underline{\underline{\zeta}}^A(z), \quad \underline{\underline{\mu}}^A(z \pm 2\Omega^A) = \underline{\underline{\mu}}^A(z) \Bigg\} \tag{3.74}$$

and

$$\underline{\underline{\epsilon}}^B(z \pm 2\Omega^B) = \underline{\underline{\epsilon}}^B(z), \quad \underline{\underline{\xi}}^B(z \pm 2\Omega^B) = \underline{\underline{\xi}}^B(z)$$
$$\underline{\underline{\zeta}}^B(z \pm 2\Omega^B) = \underline{\underline{\zeta}}^B(z), \quad \underline{\underline{\mu}}^B(z \pm 2\Omega^B) = \underline{\underline{\mu}}^B(z) \Bigg\}. \tag{3.75}$$

The surface wave propagates in the xy plane in a direction at an angle $\psi \in [0°, 360°)$ with respect to the x axis. Then, the unit vector in the direction of propagation is given by

$$\underline{u}_{\text{prop}} = \underline{u}_x \cos \psi + \underline{u}_y \sin \psi \tag{3.76}$$

and, as in Sections 3.3 and 3.4, the field phasors of the surface wave vary in the xy plane as

$$\exp\left(iq\underline{u}_{\text{prop}} \cdot \underline{r}\right) = \exp\left[iq(x \cos \psi + y \sin \psi)\right], \tag{3.77}$$

where q is a complex-valued wavenumber. The phase speed of the surface wave is

$$v_p = \omega/\text{Re}\{q\} \tag{3.78}$$

and the propagation length is

$$\Delta_{\text{prop}} = 1/\text{Im}\{q\}. \tag{3.79}$$

The propagation length is defined as the distance parallel to $\underline{u}_{\text{prop}}$ over which the magnitude of the electric (or magnetic) field phasor decays by a factor of $\exp(-1)$. The attenuation rate in the direction of propagation is the reciprocal of Δ_{prop}.

In the half-space $z > 0$, the fields must decay as $z \to \infty$. Therefore, in this half-space,

$$
\left.
\begin{aligned}
\underline{E}(\underline{r}) &= \sum_{m=1}^{2} C_m^A \Big\{ t_{1m}^A(z)\underline{u}_x + t_{2m}^A(z)\underline{u}_y + \Big[v_{zx}^{ee,A}(z)t_{1m}^A(z) \\
&\quad + v_{zy}^{ee,A}(z)t_{2m}^A(z) + v_{zx}^{eh,A}(z)t_{3m}^A(z) + v_{zy}^{eh,A}(z)t_{4m}^A(z) \Big] \underline{u}_z \Big\} \\
&\quad \times \exp\left[iq(x \cos \psi + y \sin \psi)\right] \\
\underline{H}(\underline{r}) &= \sum_{m=1}^{2} C_m^A \Big\{ t_{3m}^A(z)\underline{u}_x + t_{4m}^A(z)\underline{u}_y + \Big[v_{zx}^{he,A}(z)t_{1m}^A(z) \\
&\quad + v_{zy}^{he,A}(z)t_{2m}^A(z) + v_{zx}^{hh,A}(z)t_{3m}^A(z) + v_{zy}^{hh,A}(z)t_{4m}^A(z) \Big] \underline{u}_z \Big\} \\
&\quad \times \exp\left[iq(x \cos \psi + y \sin \psi)\right]
\end{aligned}
\right\}, \tag{3.80}
$$

follow from Eqs. (3.66), (3.67), and (3.69), with the superscript $\mathcal{A}$ indicating that the
constitutive dyadics for material $\mathcal{A}$ must be used. Accordingly, the boundary value

$$
[\underline{f}(0+)] =
\begin{bmatrix}
v_{11}^{\mathcal{A}} & v_{12}^{\mathcal{A}} \\
v_{21}^{\mathcal{A}} & v_{22}^{\mathcal{A}} \\
v_{31}^{\mathcal{A}} & v_{32}^{\mathcal{A}} \\
v_{41}^{\mathcal{A}} & v_{42}^{\mathcal{A}}
\end{bmatrix}
\cdot
\begin{bmatrix}
C_1^{\mathcal{A}} \\
C_2^{\mathcal{A}}
\end{bmatrix}
\tag{3.81}
$$

emerges. Since the fields must decay as $z \to -\infty$ in the half-space $z < 0$, Eqs. (3.66),
(3.67), and (3.70) yield

$$
\left.
\begin{aligned}
\underline{E}(\underline{r}) &= \sum_{m=3}^{4} C_m^{\mathcal{B}} \left\{ t_{1m}^{\mathcal{B}}(z)\underline{u}_x + t_{2m}^{\mathcal{B}}(z)\underline{u}_y + \left[v_{zx}^{ee,\mathcal{B}}(z)t_{1m}^{\mathcal{B}}(z) \right.\right. \\
&\quad \left.\left. + v_{zy}^{ee,\mathcal{B}}(z)t_{2m}^{\mathcal{B}}(z) + v_{zx}^{eh,\mathcal{B}}(z)t_{3m}^{\mathcal{B}}(z) + v_{zy}^{eh,\mathcal{B}}(z)t_{4m}^{\mathcal{B}}(z) \right] \underline{u}_z \right\} \\
&\quad \times \exp\left[iq(x\cos\psi + y\sin\psi) \right] \\[6pt]
\underline{H}(\underline{r}) &= \sum_{m=3}^{4} C_m^{\mathcal{B}} \left\{ t_{3m}^{\mathcal{B}}(z)\underline{u}_x + t_{4m}^{\mathcal{B}}(z)\underline{u}_y + \left[v_{zx}^{he,\mathcal{B}}(z)t_{1m}^{\mathcal{B}}(z) \right.\right. \\
&\quad \left.\left. + v_{zy}^{he,\mathcal{B}}(z)t_{2m}^{\mathcal{B}}(z) + v_{zx}^{hh,\mathcal{B}}(z)t_{3m}^{\mathcal{B}}(z) + v_{zy}^{hh,\mathcal{B}}(z)t_{4m}^{\mathcal{B}}(z) \right] \underline{u}_z \right\} \\
&\quad \times \exp\left[iq(x\cos\psi + y\sin\psi) \right]
\end{aligned}
\right\}, \tag{3.82}
$$

for that half-space, with the superscript $\mathcal{B}$ indicating that the constitutive dyadics for
material $\mathcal{B}$ must be used. These equations yield the boundary value

$$
[\underline{f}(0-)] =
\begin{bmatrix}
v_{13}^{\mathcal{B}} & v_{14}^{\mathcal{B}} \\
v_{23}^{\mathcal{B}} & v_{24}^{\mathcal{B}} \\
v_{33}^{\mathcal{B}} & v_{34}^{\mathcal{B}} \\
v_{43}^{\mathcal{B}} & v_{44}^{\mathcal{B}}
\end{bmatrix}
\cdot
\begin{bmatrix}
C_3^{\mathcal{B}} \\
C_4^{\mathcal{B}}
\end{bmatrix}.
\tag{3.83}
$$

The 4×2 matrixes appearing on the right sides of Eqs. (3.81) and (3.83) comprise
two eigenvectors of the matrixes $\left[\underline{\underline{Q}}\right]^{\mathcal{A}}$ and $\left[\underline{\underline{Q}}\right]^{\mathcal{B}}$, respectively, which are defined
via Eqs. (3.56). If either or both materials are homogeneous, then the eigenvectors of
the matrixes $\left[\underline{\underline{P}}\right]^{\mathcal{A}}$ and $\left[\underline{\underline{P}}\right]^{\mathcal{B}}$, defined via Eq. (3.24), must be used as appropriate.

The tangential components of the electric and magnetic field phasors must be continuous across the interface $z = 0$, yielding the boundary condition

$$
[\underline{f}(0-)] = [\underline{f}(0+)], \tag{3.84}
$$

which can be rearranged as

$$
[\underline{\underline{Y}}] \cdot
\begin{bmatrix}
C_1^{\mathcal{A}} \\
C_2^{\mathcal{A}} \\
C_3^{\mathcal{B}} \\
C_4^{\mathcal{B}}
\end{bmatrix}
=
\begin{bmatrix}
0 \\
0 \\
0 \\
0
\end{bmatrix},
\tag{3.85}
$$

with the 4×4 matrix

$$
\left[\underline{\underline{Y}}(q) \right] =
\begin{bmatrix}
v_{11}^A(q) & v_{12}^A(q) & -v_{13}^B(q) & -v_{14}^B(q) \\
v_{21}^A(q) & v_{22}^A(q) & -v_{23}^B(q) & -v_{24}^B(q) \\
v_{31}^A(q) & v_{32}^A(q) & -v_{33}^B(q) & -v_{34}^B(q) \\
v_{41}^A(q) & v_{42}^A(q) & -v_{43}^B(q) & -v_{44}^B(q)
\end{bmatrix}
\tag{3.86}
$$

written explicitly as a function of q.

For a non-trivial solution, $\left[\underline{\underline{Y}}(q) \right]$ must be singular, so that

$$
\det \left\{ \left[\underline{\underline{Y}}(q) \right] \right\} = 0
\tag{3.87}
$$

is the dispersion equation of the surface wave. This equation has to be solved in order to determine the wavenumber q. Not only will the solutions of Eq. (3.87) depend on the constitutive dyadics of the materials $\mathcal{A}$ and $\mathcal{B}$ as well as on the angular frequency ω, but also—in general—on the direction of propagation in the xy plane as delineated by the angle ψ.

3.5.2 Computational Matters

The most convenient way to solve the dispersion equation (3.87) is to use the Newton-Raphson technique [261, Section 4.5]. Since the function

$$
\Upsilon(q) = \det \left\{ \left[\underline{\underline{Y}}(q) \right] \right\}
\tag{3.88}
$$

is not known analytically, its derivative with respect to q may be estimated using the central difference formula by

$$
\Xi(q) = \frac{\Upsilon \left(q + \delta_q q \right) - \Upsilon \left(q - \delta_q q \right)}{2 \delta_q q},
\tag{3.89}
$$

where the magnitude of the real number $\delta_q \ll 1$. Then, the Newton-Raphson algorithm

$$
q \leftarrow q - \frac{\Upsilon(q)}{\Xi(q)}
\tag{3.90}
$$

can be numerically implemented with an initial guess for the solution of Eq. (3.87). A map of $\Upsilon(q)$ versus q helps greatly in finding initial guesses.

Equation (3.88) generally works well even when only real-valued solutions of Eq. (3.87) are sought. However, sometimes, it may be convenient to use

$$
\Upsilon(q) = \left| \det \left\{ \left[\underline{\underline{Y}}(q) \right] \right\} \right|,
\tag{3.91}
$$

in order to determine real-valued solutions.

Let the solutions of Eq. (3.87) be denoted as q_1, q_2, etc. The field phasors associated with a specific solution can be found after replacing q by the numerical value of that solution in Eq. (3.85). Now, the matrix equation

$$
\begin{bmatrix}
v_{11}^A(q_\ell) & v_{12}^A(q_\ell) & -v_{13}^B(q_\ell) & -v_{14}^B(q_\ell) \\
v_{21}^A(q_\ell) & v_{22}^A(q_\ell) & -v_{23}^B(q_\ell) & -v_{24}^B(q_\ell) \\
v_{31}^A(q_\ell) & v_{32}^A(q_\ell) & -v_{33}^B(q_\ell) & -v_{34}^B(q_\ell) \\
v_{41}^A(q_\ell) & v_{42}^A(q_\ell) & -v_{43}^B(q_\ell) & -v_{44}^B(q_\ell)
\end{bmatrix}
\cdot
\begin{bmatrix}
C_1^A(q_\ell) \\
C_2^A(q_\ell) \\
C_3^B(q_\ell) \\
C_4^B(q_\ell)
\end{bmatrix}
=
\begin{bmatrix}
0 \\ 0 \\ 0 \\ 0
\end{bmatrix}
, \quad \ell \in \{1, 2, \ldots\},
$$

(3.92)

contains only three linearly independent algebraic equations, which means that only three of the four coefficients in the column vector on the left side of Eq. (3.92) can be determined. Therefore, $C_1^A(q_\ell)$ should be set equal to unity, and the remaining three coefficients then determined by the solution of any three of the four algebraic equations in Eq. (3.92). If that procedure proves infructuous because it ends up involving an indeterminate ratio, the procedure should be repeated with $C_2^A(q_\ell)$, instead of $C_1^A(q_\ell)$, set equal to unity. The spatial profiles of the field phasors associated with a surface wave with wavenumber q_ℓ then follows from Eqs. (3.80) and (3.82), and the spatial profile of the time-averaged Poynting vector

$$
\underline{P}(\underline{r}) = (1/2)\mathrm{Re}\left\{\underline{E}(\underline{r}) \times \underline{H}^*(\underline{r})\right\}
$$

(3.93)

can be ascertained thereafter.

3.6 Modified Canonical Boundary-Value Problem

Whereas the materials $\mathcal{A}$ and $\mathcal{B}$ are in contact in Section 3.5.1, a modification of the canonical boundary-value problem [182, 297, 298] involves the interposition of a slab of a third material—or, even, vacuum—between the half-spaces filled with $\mathcal{A}$ and $\mathcal{B}$. Depending on the separation d between the two half-spaces, the modified structure can guide several different types of waves: (i) each of the two interfaces can guide surface waves independently of the other, (ii) the two interfaces guide coupled waves, and (iii) the interposing slab guides bulk modes or waveguide modes [299].

Suppose that the material $\mathcal{A}$ occupies the region $z > d$ and the material $\mathcal{B}$ fills the region $z < 0$. The constitutive relations (3.73) then have to be stated in a slightly modified form as

$$
\left.
\begin{aligned}
\underline{D}(\underline{r}) &= \begin{cases} \underline{\underline{\epsilon}}^A(z-d) \bullet \underline{E}(\underline{r}) + \underline{\underline{\xi}}^A(z-d) \bullet \underline{H}(\underline{r}), & z > d \\ \underline{\underline{\epsilon}}^B(z) \bullet \underline{E}(\underline{r}) + \underline{\underline{\xi}}^B(z) \bullet \underline{H}(\underline{r}), & z < 0 \end{cases} \\
\underline{B}(\underline{r}) &= \begin{cases} \underline{\underline{\zeta}}^A(z-d) \bullet \underline{E}(\underline{r}) + \underline{\underline{\mu}}^A(z-d) \bullet \underline{H}(\underline{r}), & z > d \\ \underline{\underline{\zeta}}^B(z) \bullet \underline{E}(\underline{r}) + \underline{\underline{\mu}}^B(z) \bullet \underline{H}(\underline{r}), & z < 0 \end{cases}
\end{aligned}
\right\}.
$$

(3.94)

Accordingly, Eqs. (3.80) must be changed to

$$
\left.
\begin{aligned}
\underline{E}(\underline{r}) &= \sum_{m=1}^{2} C_m^{\mathcal{A}} \Big\{ t_{1m}^{\mathcal{A}}(z-d)\underline{u}_x + t_{2m}^{\mathcal{A}}(z-d)\underline{u}_y + \Big[v_{zx}^{ee,\mathcal{A}}(z-d)t_{1m}^{\mathcal{A}}(z-d) \\
&\quad + v_{zy}^{ee,\mathcal{A}}(z-d)t_{2m}^{\mathcal{A}}(z-d) + v_{zx}^{eh,\mathcal{A}}(z-d)t_{3m}^{\mathcal{A}}(z-d) \\
&\quad + v_{zy}^{eh,\mathcal{A}}(z-d)t_{4m}^{\mathcal{A}}(z-d) \Big] \underline{u}_z \Big\} \exp\big[iq(x\cos\psi + y\sin\psi) \big] \\[2mm]
\underline{H}(\underline{r}) &= \sum_{m=1}^{2} C_m^{\mathcal{A}} \Big\{ t_{3m}^{\mathcal{A}}(z-d)\underline{u}_x + t_{4m}^{\mathcal{A}}(z-d)\underline{u}_y + \Big[v_{zx}^{he,\mathcal{A}}(z-d)t_{1m}^{\mathcal{A}}(z-d) \\
&\quad + v_{zy}^{he,\mathcal{A}}(z-d)t_{2m}^{\mathcal{A}}(z-d) + v_{zx}^{hh,\mathcal{A}}(z-d)t_{3m}^{\mathcal{A}}(z-d) \\
&\quad + v_{zy}^{hh,\mathcal{A}}(z-d)t_{4m}^{\mathcal{A}}(z-d) \Big] \underline{u}_z \Big\} \exp\big[iq(x\cos\psi + y\sin\psi) \big]
\end{aligned}
\right\}
$$

$$(3.95)$$

for $z \geqslant d$, so that one of the two needed boundary values is

$$
[\,\underline{f}(d+)\,] =
\begin{bmatrix}
v_{11}^{\mathcal{A}} & v_{12}^{\mathcal{A}} \\
v_{21}^{\mathcal{A}} & v_{22}^{\mathcal{A}} \\
v_{31}^{\mathcal{A}} & v_{32}^{\mathcal{A}} \\
v_{41}^{\mathcal{A}} & v_{42}^{\mathcal{A}}
\end{bmatrix}
\cdot
\begin{bmatrix}
C_1^{\mathcal{A}} \\
C_2^{\mathcal{A}}
\end{bmatrix}.
$$

$$(3.96)$$

As Eqs. (3.82) still hold for $z \leqslant 0$, the other boundary value $\big[\,\underline{f}(0-)\,\big]$ is still given by Eq. (3.83).

In order to connect $\big[\,\underline{f}(d+)\,\big]$ and $\big[\,\underline{f}(0-)\,\big]$, the material occupying the region $0 < z < d$ has to be specified. Suppose that this region is occupied by a cascade of slabs of three homogeneous materials—labeled I, II, and III, and of thicknesses d_I, d_{II}, and d_{III} —as shown in Figure 3.2. The applicable constitutive relations can

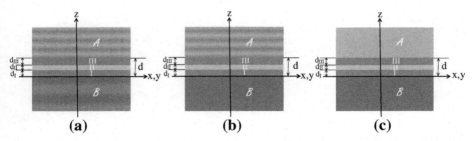

Figure 3.2 Schematics of the modified canonical boundary-value problem for wave propagation guided by two dissimilar materials $\mathcal{A}$ and $\mathcal{B}$ separated by a cascade of three slabs of homogeneous materials I, II, and III. (a) Both materials $\mathcal{A}$ and $\mathcal{B}$ are periodically nonhomogeneous in the direction normal to the interface. (b) $\mathcal{A}$ is periodically nonhomogeneous but $\mathcal{B}$ is homogeneous. (c) Both $\mathcal{A}$ and $\mathcal{B}$ are homogeneous.

be stated as

$$\underline{D}(\underline{r}) = \begin{cases} \underline{\underline{\epsilon}}^I \cdot \underline{E}(\underline{r}) + \underline{\underline{\xi}}^I \cdot \underline{H}(\underline{r}), & 0 < z < d_I \\ \underline{\underline{\epsilon}}^{II} \cdot \underline{E}(\underline{r}) + \underline{\underline{\xi}}^{II} \cdot \underline{H}(\underline{r}), & d_I < z < d_I + d_{II} \\ \underline{\underline{\epsilon}}^{III} \cdot \underline{E}(\underline{r}) + \underline{\underline{\xi}}^{III} \cdot \underline{H}(\underline{r}), & d_I + d_{II} < z < d_I + d_{II} + d_{III} \end{cases}$$

$$\underline{B}(\underline{r}) = \begin{cases} \underline{\underline{\zeta}}^I \cdot \underline{E}(\underline{r}) + \underline{\underline{\mu}}^I \cdot \underline{H}(\underline{r}), & 0 < z < d_I \\ \underline{\underline{\zeta}}^{II} \cdot \underline{E}(\underline{r}) + \underline{\underline{\mu}}^{II} \cdot \underline{H}(\underline{r}), & d_I < z < d_I + d_{II} \\ \underline{\underline{\zeta}}^{III} \cdot \underline{E}(\underline{r}) + \underline{\underline{\mu}}^{III} \cdot \underline{H}(\underline{r}), & d_I + d_{II} < z < d_I + d_{II} + d_{III} \end{cases}$$

$$(3.97)$$

with $d = d_I + d_{II} + d_{III}$. Then, the application of Eq. (3.28) to each of the three slabs in succession yields

$$\begin{aligned} [\underline{f}(d_I-)] &= \exp\left\{i\,[\underline{\underline{P}}]^I\,d_I\right\} \cdot [\underline{f}(0+)] \\ [\underline{f}(d_I + d_{II}-)] &= \exp\left\{i\,[\underline{\underline{P}}]^{II}\,d_{II}\right\} \cdot [\underline{f}(d_I+)] \\ [\underline{f}(d_I + d_{II} + d_{III}-)] &= \exp\left\{i\,[\underline{\underline{P}}]^{III}\,d_{III}\right\} \cdot [\underline{f}(d_I + d_{II}+)] \end{aligned} \right\}. \quad (3.98)$$

As the tangential components of the electric and magnetic field phasors must be continuous across both bimaterial surfaces within the region $0 < z < d$, the foregoing equations lead to

$$\begin{aligned} [\underline{f}(d-)] &= \exp\left\{i\,[\underline{\underline{P}}]^{III}\,d_{III}\right\} \cdot \exp\left\{i\,[\underline{\underline{P}}]^{II}\,d_{II}\right\} \\ &\quad \cdot \exp\left\{i\,[\underline{\underline{P}}]^I\,d_I\right\} \cdot [\underline{f}(0+)]\,. \end{aligned} \quad (3.99)$$

Finally, enforcement of the boundary conditions across the interfaces $z = 0$ and $z = d$ yields

$$\begin{bmatrix} v_{11}^A & v_{12}^A \\ v_{21}^A & v_{22}^A \\ v_{31}^A & v_{32}^A \\ v_{41}^A & v_{42}^A \end{bmatrix} \cdot \begin{bmatrix} C_1^A \\ C_2^A \end{bmatrix} = \exp\left\{i\,[\underline{\underline{P}}]^{III}\,d_{III}\right\} \cdot \exp\left\{i\,[\underline{\underline{P}}]^{II}\,d_{II}\right\}$$

$$\cdot \exp\left\{i\,[\underline{\underline{P}}]^I\,d_I\right\} \cdot \begin{bmatrix} v_{13}^B & v_{14}^B \\ v_{23}^B & v_{24}^B \\ v_{33}^B & v_{34}^B \\ v_{43}^B & v_{44}^B \end{bmatrix} \cdot \begin{bmatrix} C_3^B \\ C_4^B \end{bmatrix}.$$

$$(3.100)$$

This equation can be rearranged as

$$[\underline{\underline{Y}}] \cdot \begin{bmatrix} C_1^A \\ C_2^A \\ C_3^B \\ C_4^B \end{bmatrix} = \begin{bmatrix} 0 \\ 0 \\ 0 \\ 0 \end{bmatrix}, \quad (3.101)$$

but now the matrix $\left[\underline{\underline{Y}}\right]$ is different from the right side of Eq. (3.86). The dispersion equation

$$\det\left\{\left[\underline{\underline{Y}}(q)\right]\right\} = 0 \tag{3.102}$$

can then be solved numerically, using the Newton-Raphson technique described in Section 3.5.2.

3.7 Prism-Coupled Configuration

Canonical boundary-value problems are the playthings of theorists. Although the solutions of these problems are essential for understanding the characteristics of surface waves, exploitation of surface waves is done in a practical setting in which device dimensions cannot be infinite. The most common practical configuration for exciting SPP waves is the Turbadar-Kretschmann-Raether configuration of Figure 1.2a., the Turbadar-Otto configuration of Figure 1.2b. running a somewhat distant second. Both of these prism-coupled configurations can be theoretically treated in the same way, as also the prism-coupled configuration—shown in Figure 3.3—that can be used for the first reported observation [41] of a Dyakonov wave.

The essential part of the boundary-value problem for the prism-coupled configuration is depicted in Figure 3.4. The half-space $z < 0$ is occupied by a homogeneous, isotropic, dielectric material described by the refractive index n_1. Dissipation in this material is considered to be negligible so that n_1 is real valued and positive. The region $0 < z < d_A$ is filled with a nonhomogeneous bianisotropic material with constitutive relations

$$\left.\begin{array}{l}\underline{D}(\underline{r}) = \underline{\underline{\epsilon}}^{A}(z) \cdot \underline{E}(\underline{r}) + \underline{\underline{\xi}}^{A}(z) \cdot \underline{H}(\underline{r}) \\ \underline{B}(\underline{r}) = \underline{\underline{\zeta}}^{A}(z) \cdot \underline{E}(\underline{r}) + \underline{\underline{\mu}}^{A}(z) \cdot \underline{H}(\underline{r})\end{array}\right\}, \quad 0 < z < d_A, \tag{3.103}$$

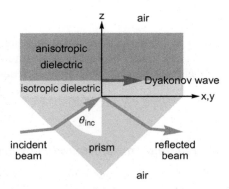

Figure 3.3 Schematic of the prism-coupled configuration used by Takayama et al. [41] for the first experimental observation of a Dyakonov wave.

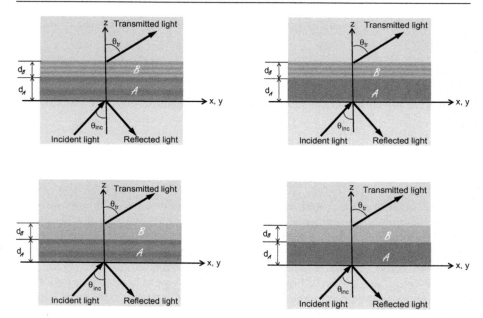

Figure 3.4 Schematics of the boundary-value problem for the prism-coupled configuration to excite surface waves guided by two dissimilar materials $\mathcal{A}$ and $\mathcal{B}$. (a) Both $\mathcal{A}$ and $\mathcal{B}$ are periodically nonhomogeneous in the direction normal to the interface. (b, c) Either $\mathcal{A}$ or $\mathcal{B}$ is periodically nonhomogeneous, but the other partnering material is homogeneous. (d) Both $\mathcal{A}$ and $\mathcal{B}$ are homogeneous. Whereas the angle $\theta_{inc} \in [0°, 90°)$, the angle θ_{tr} may be either real or complex valued.

and the region $d_A < z < d_A + d_B$ is occupied by a material with constitutive relations

$$
\left.
\begin{aligned}
\underline{D}(\underline{r}) &= \underline{\underline{\epsilon}}^{\mathcal{B}}(z) \cdot \underline{E}(\underline{r}) + \underline{\underline{\xi}}^{\mathcal{B}}(z) \cdot \underline{H}(\underline{r}) \\
\underline{B}(\underline{r}) &= \underline{\underline{\zeta}}^{\mathcal{B}}(z) \cdot \underline{E}(\underline{r}) + \underline{\underline{\mu}}^{\mathcal{B}}(z) \cdot \underline{H}(\underline{r})
\end{aligned}
\right\}, \quad d_A < z < d_A + d_B.
\tag{3.104}
$$

Finally, the half-space $z > d = d_A + d_B$ is occupied by a homogeneous, isotropic, dielectric material whose refractive index n_2 is assumed to be real valued and positive. All constitutive parameters generally depend on the angular frequency ω of the incident plane wave.

3.7.1   Incident, Reflected, and Transmitted Plane Waves

An arbitrarily polarized plane wave, propagating in the half-space $z < 0$ at an angle $\theta_{inc} \in [0°, 90°)$ to the z axis and at an angle $\psi \in [0°, 360°)$ to the x axis in the xy plane, is incident on the plane $z = 0$. The field phasors associated with the incident plane wave are represented as

$$
\left.\begin{aligned}
\underline{E}_{\text{inc}}(\underline{r}) &= \left(a_s \underline{s} + a_p \underline{p}_{\text{inc}}\right) \exp\{i[\kappa(x \cos \psi + y \sin \psi) + k_0 n_1 z \cos \theta_{\text{inc}}]\} \\
\underline{H}_{\text{inc}}(\underline{r}) &= n_1 \eta_0^{-1} \left(a_s \underline{p}_{\text{inc}} - a_p \underline{s}\right) \exp\{i[\kappa(x \cos \psi + y \sin \psi) \\
&\quad + k_0 n_1 z \cos \theta_{\text{inc}}]\} \\
&\qquad z < 0,
\end{aligned}\right\}
\tag{3.105}
$$

where $\eta_0 = \sqrt{\mu_0/\epsilon_0}$ is the intrinsic impedance of free space; the amplitudes of the s- and the p-polarized components of the incident plane wave, denoted by a_s and a_p, respectively, are assumed to be known; and

$$
\left.\begin{aligned}
\kappa &= k_0 n_1 \sin \theta_{\text{inc}} \\
\underline{s} &= -\underline{u}_x \sin \psi + \underline{u}_y \cos \psi \\
\underline{p}_{\text{inc}} &= -(\underline{u}_x \cos \psi + \underline{u}_y \sin \psi) \cos \theta_{\text{inc}} + \underline{u}_z \sin \theta_{\text{inc}}
\end{aligned}\right\}
\tag{3.106}
$$

The reflected field phasors are expressed as

$$
\left.\begin{aligned}
\underline{E}_{\text{ref}}(\underline{r}) &= (r_s \underline{s} + r_p \underline{p}_{\text{ref}}) \exp\{i[\kappa(x \cos \psi + y \sin \psi) - k_0 n_1 z \cos \theta_{\text{inc}}]\} \\
\underline{H}_{\text{ref}}(\underline{r}) &= n_1 \eta_0^{-1} \left(r_s \underline{p}_{\text{ref}} - r_p \underline{s}\right) \exp\{i[\kappa(x \cos \psi + y \sin \psi) \\
&\quad - k_0 n_1 z \cos \theta_{\text{inc}}]\} \\
&\qquad z < 0,
\end{aligned}\right\}
\tag{3.107}
$$

and the transmitted field phasors as

$$
\left.\begin{aligned}
\underline{E}_{\text{tr}}(\underline{r}) &= (t_s \underline{s} + t_p \underline{p}_{\text{tr}}) \exp\{i[\kappa(x \cos \psi + y \sin \psi) + k_0 n_2 (z - d) \cos \theta_{\text{tr}}]\} \\
\underline{H}_{\text{tr}}(\underline{r}) &= n_2 \eta_0^{-1} \left(t_s \underline{p}_{\text{tr}} - t_p \underline{s}\right) \exp\{i[\kappa(x \cos \psi + y \sin \psi) \\
&\quad + k_0 n_2 (z - d) \cos \theta_{\text{tr}}]\} \\
&\qquad z > d = d_A + d_B,
\end{aligned}\right\}
\tag{3.108}
$$

where

$$
\left.\begin{aligned}
\sin \theta_{\text{tr}} &= (n_1/n_2) \sin \theta_{\text{inc}} \\
\cos \theta_{\text{tr}} &= +\sqrt{1 - \sin \theta_{\text{tr}}^2} \\
\underline{p}_{\text{ref}} &= (\underline{u}_x \cos \psi + \underline{u}_y \sin \psi) \cos \theta_{\text{inc}} + \underline{u}_z \sin \theta_{\text{inc}} \\
\underline{p}_{\text{tr}} &= -(\underline{u}_x \cos \psi + \underline{u}_y \sin \psi) \cos \theta_{\text{tr}} + \underline{u}_z \sin \theta_{\text{tr}}
\end{aligned}\right\}
\tag{3.109}
$$

The reflection amplitudes r_s and r_p, as well as the transmission amplitudes t_s and t_p, have to be determined by the solution of a boundary-value problem. As $\sin \theta_{\text{tr}}$ can exceed unity, by virtue of the satisfaction of phase-matching conditions, $\cos \theta_{\text{tr}}$ can be complex valued—in which case, the transmitted wave is evanescent.

3.7.2   Solution of Boundary-Value Problem

In line with Eqs. (3.15), the field phasors defined in Eqs. (3.105), (3.107), and (3.108) are of the form

$$
\left.
\begin{aligned}
\underline{E}(\underline{r}) &= \underline{e}(z)\exp[i\kappa(x\cos\psi + y\sin\psi)] \\
\underline{H}(\underline{r}) &= \underline{h}(z)\exp[i\kappa(x\cos\psi + y\sin\psi)]
\end{aligned}
\right\},
\tag{3.110}
$$

for all $z \in (-\infty, \infty)$. Then the column vector $[\underline{f}(z)]$ of Eq. (3.23) can again be used profitably.

Indeed, $[\underline{f}(z)]$ satisfies the matrix ordinary differential equations

$$
\frac{d}{dz}[\underline{f}(z)] = i[\underline{P}(z)]^{\mathcal{A}} \cdot [\underline{f}(z)], \quad 0 < z < d_{\mathcal{A}},
\tag{3.111}
$$

and

$$
\frac{d}{dz}[\underline{f}(z)] = i[\underline{P}(z)]^{\mathcal{B}} \cdot [\underline{f}(z)], \quad d_{\mathcal{A}} < z < d_{\mathcal{A}} + d_{\mathcal{B}}.
\tag{3.112}
$$

The matrix $[\underline{P}(z)]^{\mathcal{A}}$ is defined by Eq. (3.24), but with q replaced by κ and ϵ_{xy} replaced by $\epsilon_{xy}^{\mathcal{A}}(z)$, etc. A similar replacement procedure yields the matrix $[\underline{P}(z)]^{\mathcal{B}}$.

The solutions of Eqs. (3.111) and (3.112) must deliver the relations

$$
\left.
\begin{aligned}
[\underline{f}(d_{\mathcal{A}}-)] &= [\underline{M}]^{\mathcal{A}} \cdot [\underline{f}(0+)] \\
[\underline{f}(d-)] &= [\underline{M}]^{\mathcal{B}} \cdot [\underline{f}(d_{\mathcal{A}}+)]
\end{aligned}
\right\}.
\tag{3.113}
$$

Computation of the 4×4 matrixes $[\underline{M}]^{\mathcal{A}}$ and $[\underline{M}]^{\mathcal{B}}$ in these relations may require the use of the piecewise-uniform-approximation method.

Let the region $0 \leqslant z \leqslant d_{\mathcal{A}}$ be divided into $N_{\mathcal{A}}$ subregions, the nth subregion being $z_{n-1} \leqslant z \leqslant z_n, n \in [1, N_{\mathcal{A}}]$. In each subregion, either $[\underline{P}(z)]^{\mathcal{A}}$ is uniform or can be approximated as being uniform. Then,

$$
[\underline{M}]^{\mathcal{A}} \cong [\underline{W}]^{(N_{\mathcal{A}})} \cdot [\underline{W}]^{(N_{\mathcal{A}}-1)} \cdot \cdots [\underline{W}]^{(2)} \cdot [\underline{W}]^{(1)},
\tag{3.114}
$$

where

$$
[\underline{W}]^{(n)} = \exp\left\{ i(z_n - z_{n-1}) \left[\underline{P}\left(\frac{z_n + z_{n-1}}{2}\right)\right]^{\mathcal{A}} \right\}, \quad n \in [1, N_{\mathcal{A}}].
\tag{3.115}
$$

The calculation of $[\underline{M}]^{\mathcal{A}}$ is very simple if material $\mathcal{A}$ is homogeneous, because then $N_{\mathcal{A}} = 1$ suffices. Similarly, if the region $d_{\mathcal{A}} \leqslant z \leqslant d_{\mathcal{A}} + d_{\mathcal{B}}$ is divided into $N_{\mathcal{B}}$ subregions, the nth subregion being $z_{n-1} \leqslant z \leqslant z_n, n \in [1, N_{\mathcal{B}}]$, and either $[\underline{P}(z)]^{\mathcal{B}}$ is uniform or can be approximated as being uniform in each of these subregions, then

$$
[\underline{M}]^{\mathcal{B}} \cong [\underline{W}]^{(N_{\mathcal{B}})} \cdot [\underline{W}]^{(N_{\mathcal{B}}-1)} \cdot \cdots [\underline{W}]^{(2)} \cdot [\underline{W}]^{(1)},
\tag{3.116}
$$

where *now*

$$\left[\underline{\underline{W}}\right]^{(n)} = \exp\left\{ i\left(z_n - z_{n-1}\right)\left[\underline{\underline{P}}\left(\frac{z_n + z_{n-1}}{2}\right)\right]^{\mathcal{B}}\right\}, \quad n \in [1, N_{\mathcal{B}}]. \quad (3.117)$$

Just $N_{\mathcal{B}} = 1$ is needed if material $\mathcal{B}$ is homogeneous.

Equations (3.114) and (3.116) indicate that there is no requirement in the present formulation that materials $\mathcal{A}$ and/or $\mathcal{B}$ be *periodically* nonhomogeneous along the z axis. Either material can be homogeneous, piecewise homogeneous, periodically nonhomogeneous, or nonhomogeneous in some other fashion—in that direction.

Equations (3.105) and (3.107) yield the boundary value

$$\left[\underline{f}(0-)\right] = \left[\underline{\underline{K}}\right]_{\text{inc}} \cdot \begin{bmatrix} a_s \\ a_p \\ r_s \\ r_p \end{bmatrix} \qquad (3.118)$$

and Eqs. (3.108) deliver the boundary value

$$\left[\underline{f}(d+)\right] = \left[\underline{\underline{K}}\right]_{\text{tr}} \cdot \begin{bmatrix} t_s \\ t_p \\ 0 \\ 0 \end{bmatrix} \qquad (3.119)$$

with

$$\left[\underline{\underline{K}}\right]_{\text{inc}} = \begin{bmatrix} -\sin\psi & -\cos\psi\cos\theta_{\text{inc}} & -\sin\psi & \cos\psi\cos\theta_{\text{inc}} \\ \cos\psi & -\sin\psi\cos\theta_{\text{inc}} & \cos\psi & \sin\psi\cos\theta_{\text{inc}} \\ -\left(\frac{n_1}{\eta_0}\right)\cos\psi\cos\theta_{\text{inc}} & \left(\frac{n_1}{\eta_0}\right)\sin\psi & \left(\frac{n_1}{\eta_0}\right)\cos\psi\cos\theta_{\text{inc}} & \left(\frac{n_1}{\eta_0}\right)\sin\psi \\ -\left(\frac{n_1}{\eta_0}\right)\sin\psi\cos\theta_{\text{inc}} & -\left(\frac{n_1}{\eta_0}\right)\cos\psi & \left(\frac{n_1}{\eta_0}\right)\sin\psi\cos\theta_{\text{inc}} & -\left(\frac{n_1}{\eta_0}\right)\cos\psi \end{bmatrix} \qquad (3.120)$$

and

$$\left[\underline{\underline{K}}\right]_{\text{tr}} = \begin{bmatrix} -\sin\psi & -\cos\psi\cos\theta_{\text{tr}} & -\sin\psi & \cos\psi\cos\theta_{\text{tr}} \\ \cos\psi & -\sin\psi\cos\theta_{\text{tr}} & \cos\psi & \sin\psi\cos\theta_{\text{tr}} \\ -\left(\frac{n_2}{\eta_0}\right)\cos\psi\cos\theta_{\text{tr}} & \left(\frac{n_2}{\eta_0}\right)\sin\psi & \left(\frac{n_2}{\eta_0}\right)\cos\psi\cos\theta_{\text{tr}} & \left(\frac{n_2}{\eta_0}\right)\sin\psi \\ -\left(\frac{n_2}{\eta_0}\right)\sin\psi\cos\theta_{\text{tr}} & -\left(\frac{n_2}{\eta_0}\right)\cos\psi & \left(\frac{n_2}{\eta_0}\right)\sin\psi\cos\theta_{\text{tr}} & -\left(\frac{n_2}{\eta_0}\right)\cos\psi \end{bmatrix}. \qquad (3.121)$$

The standard boundary conditions of electromagnetics require that

$$\left[\underline{f}(0-)\right] = \left[\underline{f}(0+)\right], \quad \left[\underline{f}(d_{\mathcal{A}}-)\right] = \left[\underline{f}(d_{\mathcal{A}}+)\right], \quad \left[\underline{f}(d-)\right] = \left[\underline{f}(d+)\right], \qquad (3.122)$$

thereby yielding the matrix equation

$$
\begin{bmatrix} t_s \\ t_p \\ 0 \\ 0 \end{bmatrix} = [\underline{\underline{K}}]_{tr}^{-1} \cdot [\underline{\underline{M}}]^{\mathcal{B}} \cdot [\underline{\underline{M}}]^{\mathcal{A}} \cdot [\underline{\underline{K}}]_{inc} \cdot \begin{bmatrix} a_s \\ a_p \\ r_s \\ r_p \end{bmatrix}.
\tag{3.123}
$$

This step completes the formulation of the boundary-value problem.

3.7.3   Linear Reflectances and Transmittances

The solution of Eq. (3.123) yields the linear reflection and transmission coefficients that appear as the elements of the 2×2 matrixes in the following relations:

$$
\left.
\begin{aligned}
\begin{bmatrix} r_s \\ r_p \end{bmatrix} &= \begin{bmatrix} r_{ss} & r_{sp} \\ r_{ps} & r_{pp} \end{bmatrix} \cdot \begin{bmatrix} a_s \\ a_p \end{bmatrix} \\
\begin{bmatrix} t_s \\ t_p \end{bmatrix} &= \begin{bmatrix} t_{ss} & t_{sp} \\ t_{ps} & t_{pp} \end{bmatrix} \cdot \begin{bmatrix} a_s \\ a_p \end{bmatrix}
\end{aligned}
\right\}.
\tag{3.124}
$$

Co-polarized coefficients have both subscripts identical, but cross-polarized coefficients do not. The square of the magnitude of a linear reflection coefficient equals the corresponding linear reflectance; thus,

$$
R_{sp} = |r_{sp}|^2
\tag{3.125}
$$

is the linear reflectance corresponding to the linear reflection coefficient r_{sp}, and so on. Linear transmittances are defined as

$$
T_{sp} = \frac{n_2}{n_1} \frac{\text{Re}\{\cos\theta_{tr}\}}{\cos\theta_{inc}} |t_{sp}|^2,
\tag{3.126}
$$

etc.

The principle of conservation of energy mandates the constraints

$$
\left.
\begin{aligned}
R_{ss} + R_{ps} + T_{ss} + T_{ps} &\leq 1 \\
R_{pp} + R_{sp} + T_{pp} + T_{sp} &\leq 1
\end{aligned}
\right\},
\tag{3.127}
$$

the inequalities turning to equalities only in the lack of dissipation in the regions occupied by materials $\mathcal{A}$ and $\mathcal{B}$.

The variations with θ_{inc} of the linear absorptances

$$
\left.
\begin{aligned}
A_s &= 1 - (R_{ss} + R_{ps} + T_{ss} + T_{ps}) \\
A_p &= 1 - (R_{pp} + R_{sp} + T_{pp} + T_{sp})
\end{aligned}
\right\}
\tag{3.128}
$$

for fixed k_0 and ψ need to be studied in order to identify the excitation of surface waves guided by the interface $z = d_{\mathcal{A}}$ of materials $\mathcal{A}$ and $\mathcal{B}$. Features that do not change with increase of $d_{\mathcal{A}}$ and $d_{\mathcal{B}}$ beyond certain thresholds are likely to indicate surface-wave propagation, and should be correlated with the solutions q of the corresponding canonical problem treated in Section 3.5.1. The thickness of a partnering material that is periodically nonhomogeneous should be increased in increments of a period, for the identification of surface waves.

3.7.4 Circular Reflectances and Transmittances

Instead of s- and p-polarized components, the arbitrarily polarized incident plane wave can be decomposed into left- and right-circularly polarized components. Thus, Eqs. (3.105) may be replaced by

$$
\left.
\begin{aligned}
\underline{E}_{\text{inc}}(\underline{r}) &= \left(a_L \frac{i\underline{s} - \underline{p}_{\text{inc}}}{\sqrt{2}} - a_R \frac{i\underline{s} + \underline{p}_{\text{inc}}}{\sqrt{2}}\right) \exp\{i[\kappa(x\cos\psi + y\sin\psi) \\
&\quad + k_0 n_1 z \cos\theta_{\text{inc}}]\} \\
\underline{H}_{\text{inc}}(\underline{r}) &= -in_1\eta_0^{-1}\left(a_L \frac{i\underline{s} - \underline{p}_{\text{inc}}}{\sqrt{2}} + a_R \frac{i\underline{s} + \underline{p}_{\text{inc}}}{\sqrt{2}}\right) \exp\{i[\kappa(x\cos\psi + y\sin\psi) \\
&\quad + k_0 n_1 z \cos\theta_{\text{inc}}]\} \\
&\quad\quad z < 0,
\end{aligned}
\right\}
\tag{3.129}
$$

where $a_L = -(ia_s + a_p)/\sqrt{2}$ and $a_R = (ia_s - a_p)/\sqrt{2}$ are the amplitudes of the left- and right-circularly polarized components, respectively. Likewise, Eqs. (3.107) and (3.108) may be rewritten as

$$
\left.
\begin{aligned}
\underline{E}_{\text{ref}}(\underline{r}) &= \left(-r_L \frac{i\underline{s} - \underline{p}_{\text{ref}}}{\sqrt{2}} + r_R \frac{i\underline{s} + \underline{p}_{\text{ref}}}{\sqrt{2}}\right) \exp\{i[\kappa(x\cos\psi + y\sin\psi) \\
&\quad - k_0 n_1 z \cos\theta_{\text{inc}}]\} \\
\underline{H}_{\text{ref}}(\underline{r}) &= in_1\eta_0^{-1}\left(r_L \frac{i\underline{s} - \underline{p}_{\text{ref}}}{\sqrt{2}} + r_R \frac{i\underline{s} + \underline{p}_{\text{ref}}}{\sqrt{2}}\right) \exp\{i[\kappa(x\cos\psi + y\sin\psi) \\
&\quad - k_0 n_1 z \cos\theta_{\text{inc}}]\} \\
&\quad\quad z < 0,
\end{aligned}
\right\}
\tag{3.130}
$$

and

$$
\left.
\begin{aligned}
\underline{E}_{\text{tr}}(\underline{r}) &= \left(t_L \frac{i\underline{s} - \underline{p}_{\text{tr}}}{\sqrt{2}} - t_R \frac{i\underline{s} + \underline{p}_{\text{tr}}}{\sqrt{2}}\right) \exp\{i[\kappa(x\cos\psi + y\sin\psi) \\
&\quad + k_0 n_2 (z - d)\cos\theta_{\text{tr}}]\} \\
\underline{H}_{\text{tr}}(\underline{r}) &= -in_2\eta_0^{-1}\left(t_L \frac{i\underline{s} - \underline{p}_{\text{tr}}}{\sqrt{2}} + t_R \frac{i\underline{s} + \underline{p}_{\text{tr}}}{\sqrt{2}}\right) \exp\{i[\kappa(x\cos\psi + y\sin\psi) \\
&\quad + k_0 n_2 (z - d)\cos\theta_{\text{tr}}]\} \\
&\quad\quad z > d = d_A + d_B,
\end{aligned}
\right\}
\tag{3.131}
$$

where the amplitudes $r_L = (ir_s + r_p)/\sqrt{2}, r_R = -(ir_s - r_p)/\sqrt{2}, t_L = -(it_s + t_p)/\sqrt{2}$, and $t_R = (it_s - t_p)/\sqrt{2}$ belong to circularly polarized components.

Circular reflection and transmission coefficients appear as the elements of the 2×2 matrixes in the following relations:

$$
\left.
\begin{aligned}
\begin{bmatrix} r_L \\ r_R \end{bmatrix} &= \begin{bmatrix} r_{\text{LL}} & r_{\text{LR}} \\ r_{\text{RL}} & r_{\text{RR}} \end{bmatrix} \cdot \begin{bmatrix} a_L \\ a_R \end{bmatrix} \\
\begin{bmatrix} t_L \\ t_R \end{bmatrix} &= \begin{bmatrix} t_{\text{LL}} & t_{\text{LR}} \\ t_{\text{RL}} & t_{\text{RR}} \end{bmatrix} \cdot \begin{bmatrix} a_L \\ a_R \end{bmatrix}
\end{aligned}
\right\}.
\tag{3.132}
$$

Again, co-polarized coefficients have both subscripts identical, but cross-polarized coefficients do not. The relationships between the linear and circular coefficients are as follows:

$$r_{ss} = -\frac{(r_{LL} + r_{RR}) - (r_{LR} + r_{RL})}{2}$$

$$r_{sp} = i\frac{(r_{LL} - r_{RR}) + (r_{LR} - r_{RL})}{2}$$

$$r_{ps} = -i\frac{(r_{LL} - r_{RR}) - (r_{LR} - r_{RL})}{2}$$

$$r_{pp} = -\frac{(r_{LL} + r_{RR}) + (r_{LR} + r_{RL})}{2}$$

$$t_{ss} = \frac{(t_{LL} + t_{RR}) - (t_{LR} + t_{RL})}{2}$$

$$t_{sp} = -i\frac{(t_{LL} - t_{RR}) + (t_{LR} - t_{RL})}{2}$$

$$t_{ps} = i\frac{(t_{LL} - t_{RR}) - (t_{LR} - t_{RL})}{2}$$

$$t_{pp} = \frac{(t_{LL} + t_{RR}) + (t_{LR} + t_{RL})}{2}$$

$$r_{LL} = -\frac{(r_{ss} + r_{pp}) + i(r_{sp} - r_{ps})}{2}$$

$$r_{LR} = \frac{(r_{ss} - r_{pp}) - i(r_{sp} + r_{ps})}{2}$$

$$r_{RL} = \frac{(r_{ss} - r_{pp}) + i(r_{sp} + r_{ps})}{2}$$

$$r_{RR} = -\frac{(r_{ss} + r_{pp}) - i(r_{sp} - r_{ps})}{2}$$

$$t_{LL} = \frac{(t_{ss} + t_{pp}) + i(t_{sp} - t_{ps})}{2}$$

$$t_{LR} = -\frac{(t_{ss} - t_{pp}) - i(t_{sp} + t_{ps})}{2}$$

$$t_{RL} = -\frac{(t_{ss} - t_{pp}) + i(t_{sp} + t_{ps})}{2}$$

$$t_{RR} = \frac{(t_{ss} + t_{pp}) - i(t_{sp} - t_{ps})}{2}$$

$$(3.133)$$

Circular reflectances are defined as

$$R_{LR} = |r_{LR}|^2, \tag{3.134}$$

etc., and circular transmittances as

$$T_{LR} = \frac{n_2}{n_1}\frac{\text{Re}\{\cos\theta_{tr}\}}{\cos\theta_{inc}}|t_{LR}|^2, \tag{3.135}$$

etc. The principle of conservation of energy requires satisfaction of the constraints

$$\left.\begin{array}{l} R_{LL} + R_{RL} + T_{LL} + T_{RL} \leqslant 1 \\ R_{RR} + R_{LR} + T_{RR} + T_{LR} \leqslant 1 \end{array}\right\}, \tag{3.136}$$

the inequalities turning to equalities only in the absence of dissipation.

The variations with θ_{inc} of the circular absorptances

$$\left.\begin{array}{l} A_L = 1 - (R_{LL} + R_{RL} + T_{LL} + T_{RL}) \\ A_R = 1 - (R_{RR} + R_{LR} + T_{RR} + T_{LR}) \end{array}\right\} \tag{3.137}$$

for fixed k_0 and ψ need to be studied in order to identify the excitation of surface waves guided by the interface $z = d_A$ of materials A and B. Certain features may not change with increase of d_A and d_B beyond certain thresholds and are likely to indicate surface-wave propagation; the thickness of a partnering material that is periodically nonhomogeneous should be increased in increments of a period, in order to identify surface waves. These features should be correlated with the solutions q of the corresponding canonical boundary-value problem treated in Section 3.5.1.

3.8 Grating-Coupled Configuration

The geometry of the grating-coupled configuration is almost the same as of the prism-coupled configuration depicted in Figure 3.4, except that the plane $z = d_A$ is replaced by the periodically undulating surface $z = d_A + g(x, y)$, where $g(x, y)$ is the grating function.

The half-space $z < 0$ is occupied by a homogeneous, isotropic, dielectric material described by the refractive index n_1, which is real valued and positive. The region $0 < z < d_A + g(x, y)$ is filled with a nonhomogeneous bianisotropic material with the constitutive relations stated in Eqs. (3.103), and the region $d_A + g(x, y) < z < d_A + d_B$ is occupied by a material with constitutive relations stated in Eqs. (3.104). Either of these two materials can be homogeneous, piecewise homogeneous, periodically nonhomogeneous, or nonhomogeneous in some other fashion along the z axis. Finally, the half-space $z > d = d_A + d_B$ is occupied by a homogeneous, isotropic, dielectric material whose refractive index n_2 is assumed to be real valued and positive.

As shown in Figure 3.5, the undulating surface $z = d_A + g(x, y)$ can be singly periodic with

$$g(x, y) \equiv g(x) = g(x \pm L_x), \tag{3.138}$$

say, where L_x is the period along the x axis. Alternatively, the undulating surface can be doubly periodic with

$$g(x, y) = g(x \pm L_x, y) = g(x, y \pm L_y), \tag{3.139}$$

where L_x and L_y are the periods along the x and y axes, respectively. In either case, the restrictions

$$\left.\begin{array}{l} g(x, y) < d_B \\ g(x, y) > -d_A \end{array}\right\}, \quad \forall x \in (-\infty, \infty), \quad \forall y \in (-\infty, \infty), \tag{3.140}$$

must hold. If the maximum value of $g(x, y)$ is denoted by $g_{max} > 0$ and the minimum value by $g_{min} < 0$, the trough-to-crest height of the surface-relief grating is $L_g = g_{max} - g_{min}$. Typically, L_g is a small fraction of the free-space wavelength.

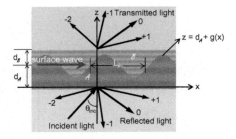

Figure 3.5 Schematic of the boundary-value problem for the grating-coupled configuration to excite surface waves guided by the periodically corrugated interface $z = d_A + g(x)$ of two dissimilar materials $\mathcal{A}$ and $\mathcal{B}$. The reflected field phasors comprise specular components (identified by 0) and nonspecular components (identified by ± 1, etc.), and the transmitted field phasors may also have specular and nonspecular components. The trough-to-crest height of the periodically corrugated surface is typically a small fraction of the free-space wavelength.

The surface-relief grating may be a simple one, exemplified by a sinusoidal grating [260], or it can be a compound surface-relief grating, each period of which comprises several periods each of two or more simple surface-relief gratings [85]. A compound grating can help excite SPP waves over a range of free-space wavelengths and for several different incidence conditions. A compound grating can also help excite all SPP waves predicted by the solution of the corresponding canonical boundary-value problem [86].

3.8.1 Incident Plane Wave

An arbitrarily polarized plane wave, propagating in the half-space $z < 0$ at an angle $\theta_{\text{inc}} \in [0°, 90°)$ to the z axis and at an angle $\psi \in [0°, 360°)$ to the x axis in the xy plane, is incident on the plane $z = 0$. The field phasors associated with the incident plane wave are represented by Eqs. (3.105). In view of the periodic undulations of the interface of materials $\mathcal{A}$ and $\mathcal{B}$, it is better to write the field phasors equivalently as a sum of linear Floquet harmonics as

$$
\left.
\begin{aligned}
\underline{E}_{\text{inc}}(\underline{r}) &= \sum_{m \in \mathbb{Z}} \sum_{n \in \mathbb{Z}} \left\{ \left(a_s^{(m,n)} \underline{s}^{(m,n)} + a_p^{(m,n)} \underline{p}_{\text{inc}}^{(m,n)} \right) \right. \\
&\qquad\qquad \left. \exp\left[i \left(\underline{\kappa}^{(m,n)} + \alpha_1^{(m,n)} \underline{u}_z \right) \cdot \underline{r} \right] \right\} \\
\underline{H}_{\text{inc}}(\underline{r}) &= n_1 \eta_0^{-1} \sum_{m \in \mathbb{Z}} \sum_{n \in \mathbb{Z}} \left\{ \left(a_s^{(m,n)} \underline{p}_{\text{inc}}^{(m,n)} - a_p^{(m,n)} \underline{s}^{(m,n)} \right) \right. \\
&\qquad\qquad \left. \exp\left[i \left(\underline{\kappa}^{(m,n)} + \alpha_1^{(m,n)} \underline{u}_z \right) \cdot \underline{r} \right] \right\}
\end{aligned}
\right\},
$$
$$
z < 0, \tag{3.141}
$$

where $\mathbb{Z} = \{0, \pm 1, \pm 2, \ldots\}$ is the set of all signed integers and zero, the amplitudes

$$
\left.
\begin{aligned}
a_s^{(m,n)} &= a_s^{(0,0)} \delta_{m0} \delta_{n0} \\
a_p^{(m,n)} &= a_p^{(0,0)} \delta_{m0} \delta_{n0}
\end{aligned}
\right\} \tag{3.142}
$$

involve the Kronecker delta (2.92), and

$$
\left.
\begin{aligned}
\underline{\kappa}^{(m,n)} &= k_x^{(m)} \underline{u}_x + k_y^{(n)} \underline{u}_y \\
k_x^{(m)} &= k_0 n_1 \sin \theta_{\text{inc}} \cos \psi + m(2\pi/L_x) \\
k_y^{(n)} &= k_0 n_1 \sin \theta_{\text{inc}} \sin \psi + n(2\pi/L_y) \\
k_{xy}^{(m,n)} &= +\sqrt{\underline{\kappa}^{(m,n)} \cdot \underline{\kappa}^{(m,n)}} \\
\alpha_1^{(m,n)} &= +\sqrt{(k_0 n_1)^2 - \underline{\kappa}^{(m,n)} \cdot \underline{\kappa}^{(m,n)}} \\
\underline{s}^{(m,n)} &= -\frac{k_y^{(n)}}{k_{xy}^{(m,n)}} \underline{u}_x + \frac{k_x^{(m)}}{k_{xy}^{(m,n)}} \underline{u}_y \\
\underline{p}_{\text{inc}}^{(m,n)} &= -\left(\frac{k_x^{(m)}}{k_{xy}^{(m,n)}} \underline{u}_x + \frac{k_y^{(n)}}{k_{xy}^{(m,n)}} \underline{u}_y \right) \frac{\alpha_1^{(m,n)}}{k_0 n_1} + \frac{k_{xy}^{(m,n)}}{k_0 n_1} \underline{u}_z
\end{aligned}
\right\}. \tag{3.143}
$$

Each index pair (m, n) on the right sides of Eqs. (3.141) identifies a Floquet harmonic. The quantity $\alpha_1^{(m,n)}$ is either positive real or positive imaginary, depending upon whether $k_0 n_1$ is greater or less than $k_{xy}^{(m,n)}$. If $\alpha_1^{(m,n)}$ is positive real, then the Floquet harmonic is a plane wave that can transport energy as $z \to \infty$, provided that all space were to be filled with the material of refractive index n_1. If $\alpha_1^{(m,n)}$ is positive imaginary, then the Floquet harmonic is an evanescent wave.

3.8.2   Reflected and Transmitted Field Phasors

Equations (3.107) and (3.108) are inadequate to represent the reflected and the transmitted field phasors. These phasors must also be written in terms of linear Floquet harmonics. Thus, the reflected field phasors are

$$
\left.
\begin{aligned}
\underline{E}_{\text{ref}}(\underline{r}) &= \sum_{m\in\mathbb{Z}}\sum_{n\in\mathbb{Z}} \Big\{ \big(r_s^{(m,n)} \underline{s}^{(m,n)} + r_p^{(m,n)} \underline{p}_{\text{ref}}^{(m,n)} \big) \\
&\qquad\qquad \exp\Big[i \big(\underline{\kappa}^{(m,n)} - \alpha_1^{(m,n)} \underline{u}_z \big) \cdot \underline{r} \Big] \Big\} \\
\underline{H}_{\text{ref}}(\underline{r}) &= n_1 \eta_0^{-1} \sum_{m\in\mathbb{Z}}\sum_{n\in\mathbb{Z}} \Big\{ \big(r_s^{(m,n)} \underline{p}_{\text{ref}}^{(m,n)} - r_p^{(m,n)} \underline{s}^{(m,n)} \big) \\
&\qquad\qquad \exp\Big[i \big(\underline{\kappa}^{(m,n)} - \alpha_1^{(m,n)} \underline{u}_z \big) \cdot \underline{r} \Big] \Big\}
\end{aligned}
\right\},
$$
$$
z < 0, \tag{3.144}
$$

where the reflection amplitudes $r_s^{(m,n)}$ and $r_p^{(m,n)}$ are unknown for all $m \in \mathbb{Z}$ and $n \in \mathbb{Z}$, and

$$
\underline{p}_{\text{ref}}^{(m,n)} = \left(\frac{k_x^{(m)}}{k_{xy}^{(m,n)}} \underline{u}_x + \frac{k_y^{(n)}}{k_{xy}^{(m,n)}} \underline{u}_y \right) \frac{\alpha_1^{(m,n)}}{k_0 n_1} + \frac{k_{xy}^{(m,n)}}{k_0 n_1} \underline{u}_z. \tag{3.145}
$$

If $\alpha_1^{(m,n)}$ is positive real, then the corresponding Floquet harmonic in Eqs. (3.144) is a plane wave that transports energy as $z \to -\infty$; if not, then the Floquet harmonic is an evanescent wave.

The transmitted field phasors are given by

$$
\left.
\begin{aligned}
\underline{E}_{\text{tr}}(\underline{r}) &= \sum_{m\in\mathbb{Z}}\sum_{n\in\mathbb{Z}} \Big\{ \big(t_s^{(m,n)} \underline{s}^{(m,n)} + t_p^{(m,n)} \underline{p}_{\text{tr}}^{(m,n)} \big) \\
&\qquad \exp\Big[i \big(\underline{\kappa}^{(m,n)} + \alpha_2^{(m,n)} \underline{u}_z \big) \cdot \big(\underline{r} - d\underline{u}_z \big) \Big] \Big\} \\
\underline{H}_{\text{tr}}(\underline{r}) &= n_2 \eta_0^{-1} \sum_{m\in\mathbb{Z}}\sum_{n\in\mathbb{Z}} \Big\{ \big(t_s^{(m,n)} \underline{p}_{\text{tr}}^{(m,n)} - t_p^{(m,n)} \underline{s}^{(m,n)} \big) \\
&\qquad \exp\Big[i \big(\underline{\kappa}^{(m,n)} + \alpha_2^{(m,n)} \underline{u}_z \big) \cdot \big(\underline{r} - d\underline{u}_z \big) \Big] \Big\}
\end{aligned}
\right\},
$$
$$
z > d = d_A + d_B, \tag{3.146}
$$

where the transmission amplitudes $t_s^{(m,n)}$ and $t_p^{(m,n)}$ are also unknown for all $m \in \mathbb{Z}$ and $n \in \mathbb{Z}$, and

$$
\left.
\begin{aligned}
\alpha_2^{(m,n)} &= +\sqrt{\big(k_0 n_2\big)^2 - \underline{\kappa}^{(m,n)} \cdot \underline{\kappa}^{(m,n)}} \\
\underline{p}_{\text{tr}}^{(m,n)} &= -\left(\frac{k_x^{(m)}}{k_{xy}^{(m,n)}} \underline{u}_x + \frac{k_y^{(n)}}{k_{xy}^{(m,n)}} \underline{u}_y \right) \frac{\alpha_2^{(m,n)}}{k_0 n_2} + \frac{k_{xy}^{(m,n)}}{k_0 n_2} \underline{u}_z
\end{aligned}
\right\}. \tag{3.147}
$$

The quantity $\alpha_2^{(m,n)}$ is either positive real or positive imaginary, depending upon whether $k_0 n_2$ is greater or less than $k_{xy}^{(m,n)}$. If $\alpha_2^{(m,n)}$ is positive real, then the corresponding Floquet harmonic in Eqs. (3.146) is a plane wave that transports energy as $z \to \infty$; if not, then the Floquet harmonic is an evanescent wave.

The specular terms in sums on the right sides of Eqs. (3.144) and (3.146) are the ones for which $m = n = 0$. The remaining terms in these sums are nonspecular terms which arise due to the periodic nature of the function $g(x, y)$. If this function does not vary with x, then $m \in \{0\}$. Likewise, $n \in \{0\}$ if $g(x, y)$ is invariant with respect to y. In the simplest possible case, $g(x, y) = 0$ and all nonspecular terms can be discarded.

Equations (3.141), (3.144), and (3.146) may be considered as Fourier series of $\underline{E}_{\text{inc}}(\underline{r}) \exp[-i k_0 n_1 \sin \theta_{\text{inc}} (x \cos \psi + y \sin \psi)]$, etc., with respect to x and y. These representations constitute an essential feature of the rigorous coupled-wave approach (RCWA) employed in Section 3.8.5 to determine the reflection and transmission amplitudes.

3.8.3 Linear Reflectances and Transmittances

Reflection and transmission coefficients of order (m, n) are defined as the elements in the 2×2 matrices appearing in the following relations:

$$\left.\begin{array}{c} \begin{bmatrix} r_s^{(m,n)} \\ r_p^{(m,n)} \end{bmatrix} = \begin{bmatrix} r_{ss}^{(m,n)} & r_{sp}^{(m,n)} \\ r_{ps}^{(m,n)} & r_{pp}^{(m,n)} \end{bmatrix} \cdot \begin{bmatrix} a_s^{(0,0)} \\ a_p^{(0,0)} \end{bmatrix} \\[2em] \begin{bmatrix} t_s^{(m,n)} \\ t_p^{(m,n)} \end{bmatrix} = \begin{bmatrix} t_{ss}^{(m,n)} & t_{sp}^{(m,n)} \\ t_{ps}^{(m,n)} & t_{pp}^{(m,n)} \end{bmatrix} \cdot \begin{bmatrix} a_s^{(0,0)} \\ a_p^{(0,0)} \end{bmatrix} \end{array}\right\} . \tag{3.148}$$

Only the coefficients of order $(0, 0)$ are classified as specular, whereas all other coefficients are nonspecular.

Linear reflectances are defined as

$$R_{sp}^{(m,n)} = \frac{\text{Re}\left\{\alpha_1^{(m,n)}\right\}}{\alpha_1^{(0,0)}} |r_{sp}^{(m,n)}|^2, \tag{3.149}$$

etc., and linear transmittances as

$$T_{sp}^{(m,n)} = \frac{\text{Re}\left\{\alpha_2^{(m,n)}\right\}}{\alpha_1^{(0,0)}} |t_{sp}^{(m,n)}|^2, \tag{3.150}$$

etc. Again, the adjective *specular* may be attached only to reflectances and transmittances of order $(0, 0)$, all others being of the nonspecular kind.

The principle of conservation of energy mandates that the linear absorptances

$$\left.\begin{array}{l} A_s = 1 - \displaystyle\sum_{m \in \mathbb{Z}} \sum_{n \in \mathbb{Z}} \left(R_{ss}^{(m,n)} + R_{ps}^{(m,n)} + T_{ss}^{(m,n)} + T_{ps}^{(m,n)} \right) \\[2em] A_p = 1 - \displaystyle\sum_{m \in \mathbb{Z}} \sum_{n \in \mathbb{Z}} \left(R_{pp}^{(m,n)} + R_{sp}^{(m,n)} + T_{pp}^{(m,n)} + T_{sp}^{(m,n)} \right) \end{array}\right\} \tag{3.151}$$

can neither be negative nor exceed unity, both materials $\mathcal{A}$ and $\mathcal{B}$ being passive [17, Section 1.7.2.2].

3.8.4   Circular Reflectances and Transmittances

As in Section 3.7.4, the incident, reflected, and transmitted field phasors can be represented using circularly polarized Floquet harmonics instead of the linearly polarized ones.

The counterparts of Eqs. (3.141) are

$$
\begin{aligned}
\underline{E}_{\text{inc}}(\underline{r}) &= \sum_{m\in\mathbb{Z}}\sum_{n\in\mathbb{Z}}\left\{\left(a_L^{(m,n)}\frac{i\underline{s}^{(m,n)}-\underline{p}_{\text{inc}}^{(m,n)}}{\sqrt{2}}-a_R^{(m,n)}\frac{i\underline{s}^{(m,n)}+\underline{p}_{\text{inc}}^{(m,n)}}{\sqrt{2}}\right)\right. \\
&\qquad \left. \exp\left[i\left(\underline{\kappa}^{(m,n)}+\alpha_1^{(m,n)}\underline{u}_z\right)\cdot\underline{r}\right]\right\} \\
\underline{H}_{\text{inc}}(\underline{r}) &= -in_1\eta_0^{-1}\sum_{m\in\mathbb{Z}}\sum_{n\in\mathbb{Z}}\left\{\left(a_L^{(m,n)}\frac{i\underline{s}^{(m,n)}-\underline{p}_{\text{inc}}^{(m,n)}}{\sqrt{2}}+a_R^{(m,n)}\frac{i\underline{s}^{(m,n)}+\underline{p}_{\text{inc}}^{(m,n)}}{\sqrt{2}}\right)\right. \\
&\qquad \left. \exp\left[i\left(\underline{\kappa}^{(m,n)}+\alpha_1^{(m,n)}\underline{u}_z\right)\cdot\underline{r}\right]\right\}
\end{aligned}
$$

$$z < 0. \tag{3.152}$$

The reflected field phasors must be written as

$$
\begin{aligned}
\underline{E}_{\text{ref}}(\underline{r}) &= \sum_{m\in\mathbb{Z}}\sum_{n\in\mathbb{Z}}\left\{\left(-r_L^{(m,n)}\frac{i\underline{s}^{(m,n)}-\underline{p}_{\text{ref}}^{(m,n)}}{\sqrt{2}}+r_R^{(m,n)}\frac{i\underline{s}^{(m,n)}+\underline{p}_{\text{ref}}^{(m,n)}}{\sqrt{2}}\right)\right. \\
&\qquad \left. \exp\left[i\left(\underline{\kappa}^{(m,n)}-\alpha_1^{(m,n)}\underline{u}_z\right)\cdot\underline{r}\right]\right\} \\
\underline{H}_{\text{ref}}(\underline{r}) &= in_1\eta_0^{-1}\sum_{m\in\mathbb{Z}}\sum_{n\in\mathbb{Z}}\left\{\left(r_L^{(m,n)}\frac{i\underline{s}^{(m,n)}-\underline{p}_{\text{ref}}^{(m,n)}}{\sqrt{2}}+r_R^{(m,n)}\frac{i\underline{s}^{(m,n)}+\underline{p}_{\text{ref}}^{(m,n)}}{\sqrt{2}}\right)\right. \\
&\qquad \left. \exp\left[i\left(\underline{\kappa}^{(m,n)}-\alpha_1^{(m,n)}\underline{u}_z\right)\cdot\underline{r}\right]\right\}
\end{aligned}
$$

$$z < 0, \tag{3.153}$$

and the transmitted field phasors as

$$
\begin{aligned}
\underline{E}_{\text{tr}}(\underline{r}) &= \sum_{m\in\mathbb{Z}}\sum_{n\in\mathbb{Z}}\left\{\left(t_L^{(m,n)}\frac{i\underline{s}^{(m,n)}-\underline{p}_{\text{tr}}^{(m,n)}}{\sqrt{2}}-t_R^{(m,n)}\frac{i\underline{s}^{(m,n)}+\underline{p}_{\text{tr}}^{(m,n)}}{\sqrt{2}}\right)\right. \\
&\qquad \left. \exp\left[i\left(\underline{\kappa}^{(m,n)}+\alpha_2^{(m,n)}\underline{u}_z\right)\cdot\left(\underline{r}-d\underline{u}_z\right)\right]\right\} \\
\underline{H}_{\text{tr}}(\underline{r}) &= -in_2\eta_0^{-1}\sum_{m\in\mathbb{Z}}\sum_{n\in\mathbb{Z}}\left\{\left(t_L^{(m,n)}\frac{i\underline{s}^{(m,n)}-\underline{p}_{\text{tr}}^{(m,n)}}{\sqrt{2}}+t_R^{(m,n)}\frac{i\underline{s}^{(m,n)}+\underline{p}_{\text{tr}}^{(m,n)}}{\sqrt{2}}\right)\right. \\
&\qquad \left. \exp\left[i\left(\underline{\kappa}^{(m,n)}+\alpha_2^{(m,n)}\underline{u}_z\right)\cdot\left(\underline{r}-d\underline{u}_z\right)\right]\right\}
\end{aligned}
$$

$$z > d = d_A + d_B. \tag{3.154}$$

The amplitudes of the circularly polarized components are related to those of the linearly polarized components as follows:

$$
\left.
\begin{aligned}
a_L^{(m,n)} &= -\left(ia_s^{(m,n)} + a_p^{(m,n)}\right)/\sqrt{2}, \quad a_R^{(m,n)} = \left(ia_s^{(m,n)} - a_p^{(m,n)}\right)/\sqrt{2} \\
r_L^{(m,n)} &= \left(ir_s^{(m,n)} + r_p^{(m,n)}\right)/\sqrt{2}, \quad r_R^{(m,n)} = -\left(ir_s^{(m,n)} - r_p^{(m,n)}\right)/\sqrt{2} \\
t_L^{(m,n)} &= -\left(it_s^{(m,n)} + t_p^{(m,n)}\right)/\sqrt{2}, \quad t_R^{(m,n)} = \left(it_s^{(m,n)} - t_p^{(m,n)}\right)/\sqrt{2}
\end{aligned}
\right\}
$$
(3.155)

Circular reflection coefficients $\left(r_{LR}^{(m,n)}, \text{etc.}\right)$ and transmission coefficients $\left(t_{LR}^{(m,n)}, \text{etc.}\right)$ of order (m, n) appear in 2×2 matrixes in the following relations:

$$
\left.
\begin{aligned}
\begin{bmatrix} r_L^{(m,n)} \\ r_R^{(m,n)} \end{bmatrix} &= \begin{bmatrix} r_{LL}^{(m,n)} & r_{LR}^{(m,n)} \\ r_{RL}^{(m,n)} & r_{RR}^{(m,n)} \end{bmatrix} \begin{bmatrix} a_L^{(0,0)} \\ a_R^{(0,0)} \end{bmatrix} \\
\begin{bmatrix} t_L^{(m,n)} \\ t_R^{(m,n)} \end{bmatrix} &= \begin{bmatrix} t_{LL}^{(m,n)} & t_{LR}^{(m,n)} \\ t_{RL}^{(m,n)} & t_{RR}^{(m,n)} \end{bmatrix} \begin{bmatrix} a_L^{(0,0)} \\ a_R^{(0,0)} \end{bmatrix}
\end{aligned}
\right\}
$$
(3.156)

Again, co-polarized coefficients have both subscripts identical, but cross-polarized coefficients do not. Specular coefficients carry the superscript $(0, 0)$, but nonspecular coefficients do not. Equations (3.133) can be used to relate the linear and circular coefficients for each index pair (m, n).

Circular reflectances are defined as

$$
R_{LR}^{(m,n)} = \frac{\text{Re}\left\{\alpha_1^{(m,n)}\right\}}{\alpha_1^{(0,0)}} |r_{LR}^{(m,n)}|^2,
$$
(3.157)

etc., and circular transmittances as

$$
T_{LR}^{(m,n)} = \frac{\text{Re}\left\{\alpha_2^{(m,n)}\right\}}{\alpha_1^{(0,0)}} |t_{LR}^{(m,n)}|^2,
$$
(3.158)

etc. Both circular absorptances

$$
\left.
\begin{aligned}
A_L &= 1 - \sum_{m \in \mathbb{Z}} \sum_{n \in \mathbb{Z}} \left(R_{LL}^{(m,n)} + R_{RL}^{(m,n)} + T_{LL}^{(m,n)} + T_{RL}^{(m,n)}\right) \\
A_R &= 1 - \sum_{m \in \mathbb{Z}} \sum_{n \in \mathbb{Z}} \left(R_{RR}^{(m,n)} + R_{LR}^{(m,n)} + T_{RR}^{(m,n)} + T_{LR}^{(m,n)}\right)
\end{aligned}
\right\}
$$
(3.159)

can neither be negative nor exceed unity, provided that both materials $\mathcal{A}$ and $\mathcal{B}$ are passive. If dissipation in both materials is negligible, then $A_L = A_R = 0$.

3.8.5 Rigorous Coupled-Wave Approach

The calculation of the amplitudes of the Floquet harmonics in the representations of the reflected and transmitted field phasors is based on the periodicity of the grating function

$g(x, y)$. This procedure is called the rigorous coupled-wave approach [256, 300–302], which requires that all field phasors as well as all constitutive scalars and constitutive dyadics be expressed as Fourier series with respect to both x and y.

The constitutive relations of the two partnering materials are jointly written as

$$\left. \begin{array}{l} \underline{D}(\underline{r}) = \underline{\underline{\epsilon}}(x, y, z) \cdot \underline{E}(\underline{r}) + \underline{\underline{\xi}}(x, y, z) \cdot \underline{H}(\underline{r}) \\ \underline{B}(\underline{r}) = \underline{\underline{\zeta}}(x, y, z) \cdot \underline{E}(\underline{r}) + \underline{\underline{\mu}}(x, y, z) \cdot \underline{H}(\underline{r}) \end{array} \right\}, \quad z \in (0, d), \qquad (3.160)$$

where

$$\underline{\underline{\epsilon}}(x, y, z) = \begin{cases} \underline{\underline{\epsilon}}^{\mathcal{A}}, & z \in (0, d_{\mathcal{A}} + g_{\min}], \\ \underline{\underline{\epsilon}}^{\mathcal{A}} + \left(\underline{\underline{\epsilon}}^{\mathcal{B}} - \underline{\underline{\epsilon}}^{\mathcal{A}} \right) \mathcal{U}[z - g(x, y)], & z \in (d_{\mathcal{A}} + g_{\min}, d_{\mathcal{A}} + g_{\max}), \\ \underline{\underline{\epsilon}}^{\mathcal{B}}, & z \in [d_{\mathcal{A}} + g_{\max}, d_{\mathcal{A}} + d_{\mathcal{B}}), \end{cases} \qquad (3.161)$$

etc., and $\mathcal{U}(\sigma)$ is the unit step function defined by Eq. (2.102). Accordingly, we get the Fourier series

$$\underline{\underline{\epsilon}}(x, y, z) = \sum_{m \in \mathbb{Z}} \sum_{n \in \mathbb{Z}} \underline{\underline{\epsilon}}^{(m,n)}(z) \exp\left[i2\pi \left(m\frac{x}{L_x} + n\frac{y}{L_y} \right) \right], \quad z \in (0, d), \qquad (3.162)$$

where

$$\underline{\underline{\epsilon}}^{(m,n)}(z) = \begin{cases} \left(\underline{\underline{\epsilon}}^{\mathcal{B}} - \underline{\underline{\epsilon}}^{\mathcal{A}} \right) \Upsilon^{(m,n)}(z), & z \in (d_{\mathcal{A}} + g_{\min}, d_{\mathcal{A}} + g_{\max}), \\ \underline{\underline{0}}, & z \notin (d_{\mathcal{A}} + g_{\min}, d_{\mathcal{A}} + g_{\max}), \end{cases} \qquad (3.163)$$

except that

$$\underline{\underline{\epsilon}}^{(0,0)}(z) = \begin{cases} \underline{\underline{\epsilon}}^{\mathcal{A}}, & z \in (0, d_{\mathcal{A}} + g_{\min}], \\ \underline{\underline{\epsilon}}^{\mathcal{A}} + \left(\underline{\underline{\epsilon}}^{\mathcal{B}} - \underline{\underline{\epsilon}}^{\mathcal{A}} \right) \Upsilon^{(0,0)}(z), & z \in (d_{\mathcal{A}} + g_{\min}, d_{\mathcal{A}} + g_{\max}), \\ \underline{\underline{\epsilon}}^{\mathcal{B}}, & z \in [d_{\mathcal{A}} + g_{\max}, d_{\mathcal{A}} + d_{\mathcal{B}}), \end{cases} \qquad (3.164)$$

and

$$\Upsilon^{(m,n)}(z) = \frac{1}{L_x} \frac{1}{L_y} \int_0^{L_x} \int_0^{L_y} \mathcal{U}[z - g(x, y)] \exp\left[-i2\pi \left(m\frac{x}{L_x} + n\frac{y}{L_y} \right) \right] dx \, dy. \qquad (3.165)$$

The Fourier series of the other three constitutive dyadics in Eqs. (3.160) are similar to that of $\underline{\underline{\epsilon}}(x, y, z)$.

The field phasors in the region occupied by either material $\mathcal{A}$ or material $\mathcal{B}$ are expressed in terms of Fourier series too, just like the reflected and transmitted field phasors; thus,

$$\left. \begin{array}{l} \underline{E}(\underline{r}) = \sum_{m \in \mathbb{Z}} \sum_{n \in \mathbb{Z}} \left\{ \underline{e}^{(m,n)}(z) \exp\left[i\underline{\kappa}^{(m,n)} \cdot \underline{r} \right] \right\} \\ \underline{H}(\underline{r}) = \sum_{m \in \mathbb{Z}} \sum_{n \in \mathbb{Z}} \left\{ \underline{h}^{(m,n)}(z) \exp\left[i\underline{\kappa}^{(m,n)} \cdot \underline{r} \right] \right\} \end{array} \right\}, \quad z \in (0, d), \qquad (3.166)$$

where the vector functions $\underline{e}^{(m,n)}(z)$ and $\underline{h}^{(m,n)}(z)$ remain to be determined for all index pairs (m, n).

The numbers of terms in the sums on the right sides of Eqs. (3.141), (3.144), (3.152), (3.153), (3.154), (3.162), and (3.166) are infinite. For computational purposes, the number of terms in a series must be finite. Accordingly, we restrict the index $m \in [-M_t, M_t]$ and the index $n \in [-N_t, N_t]$ in these equations, with $M_t \geqslant 0$ and $N_t \geqslant 0$ being sufficiently large. Then every field phasor and every constitutive dyadic is represented by a series containing $(2M_t+1)(2N_t+1)$ terms. Thus, for example, Eqs. (3.144) are rewritten as

$$
\left.
\begin{aligned}
\underline{E}_{\text{ref}}(\underline{r}) &= \sum_{m=-M_t}^{M_t} \sum_{n=-N_t}^{N_t} \left\{ \left(r_s^{(m,n)} \underline{s}^{(m,n)} + r_p^{(m,n)} \underline{p}_{\text{ref}}^{(m,n)} \right) \right. \\
&\qquad \left. \times \exp\left[i\left(\underline{\kappa}^{(m,n)} - \alpha_1^{(m,n)} \underline{u}_z \right) \cdot \underline{r} \right] \right\} \\
\underline{H}_{\text{ref}}(\underline{r}) &= n_1 \eta_0^{-1} \sum_{m=-M_t}^{M_t} \sum_{n=-N_t}^{N_t} \left\{ \left(r_s^{(m,n)} \underline{p}_{\text{ref}}^{(m,n)} - r_p^{(m,n)} \underline{s}^{(m,n)} \right) \right. \\
&\qquad \left. \times \exp\left[i\left(\underline{\kappa}^{(m,n)} - \alpha_1^{(m,n)} \underline{u}_z \right) \cdot \underline{r} \right] \right\}
\end{aligned}
\right\},
$$
$$
z < 0. \qquad (3.167)
$$

Furthermore, a superindex

$$
\tau = m(2N_t + 1) + n, \quad m \in [-M_t, M_t], \quad n \in [-N_t, N_t] \qquad (3.168)
$$

is defined for convenience; $\tau \in [-\tau_t, \tau_t]$, where $\tau_t = 2M_t N_t + M_t + N_t$. Then, Eqs. (3.167) can be recast as

$$
\left.
\begin{aligned}
\underline{E}_{\text{ref}}(\underline{r}) &= \sum_{\tau=-\tau_t}^{\tau_t} \left\{ \left(r_s^{(\tau)} \underline{s}^{(\tau)} + r_p^{(\tau)} \underline{p}_{\text{ref}}^{(\tau)} \right) \exp\left[i\left(\underline{\kappa}^{(\tau)} - \alpha_1^{(\tau)} \underline{u}_z \right) \cdot \underline{r} \right] \right\} \\
\underline{H}_{\text{ref}}(\underline{r}) &= n_1 \eta_0^{-1} \sum_{\tau=-\tau_t}^{\tau_t} \left\{ \left(r_s^{(\tau)} \underline{p}_{\text{ref}}^{(\tau)} - r_p^{(\tau)} \underline{s}^{(\tau)} \right) \exp\left[i\left(\underline{\kappa}^{(\tau)} - \alpha_1^{(\tau)} \underline{u}_z \right) \cdot \underline{r} \right] \right\}
\end{aligned}
\right\},
$$
$$
z < 0, \qquad (3.169)
$$

wherein the superscript $^{(\tau)}$ stands for $^{(m,n)}$.

For use in the region $z \in [0, d]$, the column $(2\tau_t + 1)$-vectors[1]

$$
[\breve{\underline{e}}_\sigma(z)] = [\breve{e}_\sigma^{(-\tau_t)}(z), \breve{e}_\sigma^{(-\tau_t+1)}(z), \ldots, \breve{e}_\sigma^{(\tau_t-1)}(z), \breve{e}_\sigma^{(\tau_t)}(z)]^T, \quad \sigma \in \{x, y, z\} \qquad (3.170)
$$

and

$$
[\breve{\underline{h}}_\sigma(z)] = [\breve{h}_\sigma^{(-\tau_t)}(z), \breve{h}_\sigma^{(-\tau_t+1)}(z), \ldots, \breve{h}_\sigma^{(\tau_t-1)}(z), \breve{h}_\sigma^{(\tau_t)}(z)]^T, \quad \sigma \in \{x, y, z\} \qquad (3.171)
$$

[1] The symbol $\breve{}$ identifies quantities associated with the RCWA.

are defined, with

$$\check{e}_\sigma^{(\tau)}(z) = e_\sigma^{(m,n)}(z), \qquad \check{h}_\sigma^{(\tau)}(z) = h_\sigma^{(m,n)}(z). \tag{3.172}$$

Along with these column vectors, $(2\tau_t + 1) \times (2\tau_t + 1)$ constitutive matrixes such as

$$\left[\underline{\check{\epsilon}}_{xz}(z)\right] = \begin{bmatrix} \check{\epsilon}_{xz}^{(-\tau_t,-\tau_t)}(z) & \check{\epsilon}_{xz}^{(-\tau_t,-\tau_t+1)}(z) & \cdots & \check{\epsilon}_{xz}^{(-\tau_t,\tau_t-1)}(z) & \check{\epsilon}_{xz}^{(-\tau_t,\tau_t)}(z) \\ \check{\epsilon}_{xz}^{(-\tau_t+1,-\tau_t)}(z) & \check{\epsilon}_{xz}^{(-\tau_t+1,-\tau_t+1)}(z) & \cdots & \check{\epsilon}_{xz}^{(-\tau_t+1,\tau_t-1)}(z) & \check{\epsilon}_{xz}^{(-\tau_t+1,\tau_t)}(z) \\ \cdots & \cdots & \cdots & \cdots & \cdots \\ \check{\epsilon}_{xz}^{(\tau_t-1,-\tau_t)}(z) & \check{\epsilon}_{xz}^{(\tau_t-1,-\tau_t+1)}(z) & \cdots & \check{\epsilon}_{xz}^{(\tau_t-1,\tau_t-1)}(z) & \check{\epsilon}_{xz}^{(\tau_t-1,\tau_t)}(z) \\ \check{\epsilon}_{xz}^{(\tau_t,-\tau_t)}(z) & \check{\epsilon}_{xz}^{(\tau_t,-\tau_t+1)}(z) & \cdots & \check{\epsilon}_{xz}^{(\tau_t,\tau_t-1)}(z) & \check{\epsilon}_{xz}^{(\tau_t,\tau_t)}(z) \end{bmatrix} \tag{3.173}$$

with

$$\check{\epsilon}_{xz}^{(\tau,\tau')}(z) = \epsilon_{xz}^{(m-m',n-n')}(z) \tag{3.174}$$

and

$$\tau' = m'(2N_t + 1) + n', \quad m' \in [-M_t, M_t], \quad n' \in [-N_t, N_t] \tag{3.175}$$

are defined. Finally, the $(2\tau_t + 1) \times (2\tau_t + 1)$ Fourier-wavenumber matrixes

$$\left[\underline{\check{K}}_x\right] = \mathrm{diag}\left[\check{k}_x^{(-\tau_t)}, \check{k}_x^{(-\tau_t+1)}, \ldots, \check{k}_x^{(\tau_t-1)}, \check{k}_x^{(\tau_t)}\right] \tag{3.176}$$

and

$$\left[\underline{\check{K}}_y\right] = \mathrm{diag}\left[\check{k}_y^{(-\tau_t)}, \check{k}_y^{(-\tau_t+1)}, \ldots, \check{k}_y^{(\tau_t-1)}, \check{k}_y^{(\tau_t)}\right] \tag{3.177}$$

are set up with

$$\check{k}_x^{(\tau)} = k_x^{(m)}, \qquad \check{k}_y^{(\tau)} = k_y^{(n)}. \tag{3.178}$$

Next, Eqs. (3.160)–(3.166) are substituted in Eqs. (3.7) with $\underline{J}_e(\underline{r}) = \underline{0}$ everywhere, and the orthogonality properties of the functions

$$\exp\left[i2\pi\left(m\frac{x}{L_x} + n\frac{y}{L_y}\right)\right] \tag{3.179}$$

on the rectangular region $\{|x| \leqslant L_x/2, |y| \leqslant L_y/2\}$ are exploited to obtain the four matrix ordinary differential equations

$$
\begin{aligned}
\frac{d}{dz}\left[\breve{\underline{e}}_x(z)\right] = {} & i\left[\underline{\underline{\breve{K}}}_x\right] \cdot \left[\breve{\underline{e}}_z(z)\right] + i\omega\Bigg\{ \left[\underline{\underline{\breve{\zeta}}}_{yx}\right] \cdot \left[\breve{\underline{e}}_x(z)\right] + \left[\underline{\underline{\breve{\zeta}}}_{yy}\right] \cdot \left[\breve{\underline{e}}_y(z)\right] \\
& + \left[\underline{\underline{\breve{\zeta}}}_{yz}\right] \cdot \left[\breve{\underline{e}}_z(z)\right] + \left[\underline{\underline{\breve{\mu}}}_{yx}\right] \cdot \left[\breve{\underline{h}}_x(z)\right] \\
& + \left[\underline{\underline{\breve{\mu}}}_{yy}\right] \cdot \left[\breve{\underline{h}}_y(z)\right] + \left[\underline{\underline{\breve{\mu}}}_{yz}\right] \cdot \left[\breve{\underline{h}}_z(z)\right] \Bigg\} \\[2mm]
\frac{d}{dz}\left[\breve{\underline{e}}_y(z)\right] = {} & i\left[\underline{\underline{\breve{K}}}_y\right] \cdot \left[\breve{\underline{e}}_z(z)\right] - i\omega\Bigg\{ \left[\underline{\underline{\breve{\zeta}}}_{xx}\right] \cdot \left[\breve{\underline{e}}_x(z)\right] + \left[\underline{\underline{\breve{\zeta}}}_{xy}\right] \cdot \left[\breve{\underline{e}}_y(z)\right] \\
& + \left[\underline{\underline{\breve{\zeta}}}_{xz}\right] \cdot \left[\breve{\underline{e}}_z(z)\right] + \left[\underline{\underline{\breve{\mu}}}_{xx}\right] \cdot \left[\breve{\underline{h}}_x(z)\right] \\
& + \left[\underline{\underline{\breve{\mu}}}_{xy}\right] \cdot \left[\breve{\underline{h}}_y(z)\right] + \left[\underline{\underline{\breve{\mu}}}_{xz}\right] \cdot \left[\breve{\underline{h}}_z(z)\right] \Bigg\} \\[2mm]
\frac{d}{dz}\left[\breve{\underline{h}}_x(z)\right] = {} & i\left[\underline{\underline{\breve{K}}}_x\right] \cdot \left[\breve{\underline{h}}_z(z)\right] - i\omega\Bigg\{ \left[\underline{\underline{\breve{\epsilon}}}_{yx}\right] \cdot \left[\breve{\underline{e}}_x(z)\right] + \left[\underline{\underline{\breve{\epsilon}}}_{yy}\right] \cdot \left[\breve{\underline{e}}_y(z)\right] \\
& + \left[\underline{\underline{\breve{\epsilon}}}_{yz}\right] \cdot \left[\breve{\underline{e}}_z(z)\right] + \left[\underline{\underline{\breve{\xi}}}_{yx}\right] \cdot \left[\breve{\underline{h}}_x(z)\right] \\
& + \left[\underline{\underline{\breve{\xi}}}_{yy}\right] \cdot \left[\breve{\underline{h}}_y(z)\right] + \left[\underline{\underline{\breve{xi}}}_{yz}\right] \cdot \left[\breve{\underline{h}}_z(z)\right] \Bigg\} \\[2mm]
\frac{d}{dz}\left[\breve{\underline{h}}_y(z)\right] = {} & i\left[\underline{\underline{\breve{K}}}_y\right] \cdot \left[\breve{\underline{h}}_z(z)\right] + i\omega\Bigg\{ \left[\underline{\underline{\breve{\epsilon}}}_{xx}\right] \cdot \left[\breve{\underline{e}}_x(z)\right] + \left[\underline{\underline{\breve{\epsilon}}}_{xy}\right] \cdot \left[\breve{\underline{e}}_y(z)\right] \\
& + \left[\underline{\underline{\breve{\epsilon}}}_{xz}\right] \cdot \left[\breve{\underline{e}}_z(z)\right] + \left[\underline{\underline{\breve{\xi}}}_{xx}\right] \cdot \left[\breve{\underline{h}}_x(z)\right] \\
& + \left[\underline{\underline{\breve{\xi}}}_{xy}\right] \cdot \left[\breve{\underline{h}}_y(z)\right] + \left[\underline{\underline{\breve{\xi}}}_{xz}\right] \cdot \left[\breve{\underline{h}}_z(z)\right] \Bigg\}
\end{aligned}
$$

$$(3.180)$$

and the two algebraic equations

$$
\begin{aligned}
\left[\underline{\underline{\breve{K}}}_x\right] \cdot \left[\breve{\underline{e}}_y\right] - \left[\underline{\underline{\breve{K}}}_y\right] \cdot \left[\breve{\underline{e}}_x\right] = {} & \omega\Bigg\{ \left[\underline{\underline{\breve{\zeta}}}_{zx}\right] \cdot \left[\breve{\underline{e}}_x(z)\right] + \left[\underline{\underline{\breve{\zeta}}}_{zy}\right] \cdot \left[\breve{\underline{e}}_y(z)\right] \\
& + \left[\underline{\underline{\breve{\zeta}}}_{zz}\right] \cdot \left[\breve{\underline{e}}_z(z)\right] + \left[\underline{\underline{\breve{\mu}}}_{zx}\right] \cdot \left[\breve{\underline{h}}_x(z)\right] \\
& + \left[\underline{\underline{\breve{\mu}}}_{zy}\right] \cdot \left[\breve{\underline{h}}_y(z)\right] + \left[\underline{\underline{\breve{\mu}}}_{zz}\right] \cdot \left[\breve{\underline{h}}_z(z)\right] \Bigg\} \\[2mm]
\left[\underline{\underline{\breve{K}}}_x\right] \cdot \left[\breve{\underline{h}}_y\right] - \left[\underline{\underline{\breve{K}}}_y\right] \cdot \left[\breve{\underline{h}}_x\right] = {} & -\omega\Bigg\{ \left[\underline{\underline{\breve{\epsilon}}}_{zx}\right] \cdot \left[\breve{\underline{e}}_x(z)\right] + \left[\underline{\underline{\breve{\epsilon}}}_{zy}\right] \cdot \left[\breve{\underline{e}}_y(z)\right] \\
& + \left[\underline{\underline{\breve{\epsilon}}}_{zz}\right] \cdot \left[\breve{\underline{e}}_z(z)\right] + \left[\underline{\underline{\breve{\xi}}}_{zx}\right] \cdot \left[\breve{\underline{h}}_x(z)\right] \\
& + \left[\underline{\underline{\breve{\xi}}}_{zy}\right] \cdot \left[\breve{\underline{h}}_y(z)\right] + \left[\underline{\underline{\breve{\xi}}}_{zz}\right] \cdot \left[\breve{\underline{h}}_z(z)\right] \Bigg\}
\end{aligned}
$$

$$(3.181)$$

With the assumption that the column vectors $\left[\breve{\underline{e}}_z(z)\right]$ and $\left[\breve{\underline{h}}_z(z)\right]$ can be eliminated by solving Eqs. (3.181), Eqs. (3.180) are compactly written as a single matrix ordinary differential equation

$$\frac{d}{dz}\left[\breve{\underline{f}}(z)\right] = i \left[\underline{\underline{\breve{P}}}(z)\right] \cdot \left[\breve{\underline{f}}(z)\right], \quad z \in (0, d), \tag{3.182}$$

where the column $4(2\tau_t + 1)$-vector

$$\left[\breve{\underline{f}}(z)\right] = \begin{bmatrix} \left[\breve{\underline{e}}_x(z)\right] \\ \left[\breve{\underline{e}}_y(z)\right] \\ \left[\breve{\underline{h}}_x(z)\right] \\ \left[\breve{\underline{h}}_y(z)\right] \end{bmatrix}, \tag{3.183}$$

but the $4(2\tau_t+1) \times 4(2\tau_t+1)$ matrix $\left[\underline{\underline{\breve{P}}}(z)\right]$ is too cumbersome for reproduction here.

The boundary values of $\left[\breve{\underline{f}}(z)\right]$ follow from the expansions of the incident, reflected, and transmitted field phasors. In the matrix notation used for RCWA, the boundary value

$$\left[\breve{\underline{f}}(0-)\right] =$$

$$\begin{bmatrix} \left[\underline{\breve{s}}_x\right] & \left[\underline{\breve{p}}_x^{\text{inc}}\right] & \left[\underline{\breve{s}}_x\right] & \left[\underline{\breve{p}}_x^{\text{ref}}\right] \\ \left[\underline{\breve{s}}_y\right] & \left[\underline{\breve{p}}_y^{\text{inc}}\right] & \left[\underline{\breve{s}}_y\right] & \left[\underline{\breve{p}}_y^{\text{ref}}\right] \\ n_1\eta_0^{-1}\left[\underline{\breve{p}}_x^{\text{inc}}\right] & -n_1\eta_0^{-1}\left[\underline{\breve{s}}_x\right] & n_1\eta_0^{-1}\left[\underline{\breve{p}}_x^{\text{ref}}\right] & -n_1\eta_0^{-1}\left[\underline{\breve{s}}_x\right] \\ n_1\eta_0^{-1}\left[\underline{\breve{p}}_y^{\text{inc}}\right] & -n_1\eta_0^{-1}\left[\underline{\breve{s}}_y\right] & n_1\eta_0^{-1}\left[\underline{\breve{p}}_y^{\text{ref}}\right] & -n_1\eta_0^{-1}\left[\underline{\breve{s}}_y\right] \end{bmatrix} \cdot \begin{bmatrix} \left[\underline{\breve{A}}\right] \\ \left[\underline{\breve{R}}\right] \end{bmatrix}, \tag{3.184}$$

where the column $2(2\tau_t + 1)$-vector

$$\left[\underline{\breve{A}}\right] = \left[a_s^{(-\tau_t)}, a_s^{(-\tau_t+1)}, \dots, a_s^{(\tau_t-1)}, a_s^{(\tau_t)}, a_p^{(-\tau_t)}, a_p^{(-\tau_t+1)}, \dots, a_p^{(\tau_t-1)}, a_p^{(\tau_t)}\right]^T \tag{3.185}$$

comprises known quantities, but the column $2(2\tau_t + 1)$-vector

$$\left[\underline{\breve{R}}\right] = \left[r_s^{(-\tau_t)}, r_s^{(-\tau_t+1)}, \dots, r_s^{(\tau_t-1)}, r_s^{(\tau_t)}, r_p^{(-\tau_t)}, r_p^{(-\tau_t+1)}, \dots, r_p^{(\tau_t-1)}, r_p^{(\tau_t)}\right]^T \tag{3.186}$$

contains the unknown reflection amplitudes. In addition, the following six diagonal $(2\tau_t + 1) \times (2\tau_t + 1)$ matrixes enter Eq. (3.184):

$$\left[\underline{\underline{\breve{s}}}_x\right] = \begin{bmatrix} \underline{u}_x \cdot \underline{s}^{(-\tau_t)} & 0 & \ldots & 0 & 0 \\ 0 & \underline{u}_x \cdot \underline{s}^{(-\tau_t+1)} & \ldots & 0 & 0 \\ \ldots & \ldots & \ldots & \ldots & \ldots \\ 0 & 0 & \ldots & \underline{u}_x \cdot \underline{s}^{(\tau_t-1)} & 0 \\ 0 & 0 & \ldots & 0 & \underline{u}_x \cdot \underline{s}^{(\tau_t)} \end{bmatrix}, \quad (3.187)$$

$$\left[\underline{\underline{\breve{s}}}_y\right] = \begin{bmatrix} \underline{u}_y \cdot \underline{s}^{(-\tau_t)} & 0 & \ldots & 0 & 0 \\ 0 & \underline{u}_y \cdot \underline{s}^{(-\tau_t+1)} & \ldots & 0 & 0 \\ \ldots & \ldots & \ldots & \ldots & \ldots \\ 0 & 0 & \ldots & \underline{u}_y \cdot \underline{s}^{(\tau_t-1)} & 0 \\ 0 & 0 & \ldots & 0 & \underline{u}_y \cdot \underline{s}^{(\tau_t)} \end{bmatrix}, \quad (3.188)$$

$$\left[\underline{\underline{\breve{p}}}_x^{\text{inc}}\right] = \begin{bmatrix} \underline{u}_x \cdot \underline{p}_{\text{inc}}^{(-\tau_t)} & 0 & \ldots & 0 & 0 \\ 0 & \underline{u}_x \cdot \underline{p}_{\text{inc}}^{(-\tau_t+1)} & \ldots & 0 & 0 \\ \ldots & \ldots & \ldots & \ldots & \ldots \\ 0 & 0 & \ldots & \underline{u}_x \cdot \underline{p}_{\text{inc}}^{(\tau_t-1)} & 0 \\ 0 & 0 & \ldots & 0 & \underline{u}_x \cdot \underline{p}_{\text{inc}}^{(\tau_t)} \end{bmatrix}, \quad (3.189)$$

$$\left[\underline{\underline{\breve{p}}}_y^{\text{inc}}\right] = \begin{bmatrix} \underline{u}_y \cdot \underline{p}_{\text{inc}}^{(-\tau_t)} & 0 & \ldots & 0 & 0 \\ 0 & \underline{u}_y \cdot \underline{p}_{\text{inc}}^{(-\tau_t+1)} & \ldots & 0 & 0 \\ \ldots & \ldots & \ldots & \ldots & \ldots \\ 0 & 0 & \ldots & \underline{u}_y \cdot \underline{p}_{\text{inc}}^{(\tau_t-1)} & 0 \\ 0 & 0 & \ldots & 0 & \underline{u}_y \cdot \underline{p}_{\text{inc}}^{(\tau_t)} \end{bmatrix}, \quad (3.190)$$

$$\left[\underline{\underline{\breve{p}}}_x^{\text{ref}}\right] = \begin{bmatrix} \underline{u}_x \cdot \underline{p}_{\text{ref}}^{(-\tau_t)} & 0 & \ldots & 0 & 0 \\ 0 & \underline{u}_x \cdot \underline{p}_{\text{ref}}^{(-\tau_t+1)} & \ldots & 0 & 0 \\ \ldots & \ldots & \ldots & \ldots & \ldots \\ 0 & 0 & \ldots & \underline{u}_x \cdot \underline{p}_{\text{ref}}^{(\tau_t-1)} & 0 \\ 0 & 0 & \ldots & 0 & \underline{u}_x \cdot \underline{p}_{\text{ref}}^{(\tau_t)} \end{bmatrix}, \quad (3.191)$$

$$\left[\underline{\underline{\breve{p}}}_y^{\text{ref}}\right] = \begin{bmatrix} \underline{u}_y \cdot \underline{p}_{\text{ref}}^{(-\tau_t)} & 0 & \ldots & 0 & 0 \\ 0 & \underline{u}_y \cdot \underline{p}_{\text{ref}}^{(-\tau_t+1)} & \ldots & 0 & 0 \\ \ldots & \ldots & \ldots & \ldots & \ldots \\ 0 & 0 & \ldots & \underline{u}_y \cdot \underline{p}_{\text{ref}}^{(\tau_t-1)} & 0 \\ 0 & 0 & \ldots & 0 & \underline{u}_y \cdot \underline{p}_{\text{ref}}^{(\tau_t)} \end{bmatrix}. \quad (3.192)$$

The other boundary value

$$\left[\underline{\breve{f}}(d+)\right] = \begin{bmatrix} \left[\underline{\underline{\breve{s}}}_x\right] & \left[\underline{\underline{\breve{p}}}_x^{\text{tr}}\right] \\ \left[\underline{\underline{\breve{s}}}_y\right] & \left[\underline{\underline{\breve{p}}}_y^{\text{tr}}\right] \\ n_2\eta_0^{-1}\left[\underline{\underline{\breve{p}}}_x^{\text{tr}}\right] & -n_2\eta_0^{-1}\left[\underline{\underline{\breve{s}}}_x\right] \\ n_2\eta_0^{-1}\left[\underline{\underline{\breve{p}}}_y^{\text{tr}}\right] & -n_2\eta_0^{-1}\left[\underline{\underline{\breve{s}}}_y\right] \end{bmatrix} \cdot \left[\underline{\breve{T}}\right] \quad (3.193)$$

contains the column $2(2\tau_t + 1)$-vector

$$\left[\check{\underline{T}}\right] = \left[t_s^{(-\tau_t)}, t_s^{(-\tau_t+1)}, \ldots, t_s^{(\tau_t-1)}, t_s^{(\tau_t)}, t_p^{(-\tau_t)}, t_p^{(-\tau_t+1)}, \ldots, t_p^{(\tau_t-1)}, t_p^{(\tau_t)}\right]^T$$

(3.194)

comprising the unknown transmission amplitudes and employs the diagonal matrixes

$$\left[\check{\underline{\underline{p}}}_x^{\mathrm{tr}}\right] = \begin{bmatrix} \underline{u}_x \bullet \underline{p}_{\mathrm{tr}}^{(-\tau_t)} & 0 & \cdots & 0 & 0 \\ 0 & \underline{u}_x \bullet \underline{p}_{\mathrm{tr}}^{(-\tau_t+1)} & \cdots & 0 & 0 \\ \cdots & \cdots & \cdots & \cdots & \cdots \\ 0 & 0 & \cdots & \underline{u}_x \bullet \underline{p}_{\mathrm{tr}}^{(\tau_t-1)} & 0 \\ 0 & 0 & \cdots & 0 & \underline{u}_x \bullet \underline{p}_{\mathrm{tr}}^{(\tau_t)} \end{bmatrix}$$

(3.195)

and

$$\left[\check{\underline{\underline{p}}}_y^{\mathrm{tr}}\right] = \begin{bmatrix} \underline{u}_y \bullet \underline{p}_{\mathrm{tr}}^{(-\tau_t)} & 0 & \cdots & 0 & 0 \\ 0 & \underline{u}_y \bullet \underline{p}_{\mathrm{tr}}^{(-\tau_t+1)} & \cdots & 0 & 0 \\ \cdots & \cdots & \cdots & \cdots & \cdots \\ 0 & 0 & \cdots & \underline{u}_y \bullet \underline{p}_{\mathrm{tr}}^{(\tau_t-1)} & 0 \\ 0 & 0 & \cdots & 0 & \underline{u}_y \bullet \underline{p}_{\mathrm{tr}}^{(\tau_t)} \end{bmatrix}.$$

(3.196)

For convenience, the boundary values $\left[\check{\underline{f}}(0-)\right]$ and $\left[\check{\underline{f}}(d+)\right]$ are recast compactly as

$$\left[\check{\underline{f}}(0-)\right] = \begin{bmatrix} \left[\check{\underline{\underline{Y}}}_e^{\mathrm{inc}}\right] & \left[\check{\underline{\underline{Y}}}_e^{\mathrm{ref}}\right] \\ \left[\check{\underline{\underline{Y}}}_h^{\mathrm{inc}}\right] & \left[\check{\underline{\underline{Y}}}_h^{\mathrm{ref}}\right] \end{bmatrix} \bullet \begin{bmatrix} \left[\check{\underline{A}}\right] \\ \left[\check{\underline{R}}\right] \end{bmatrix}$$

(3.197)

and

$$\left[\check{\underline{f}}(d+)\right] = \begin{bmatrix} \left[\check{\underline{\underline{Y}}}_e^{\mathrm{tr}}\right] \\ \left[\check{\underline{\underline{Y}}}_h^{\mathrm{tr}}\right] \end{bmatrix} \bullet \left[\check{\underline{T}}\right].$$

(3.198)

The $2(2\tau_t + 1) \times 2(2\tau_t + 1)$ matrixes $\left[\check{\underline{\underline{Y}}}_e^{\mathrm{inc}}\right]$, etc., can be synthesized from Eqs. (3.184) and (3.193) by inspection; for instance,

$$\left[\check{\underline{\underline{Y}}}_e^{\mathrm{inc}}\right] = \begin{bmatrix} \left[\underline{\underline{s}}_x\right] & \left[\check{\underline{\underline{p}}}_x^{\mathrm{inc}}\right] \\ \left[\underline{\underline{s}}_y\right] & \left[\check{\underline{\underline{p}}}_y^{\mathrm{inc}}\right] \end{bmatrix}, \qquad \left[\check{\underline{\underline{Y}}}_h^{\mathrm{inc}}\right] = n_1 \eta_0^{-1} \begin{bmatrix} \left[\check{\underline{\underline{p}}}_x^{\mathrm{inc}}\right] & -\left[\underline{\underline{s}}_x\right] \\ \left[\check{\underline{\underline{p}}}_y^{\mathrm{inc}}\right] & -\left[\underline{\underline{s}}_y\right] \end{bmatrix}.$$

(3.199)

As in Section 3.7.2, the piecewise-uniform-approximation method is very suitable to solve Eq. (3.182) numerically. Let the region $0 \leqslant z \leqslant d$ be divided into N_s subregions, the ℓth subregion being bounded by the planes $z = z_{\ell-1}$ and $z = z_\ell$, $\ell \in [1, N_s]$, where $z_0 = 0$ and $z_{N_s} = d$. In the ℓth subregion, the matrix $\left[\underline{\underline{\breve{P}}}(z)\right]$ is approximated by the uniform matrix

$$\left[\underline{\underline{\breve{P}}}\right]^{(\ell)} = \left[\underline{\underline{\breve{P}}}\left(\frac{z_\ell + z_{\ell-1}}{2}\right)\right], \qquad \ell \in [1, N_s]. \tag{3.200}$$

Then, provided that every subregion is sufficiently thin,

$$\left[\underline{\breve{f}}(d-)\right] \simeq \left[\underline{\underline{\breve{W}}}\right]^{(N_s)} \cdot \left[\underline{\underline{\breve{W}}}\right]^{(N_s-1)} \cdot \ldots \cdot \left[\underline{\underline{\breve{W}}}\right]^{(2)} \cdot \left[\underline{\underline{\breve{W}}}\right]^{(1)} \cdot \left[\underline{\breve{f}}(0+)\right], \tag{3.201}$$

where

$$\left[\underline{\underline{\breve{W}}}\right]^{(\ell)} = \exp\left\{i(z_\ell - z_{\ell-1})\left[\underline{\underline{\breve{P}}}\right]^{(\ell)}\right\}, \qquad \ell \in [1, N_s]. \tag{3.202}$$

The usual boundary conditions being

$$\left[\underline{\breve{f}}(0-)\right] = \left[\underline{\breve{f}}(0+)\right], \qquad \left[\underline{\breve{f}}(d-)\right] = \left[\underline{\breve{f}}(d+)\right], \tag{3.203}$$

Equations (3.197), (3.198), and (3.201) lead to the algebraic equation

$$\left[\begin{bmatrix}\underline{\underline{\breve{Y}}}_e^{tr}\end{bmatrix}\\\begin{bmatrix}\underline{\underline{\breve{Y}}}_h^{tr}\end{bmatrix}\right] \cdot \left[\underline{\breve{T}}\right] = \left[\underline{\underline{\breve{W}}}\right]^{(N_s)} \cdot \left[\underline{\underline{\breve{W}}}\right]^{(N_s-1)} \cdot \ldots$$

$$\cdot \left[\underline{\underline{\breve{W}}}\right]^{(2)} \cdot \left[\underline{\underline{\breve{W}}}\right]^{(1)} \cdot \left[\begin{bmatrix}\underline{\underline{\breve{Y}}}_e^{inc}\end{bmatrix} & \begin{bmatrix}\underline{\underline{\breve{Y}}}_e^{ref}\end{bmatrix}\\\begin{bmatrix}\underline{\underline{\breve{Y}}}_h^{inc}\end{bmatrix} & \begin{bmatrix}\underline{\underline{\breve{Y}}}_h^{ref}\end{bmatrix}\end{bmatrix} \cdot \left[\begin{bmatrix}\underline{\breve{A}}\end{bmatrix}\\\begin{bmatrix}\underline{\breve{R}}\end{bmatrix}\right], \tag{3.204}$$

which may be solved for $\left[\underline{\breve{R}}\right]$ and $\left[\underline{\breve{T}}\right]$ using standard matrix techniques [261].

3.8.6   Stable RCWA Algorithm

The application of a standard matrix-inversion technique such as the Gauss elimination technique [261] to solve Eq. (3.204) is prone to numerical problems, particularly when either d is electrically large and/or the incidence is highly oblique. As stated in Section 2.3.4.5, a stable algorithm is available to overcome these problems. It requires that $\left[\underline{\underline{\breve{P}}}\right]^{(\ell)}$ be diagonalizable for all $\ell \in [1, N_s]$. Then,

$$\left[\underline{\underline{\breve{P}}}\right]^{(\ell)} = \left[\underline{\underline{\breve{V}}}\right]^{(\ell)} \cdot \left[\underline{\underline{\breve{G}}}\right]^{(\ell)} \cdot \left(\left[\underline{\underline{\breve{V}}}\right]^{(\ell)}\right)^{-1}, \qquad \ell \in [1, N_s], \tag{3.205}$$

where the diagonal matrix $\left[\underline{\underline{\breve{G}}}\right]^{(\ell)}$ contains the eigenvalues of $\left[\underline{\underline{\breve{P}}}\right]^{(\ell)}$ in decreasing order of the magnitude of the imaginary part, and $\left[\underline{\underline{\breve{V}}}\right]^{(\ell)}$ is a square matrix comprising the eigenvectors of $\left[\underline{\underline{\check{P}}}\right]^{(\ell)}$ as its columns, arranged so that each eigenvector is in the same position as the corresponding eigenvalue in $\left[\underline{\underline{\check{G}}}\right]^{(\ell)}$. Accordingly, Eq. (3.182) yields

$$\left[\underline{\check{f}}(z_{\ell-1})\right] = \left[\underline{\underline{\check{V}}}\right]^{(\ell)} \cdot \exp\left\{-i(z_\ell - z_{\ell-1})\left[\underline{\underline{\check{G}}}\right]^{(\ell)}\right\} \cdot \left(\left[\underline{\underline{\check{V}}}\right]^{(\ell)}\right)^{-1} \cdot \left[\underline{\check{f}}(z_\ell)\right].$$
(3.206)

A set of auxiliary column $2(2\tau_t + 1)$-vectors $\left[\underline{\check{T}}\right]^{(\ell)}$ and auxiliary transmission matrixes $\left[\underline{\underline{\check{Z}}}\right]^{(\ell)}$ of size $4(2\tau_t+1)\times 2(2\tau_t+1)$ are postulated to satisfy the relations [258]

$$\left[\underline{\check{f}}(z_\ell)\right] = \left[\underline{\underline{\check{Z}}}\right]^{(\ell)} \cdot \left[\underline{\check{T}}\right]^{(\ell)}, \qquad \ell \in [0, N_s],$$
(3.207)

where

$$\left[\underline{\check{T}}\right]^{(N_s)} = \left[\underline{\check{T}}\right], \qquad \left[\underline{\underline{\check{Z}}}\right]^{(N_s)} = \left[\begin{array}{c}\left[\underline{\underline{\check{Y}}}_e^{\mathrm{tr}}\right] \\ \left[\underline{\underline{\check{Y}}}_h^{\mathrm{tr}}\right]\end{array}\right].$$
(3.208)

Equation (3.207) is then substituted in Eq. (3.206) to obtain the relation

$$\left[\underline{\underline{\check{Z}}}\right]^{(\ell-1)} \cdot \left[\underline{\check{T}}\right]^{(\ell-1)} = \left[\underline{\underline{\check{V}}}\right]^{(\ell)}$$

$$\cdot \left[\begin{array}{cc} \exp\left\{-i(z_\ell - z_{\ell-1})\left[\underline{\underline{\check{G}}}_{\mathrm{upper}}\right]^{(\ell)}\right\} & \left[\underline{\underline{o}}\right] \\ \left[\underline{\underline{o}}\right] & \exp\left\{-i(z_\ell - z_{\ell-1})\left[\underline{\underline{\check{G}}}_{\mathrm{lower}}\right]^{(\ell)}\right\} \end{array}\right]$$

$$\cdot \left(\left[\underline{\underline{\check{V}}}\right]^{(\ell)}\right)^{-1} \cdot \left[\underline{\underline{\check{Z}}}\right]^{(\ell)} \cdot \left[\underline{\check{T}}\right]^{(\ell)}, \qquad \ell \in [1, N_s],$$
(3.209)

where $\left[\underline{\underline{o}}\right]$ is the $2(2\tau_t + 1) \times 2(2\tau_t + 1)$ null matrix. The $2(2\tau_t + 1) \times 2(2\tau_t + 1)$ matrixes $\left[\underline{\underline{\check{G}}}_{\mathrm{upper}}\right]^{(\ell)}$ and $\left[\underline{\underline{\check{G}}}_{\mathrm{lower}}\right]^{(\ell)}$ are the upper and lower diagonal submatrixes of the $4(2\tau_t + 1) \times 4(2\tau_t + 1)$ matrix $\left[\underline{\underline{\check{G}}}\right]^{(\ell)}$, respectively.

Next, the $2(2\tau_t+1)\times 2(2\tau_t+1)$ matrixes $\left[\underline{\underline{\check{X}}}_{\mathrm{upper}}\right]^{(\ell)}$ and $\left[\underline{\underline{\check{X}}}_{\mathrm{lower}}\right]^{(\ell)}$ are defined via

$$\left[\begin{array}{c}\left[\underline{\underline{\check{X}}}_{\mathrm{upper}}\right]^{(\ell)} \\ \left[\underline{\underline{\check{X}}}_{\mathrm{lower}}\right]^{(\ell)}\end{array}\right] = \left(\left[\underline{\underline{\check{V}}}\right]^{(\ell)}\right)^{-1} \cdot \left[\underline{\underline{\check{Z}}}\right]^{(\ell)}.$$
(3.210)

Furthermore, the recurrence relation

$$\left[\underline{\check{T}}\right]^{(\ell-1)} = \exp\left\{-i(z_\ell - z_{\ell-1})\left[\underline{\underline{\check{G}}}_{\mathrm{upper}}\right]^{(\ell)}\right\} \cdot \left[\underline{\underline{\check{X}}}_{\mathrm{upper}}\right]^{(\ell)} \cdot \left[\underline{\check{T}}\right]^{(\ell)}$$
(3.211)

is postulated. Now, substitution of Eq. (3.211) in Eq. (3.209) leads to the relation

$$\left[\underline{\underline{\breve{Z}}}\right]^{(\ell-1)} = \left[\underline{\underline{\breve{V}}}\right]^{(\ell)} \cdot \left[\begin{array}{c} [\underline{\breve{1}}] \\ [\underline{\breve{U}}]^{(\ell)} \end{array}\right], \qquad \ell \in [1, N_s], \tag{3.212}$$

where $[\underline{\breve{1}}]$ is the $2(2\tau_t + 1) \times 2(2\tau_t + 1)$ identity matrix and

$$\left[\underline{\underline{\breve{U}}}\right]^{(\ell)} = \exp\left\{-i(z_\ell - z_{\ell-1})\left[\underline{\underline{\breve{G}}}_{\text{lower}}\right]^{(\ell)}\right\} \cdot \left[\underline{\underline{\breve{X}}}_{\text{lower}}\right]^{(\ell)} \cdot \left(\left[\underline{\underline{\breve{X}}}_{\text{upper}}\right]^{(\ell)}\right)^{-1}$$

$$\cdot \exp\left\{i(z_\ell - z_{\ell-1})\left[\underline{\underline{\breve{G}}}_{\text{upper}}\right]^{(\ell)}\right\}, \qquad \ell \in [1, N_s]. \tag{3.213}$$

After using Eqs. (3.210) and (3.212) repeatedly, $\left[\underline{\underline{\breve{Z}}}\right]^{(\ell)}$ is found in terms of $\left[\underline{\underline{\breve{Z}}}\right]^{(N_s)}$ $\forall \ell \in [0, N_s - 1]$.

After it is found, $\left[\underline{\underline{\breve{Z}}}\right]^{(0)}$ is partitioned as

$$\left[\underline{\underline{\breve{Z}}}\right]^{(0)} = \left[\begin{array}{c} \left[\underline{\underline{\breve{Z}}}_{\text{upper}}\right]^{(0)} \\ \left[\underline{\underline{\breve{Z}}}_{\text{lower}}\right]^{(0)} \end{array}\right]. \tag{3.214}$$

From Eqs. (3.197) and (3.207), and after enforcing the boundary condition $\left[\underline{\breve{f}}(0-)\right] = \left[\underline{\breve{f}}(0+)\right]$, $\left[\underline{\breve{R}}\right]$ and $\left[\underline{\breve{T}}\right]^{(0)}$ are found as follows:

$$\left[\begin{array}{c} \left[\underline{\breve{T}}\right]^{(0)} \\ \left[\underline{\breve{R}}\right] \end{array}\right] = \left[\begin{array}{cc} \left[\underline{\underline{\breve{Z}}}_{\text{upper}}\right]^{(0)} & -\left[\underline{\underline{\breve{Y}}}_e^{\text{ref}}\right] \\ \left[\underline{\underline{\breve{Z}}}_{\text{lower}}\right]^{(0)} & -\left[\underline{\underline{\breve{Y}}}_h^{\text{ref}}\right] \end{array}\right]^{-1} \cdot \left[\begin{array}{c} \left[\underline{\underline{\breve{Y}}}_e^{\text{inc}}\right] \\ \left[\underline{\underline{\breve{Y}}}_h^{\text{inc}}\right] \end{array}\right] \cdot \left[\underline{\breve{A}}\right]. \tag{3.215}$$

After $\left[\underline{\breve{T}}\right]^{(0)}$ is known, $\left[\underline{\breve{T}}\right]^{(\ell)}$ is found for all $\ell \in [1, N_s]$ by reversing the sense of iterations in Eq. (3.211). Thus the stable RCWA algorithm yields both $\left[\underline{\breve{T}}\right] = \left[\underline{\breve{T}}\right]^{(N_s)}$ and $\left[\underline{\breve{R}}\right]$.

With $\left[\underline{\breve{f}}(z_0)\right] = \left[\underline{\breve{f}}(0-)\right]$ now known using Eq. (3.197), $\left[\underline{\breve{f}}(z_\ell)\right]$ can be calculated for all $\ell \in [1, N_s]$ by repeated use of Eq. (3.206). Next, with $\left[\underline{\breve{f}}(z_\ell)\right]$ known, $\left[\underline{\breve{h}}_z(z_\ell)\right]$ and $\left[\underline{\breve{e}}_z(z_\ell)\right]$ can be obtained from Eqs. (3.181). After the further use of Eqs. (3.166), the field phasors $\underline{E}(\underline{r})$ and $\underline{H}(\underline{r})$ can be mapped in the region $0 \leqslant z \leqslant d$.

The sufficiency of M_t and N_t requires convergence tests. Increasing either or both of them in unit increments, one must test that the reflectances $R_{\text{sp}}^{(m,n)}$, etc., and the transmittances $T_{\text{sp}}^{(m,n)}$, etc., converge within satisfactory tolerance limits (say, 0.1%). The principle of conservation of energy also must be satisfied simultaneously.

3.8.7 Excitation of a Surface Wave

In order to investigate the excitation of a surface wave in the grating-coupled configuration at a specified value of the free-space wavelength, some measurable intensity must be plotted as a function of the angle of incidence. This plot would contain many peaks and troughs. Some of these features would be found at about the same value of θ_{inc}, provided that average thicknesses d_A and d_B exceed certain thresholds. As mentioned in Sections 3.7.3 and 3.7.4, the thickness of a partnering material that is periodically nonhomogeneous should be increased in increments of a period, for the purpose of identifying surface waves.

For SPP waves, peaks in the plots of an absorptance versus angle of incidence should be examined, as explained in Section 2.3.4.6 and illustrated in Section 2.3.4.7. Even if both partnering materials are commonly assumed to be lossless in the spectral regime of interest, the incorporation of weak dissipation in calculations is a good strategy, as exemplified recently for the identification of Tamm waves [180].

Suppose that the value of θ_{inc} of a certain absorptance peak does not change significantly when the average thicknesses of both partnering materials are increased above some thresholds, the angle ψ being fixed. If an index pair (m, n) can be found such that

$$k_{xy}^{(m,n)} \approx \text{Re}\,\{q\}, \tag{3.216}$$

where q is a solution of Eq. (3.87) for the corresponding canonical boundary-value problem, a surface wave is excited as a Floquet harmonic of order (m, n).

4 Dyakonov Waves

4.1 Introduction

Whereas Chapter 2 is devoted to surface-wave propagation guided by the interface of an isotropic homogeneous dielectric material and an isotropic homogeneous metal, in this chapter both partnering materials are homogeneous dielectric with at least one being anisotropic. Although either or both partnering materials can even be homogeneous bianisotropic materials [37,38] and the problem of surface-wave propagation can be theoretically treated using the general methods presented in Chapter 3, a restriction to anisotropic dielectric materials in this chapter is appropriate in light of the current state of practical materials science.

The transition from isotropic dielectric materials to anisotropic dielectric materials is associated with the emergence of new phenomenons and a generally richer palette of optical response characteristics. For example, isotropic dielectric materials are unirefringent, but their anisotropic counterparts are birefringent. In a unirefringent material, the four elements of the diagonal matrix $[\underline{G}]$ of Eq. $(3.31)_2$ obey the restrictions $g_1 = g_2$, $g_3 = g_4$, and $g_1 = -g_3$; furthermore, these elements do not depend on the angle ψ of Eq. (3.15). In a birefringent material, $g_1 \neq g_2$ and/or $g_3 \neq g_4$ at the very least.

The surface wave guided by the interface of two homogeneous dielectric materials, of which at least one is anisotropic, is called a Dyakonov wave. The following rule applies: whereas such an interface *may* support the propagation of Dyakonov waves, the propagation of a Dyakonov wave is never supported if both partnering dielectric materials are isotropic. Thus, the topic of Dyakonov waves is inextricably linked to anisotropy.

Both the theory and practicality of Dyakonov waves, in the context of relatively simple anisotropic dielectric materials and beyond, are considered in this chapter. We are exclusively concerned with partnering materials which are homogeneous at macroscopic length scales, and our focus is largely on materials which exhibit negligible dissipation.

In a seminal short paper published in 1988, D'yakonov[1] presented a theoretical description of a surface wave guided by the interface of an isotropic dielectric material and a uniaxial dielectric material [35]. In contrast to the case for SPP waves, both

---

[1]The name *D'yakonov* was published in a Russian-language journal as well as in its English-language translational journal. It is nowadays commonly written as *Dyakonov* in English.

Electromagnetic Surface Waves. http://dx.doi.org/10.1016/B978-0-12-397024-4.00004-9

partnering materials involved were assumed to be lossless, both were characterized by positive-definite permittivity dyadics, and the uniaxiality of one of them was essential. All eigenvalues of a positive-definite dyadic are real and positive.

Although there were earlier studies in which surface waves guided by the interface of anisotropic/isotropic dielectric materials had been described [34,303], surface waves of this type (as well as their generalizations) have become known as *Dyakonov waves*. Since these surface waves are guided by the interface of two dielectric materials whose principal permittivity scalars are typically weakly sensitive to frequency, in the optical regime at least, these waves are sometimes referred to as being dispersionless [164,304,305].

4.2 Interface of an Anisotropic Material and an Isotropic Material

4.2.1 Interface of a Uniaxial Material and an Isotropic Material

4.2.1.1 Optic Axis in Interface Plane

The simplest interface for the propagation of a Dyakonov wave, at least from a mathematical perspective, is that of an isotropic dielectric material and a uniaxial dielectric material, with both materials assumed to be lossless. The characteristics of the Dyakonov wave can be obtained after the application of the general formulation of the canonical boundary-value problem given in Section 3.5. Although the theory of the canonical boundary-value problem for Dyakonov-wave propagation is covered in that section for even the most general of cases, it is illuminating to consider here in greater detail the simplest Dyakonov wave for analytical progress before having to resort to numerical techniques.

Following the notation of Section 3.5, let us take material $\mathcal{A}$ which occupies the half-space $z > 0$ to be a uniaxial dielectric material characterized by the permittivity dyadic

$$\underline{\underline{\epsilon}}^{\mathcal{A}} = \epsilon_s^{\mathcal{A}}\underline{\underline{I}} + \left(\epsilon_t^{\mathcal{A}} - \epsilon_s^{\mathcal{A}}\right)\underline{u}_c\underline{u}_c,\tag{4.1}$$

with the unit vector $\underline{u}_c$ pointing in the direction of the optic axis of material $\mathcal{A}$.[2] Let us further simplify matters by assuming that the optic axis of material $\mathcal{A}$ is aligned parallel to the plane of the interface. Without further loss of generality, the coordinate axes can be oriented such that $\underline{u}_c = \underline{u}_x$. Material $\mathcal{B}$ which occupies the half-space $z < 0$ is taken to be an isotropic dielectric material characterized by the permittivity dyadic

$$\underline{\underline{\epsilon}}^{\mathcal{B}} = \epsilon^{\mathcal{B}}\underline{\underline{I}}.\tag{4.2}$$

[2]The optic axis of a uniaxial dielectric material coincides with its optic ray axis, as discussed in Appendix B.

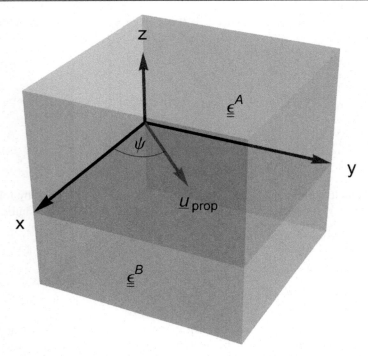

Figure 4.1 Schematic of the canonical boundary-value problem for the propagation of a Dyakonov wave guided by the planar interface of two dissimilar dielectric materials with permittivity dyadics $\underline{\underline{\epsilon}}^A$ and $\underline{\underline{\epsilon}}^B$. The direction of propagation in the interface plane is denoted by $\underline{u}_{\text{prop}}$.

Let us consider the surface wave which propagates in the xy plane (i.e. parallel to the interface) at an angle $\psi \in [0°, 360°)$ with respect to the x axis, as shown in Figure 4.1. The spatial variation of the field phasors along the direction of propagation

$$\underline{u}_{\text{prop}} = \underline{u}_x \cos \psi + \underline{u}_y \sin \psi \tag{4.3}$$

has the form

$$\exp(iq\underline{u}_{\text{prop}} \cdot \underline{r}). \tag{4.4}$$

The wavenumber q is taken to be real valued for Dyakonov waves. Complex-valued wavenumbers may exist even though both partnering materials are taken to have negligible dissipation, and the associated surface waves would then have to be called *leaky Dyakonov waves*. But, as a leaky Dyakonov wave is of little value for optical applications, it is ignored in the remainder of this chapter.

The field phasors in the half-space $z > 0$ are given by Eqs. (3.80), which may be restated here as

$$
\left.
\begin{aligned}
\underline{e}(z) &= \sum_{m=1}^{2} C_m^{\mathcal{A}} \left\{ v_{1m}^{\mathcal{A}} \underline{u}_x + v_{2m}^{\mathcal{A}} \underline{u}_y \right. \\
&\quad + \left. \left[v_{zx}^{eh,\mathcal{A}} v_{3m}^{\mathcal{A}} + v_{zy}^{eh,\mathcal{A}} v_{4m}^{\mathcal{A}} \right] \underline{u}_z \right\} \exp\left(i\alpha_m^{\mathcal{A}} z \right) \\
\underline{h}(z) &= \sum_{m=1}^{2} C_m^{\mathcal{A}} \left\{ v_{3m}^{\mathcal{A}} \underline{u}_x + v_{4m}^{\mathcal{A}} \underline{u}_y \right. \\
&\quad + \left. \left[v_{zx}^{he,\mathcal{A}} v_{1m}^{\mathcal{A}} + v_{zy}^{he,\mathcal{A}} v_{2m}^{\mathcal{A}} \right] \underline{u}_z \right\} \exp\left(i\alpha_m^{\mathcal{A}} z \right)
\end{aligned}
\right\}, \quad z > 0, \qquad (4.5)
$$

wherein explicit expressions for $v_{1m}^{\mathcal{A}}$, etc., and $v_{zx}^{eh,\mathcal{A}}$, etc., can be found by following the methodology of Chapter 3. It suffices to mention here that the wavenumbers

$$
\alpha_1^{\mathcal{A}} = \sqrt{\omega^2 \mu_0 \epsilon_s^{\mathcal{A}} - q^2} \qquad (4.6)
$$

and

$$
\alpha_2^{\mathcal{A}} = \sqrt{\omega^2 \mu_0 \epsilon_t^{\mathcal{A}} - q^2 \left(\sin^2 \psi + \frac{\epsilon_t^{\mathcal{A}}}{\epsilon_s^{\mathcal{A}}} \cos^2 \psi \right)}, \qquad (4.7)
$$

respectively, of the two eigenmodes in Eqs. (4.5) are akin to the *ordinary* and *extraordinary* waves in a uniaxial dielectric material [78] and must obey the restrictions

$$
\text{Im}\left\{ \alpha_m^{\mathcal{A}} \right\} > 0, \quad m \in [1, 2]. \qquad (4.8)
$$

Of course, the coefficients $C_1^{\mathcal{A}}$ and $C_2^{\mathcal{A}}$ are not known. Let us also note that although the symbol $g_m^{\mathcal{A}}$ could have been used in place of $\alpha_m^{\mathcal{A}}$ in accordance with Section 3.3, because the uniaxial material is homogeneous, we elected to use the latter for consistency with those types of surface waves that require at least one partnering material to be periodically nonhomogeneous normal to the interface.

A similar treatment of the field phasors in the half-space $z < 0$ shows that they are linear combinations of s- and p-polarized eigenmodes. Thus, with unknown coefficients $C_s^{\mathcal{B}}$ and $C_p^{\mathcal{B}}$, we have[3]

$$
\left.
\begin{aligned}
\underline{e}(z) &= C_s^{\mathcal{B}} \underline{u}_s + C_p^{\mathcal{B}} \left(\frac{-\alpha^{\mathcal{B}} \underline{u}_{\text{prop}} + q \underline{u}_z}{k_0 n^{\mathcal{B}}} \right) \exp\left(i\alpha^{\mathcal{B}} z \right) \\
\underline{h}(z) &= \frac{n^{\mathcal{B}}}{\eta_0} \left[C_s^{\mathcal{B}} \left(\frac{-\alpha^{\mathcal{B}} \underline{u}_{\text{prop}} + q \underline{u}_z}{k_0 n^{\mathcal{B}}} \right) - C_p^{\mathcal{B}} \underline{u}_s \right] \exp\left(i\alpha^{\mathcal{B}} z \right)
\end{aligned}
\right\}, \quad z < 0, \quad (4.9)
$$

---

[3] In Chapters 1, 2, and 6, the half-space $z < 0$ in the canonical boundary-value problem for SPP-wave propagation is occupied by a metal, and the spatial dependence of the field phasors is taken to be $\exp\left[i (q \underline{u}_{\text{prop}} - \alpha_{\text{met}} \underline{u}_z) \cdot \underline{r} \right]$ with $\text{Im}\{\alpha_{\text{met}}\} > 0$. In Chapters 4, 5, and 7, sometimes the half-space $z < 0$ in the canonical boundary-value problem for surface-wave propagation is occupied by an isotropic homogeneous dielectric material. In consonance with Chapter 3, we then adopt the spatial dependence $\exp\left[i (q \underline{u}_{\text{prop}} + \alpha^{\mathcal{B}} \underline{u}_z) \cdot \underline{r} \right]$ with $\text{Im}\{\alpha^{\mathcal{B}}\} < 0$. Both representations are identical, as can be ascertained by setting $\alpha_{\text{met}} = -\alpha^{\mathcal{B}}$.

where $\underline{u}_s = -\underline{u}_x \sin \psi + \underline{u}_y \cos \psi$, $n^B = \sqrt{\epsilon^B/\epsilon_0}$, and the wavenumber

$$\alpha^B = \sqrt{\omega^2 \mu_0 \epsilon^B - q^2} \qquad (4.10)$$

satisfies the condition

$$\text{Im}\{\alpha^B\} < 0. \qquad (4.11)$$

After following the procedure presented in Section 3.5.1, the dispersion equation (3.87) for the present problem reduces to

$$\alpha_1^A \left(\alpha^B - \alpha_1^A \right) \left[\epsilon_s^A \alpha_2^A \alpha^B - \epsilon^B \left(\alpha_1^A \right)^2 \right] \cos^2 \psi$$
$$+ \omega^2 \mu_0 \epsilon_s^A \left(\alpha^B - \alpha_2^A \right) \left(\epsilon_s^A \alpha^B - \epsilon^B \alpha_1^A \right) \sin^2 \psi = 0. \qquad (4.12)$$

When combined with Eqs. (4.6), (4.7), and (4.10), the dispersion equation simplifies to

$$\left(\text{Im}\left\{ \alpha_1^A \right\} - \text{Im}\{\alpha^B\} \right) \left(\text{Im}\left\{ \alpha_2^A \right\} - \text{Im}\{\alpha^B\} \right) \left(\epsilon^B \text{Im}\left\{ \alpha_1^A \right\} + \epsilon_s^A \text{Im}\left\{ \alpha_2^A \right\} \right)$$
$$= \omega^2 \mu_0 \left(\epsilon_t^A - \epsilon^B \right) \left(\epsilon^B - \epsilon_s^A \right) \text{Im}\left\{ \alpha_1^A \right\}. \qquad (4.13)$$

For extracting solutions q of Eq. (4.13) consistently with the inequalities (4.8) and (4.11), only two possibilities emerge: either the inequalities

$$\epsilon_s^A < \epsilon^B < \epsilon_t^A \qquad (4.14)$$

or the inequalities

$$\epsilon_s^A > \epsilon^B > \epsilon_t^A \qquad (4.15)$$

must be satisfied. However, satisfaction of the inequalities (4.15) results only in null solutions [35]. Consequently, the inequalities (4.14) must hold in order for Dyakonov waves to exist. Accordingly, only when material A is a positive uniaxial material can a Dyakonov wave be guided by the chosen interface. Positive uniaxiality requires that $\epsilon_t^A > \epsilon_s^A$, whereas negative uniaxiality requires that $\epsilon_t^A < \epsilon_s^A$, as stated in Appendix B.

The inequalities (4.8) and (4.11) are generally satisfied only for a relatively small range of propagation directions, as specified by $\psi_{min} < \psi < \psi_{max}$. The limits on ψ are determined by the zeros of α_2^A and α^B. When

$$q^2 < \omega^2 \mu_0 \epsilon^B, \qquad (4.16)$$

which corresponds to $\psi < \psi_{min}$, then α^B is real valued and the fields are not localized to the interface on the isotropic side ($z < 0$). Similarly, when

$$q^2 \left(\sin^2 \psi + \frac{\epsilon_t^A}{\epsilon_s^A} \cos^2 \psi \right) < \omega^2 \mu_0 \epsilon_t^A, \qquad (4.17)$$

then α^B is real valued and the fields are not localized to the interface on the uniaxial side ($z > 0$). As $\psi \in (\psi_{min}, \psi_{max})$ decreases towards ψ_{min}, the Dyakonov wave becomes less tightly confined to the vicinity of the interface and its mean power density shifts away from the interface in the $-z$ direction. Likewise, as $\psi \in (\psi_{min}, \psi_{max})$ increases towards ψ_{max}, the Dyakonov wave again becomes less tightly confined to the vicinity of interface, and its mean power density shifts away from the interface in the $+z$ direction. Thus, the Dyakonov wave disappears as ψ approaches either of its limiting values ψ_{min} and ψ_{max} from inside its angular existence domain (AED) (ψ_{min}, ψ_{max}).

Explicit expressions can be derived for the limits on propagation directions for Dyakonov waves guided by the interface of an isotropic dielectric material and a uniaxial dielectric material with its optic axis aligned parallel to the interface plane. The transition points $\alpha^B = 0$ and $\alpha_2^A = 0$ give rise to [35]:

$$\psi_{min} = \sin^{-1} \left(\frac{\Upsilon}{2} \left\{ 1 - \Lambda\Upsilon + \left[(1 - \Lambda\Upsilon)^2 + 4\Lambda \right]^{1/2} \right\} \right)^{1/2} \qquad (4.18)$$

and

$$\psi_{max} = \sin^{-1} \left(\frac{\Upsilon (1 + \Lambda)^3}{(1 + \Lambda)^2 (1 + \Lambda\Upsilon) - \Lambda^2 (1 - \Upsilon)^2} \right)^{1/2}, \qquad (4.19)$$

respectively, with the scalar parameters

$$\Upsilon = \frac{\epsilon^B - \epsilon_s^A}{\epsilon_t^A - \epsilon_s^A}, \quad \Lambda = \frac{\epsilon_t^A}{\epsilon_s^A} - 1. \qquad (4.20)$$

The width of the AED is the difference $\psi_{max} - \psi_{min}$. This quantity is plotted in Figure 4.2 as a function of the relative permittivity ϵ^B/ϵ_0 for three strongly anisotropic uniaxial dielectric materials. For all three uniaxial dielectric materials, the width of the AED is less than $1°$ and it rapidly vanishes in the limits $\epsilon^B \to \epsilon_s^A$ and $\epsilon^B \to \epsilon_t^A$. Also, the AED is wider for uniaxial dielectric materials which exhibit a greater degree of positive uniaxiality, as gauged by the scalar parameter Λ defined in Eq. $(4.20)_2$; the larger the value of $\Lambda > 0$, the larger is the degree of positive uniaxiality.

The angle $(\psi_{max} + \psi_{min})/2$ denoting the average direction allowed for Dyakonov-wave propagation is also plotted in Figure 4.2 as a function of the relative permittivity ϵ^B/ϵ_0. This angle tends to $0°$ as $\epsilon^B \to \epsilon_s^A+$ and tends to $90°$ as $\epsilon^B \to \epsilon_t^A-$.

From Eqs. (4.18) and (4.19), it follows that if Dyakonov-wave propagation is supported for the angular range $\psi_{min} < \psi < \psi_{max}$, then it is also supported for the angular ranges $(180° - \psi_{max}) < \psi < (180° - \psi_{min})$, $(180° + \psi_{min}) < \psi < (180° + \psi_{max})$, and $-\psi_{max} < \psi < -\psi_{min}$. Figure 4.3 illustrates all the directions for which a Dyakonov wave can be guided by the planar interface of calomel (a strongly uniaxial material with $\epsilon_t^A = 7.02\epsilon_0$ and $\epsilon_s^A = 3.88\epsilon_0$) and an isotropic dielectric material. In order to maximize the width of the AED, for this figure a value of ϵ^B was chosen which lies midway between ϵ_s^A and ϵ_t^A.

Solutions q of the dispersion equation (4.12)—equivalently, (4.13)—are plotted against $\psi \in (\psi_{min}, \psi_{max})$ in Figure 4.4 for Dyakonov waves guided by the planar interface of calomel and an isotropic dielectric material with permittivity $\epsilon^B = 4.5\epsilon_0$. A modest increase in q of about 1% is observed as the propagation angle ψ increases from ψ_{min} to ψ_{max}.

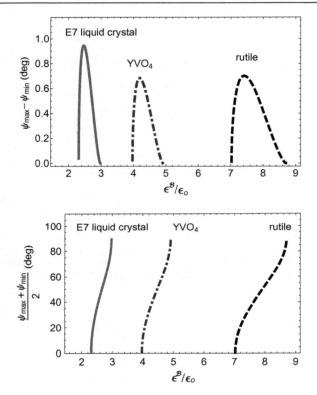

Figure 4.2 (Top) The width $\psi_{max} - \psi_{min}$ and (bottom) the central angle $(\psi_{max} + \psi_{min})/2$ of the AED as functions of the relative permittivity ϵ^{B}/ϵ_0 of the isotropic partnering material when the uniaxial dielectric partnering material is one of E7 liquid crystal $(\epsilon_t^{A} = 2.98\epsilon_0$ and $\epsilon_s^{A} = 2.31\epsilon_0)$, or YVO$_4$ $(\epsilon_t^{A} = 4.91\epsilon_0$ and $\epsilon_s^{A} = 3.97\epsilon_0)$, or rutile $(\epsilon_t^{A} = 8.70\epsilon_0$ and $\epsilon_s^{A} = 7.02\epsilon_0)$. The optic axis of the uniaxial dielectric partnering material is aligned parallel to the x axis.

Figure 4.5 presents the variations of the magnitudes of the Cartesian components of $\underline{e}(z)$, $\underline{h}(z)$, and $\underline{P}(x, y, z)$ in relation to z/λ_0 when $\psi = (\psi_{max} + \psi_{min})/2$, for the same combination of constitutive parameters as used for Figure 4.4. Most strikingly, neither do the maximum magnitudes of the field components generally occur precisely at the interface $z = 0$, nor is the energy flux associated with the Dyakonov wave centered precisely at $z = 0$. Instead the energy flux is somewhat skewed towards the half-space $z < 0$ occupied by the partnering isotropic material.

Before proceeding to more complicated cases, let us summarize the key properties of Dyakonov waves in the simplest possible case we have been considering, namely that involving the planar interface of a positive uniaxial dielectric material and an isotropic dielectric material with the optic axis of the uniaxial material aligned parallel to the interface plane:

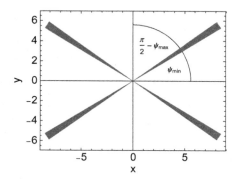

Figure 4.3 Ranges (shaded red) of angle ψ for which a Dyakonov wave is guided by the planar interface of calomel $\left(\epsilon_t^A = 7.02\epsilon_0 \text{ and } \epsilon_s^A = 3.88\epsilon_0 \right)$ and an isotropic dielectric material with permittivity $\epsilon^B = 4.5\epsilon_0$. The angular ranges are $\psi_{\min} < \psi < \psi_{\max}$, $(180° - \psi_{\max}) < \psi < (180° - \psi_{\min})$, $(180° + \psi_{\min}) < \psi < (180° + \psi_{\max})$, and $-\psi_{\max} < \psi < -\psi_{\min}$, where $\psi_{\min} = 31.88°$ and $\psi_{\max} = 36.06°$. The optic axis of calomel is aligned parallel to the x axis. (For interpretation of the references to color in this figure legend, the reader is referred to the web version of this book.)

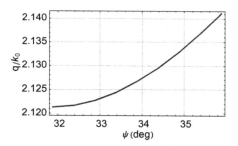

Figure 4.4 Solutions q with respect to ψ for Dyakonov waves guided by the planar interface of calomel $\left(\epsilon_t^A = 7.02\epsilon_0 \text{ and } \epsilon_s^A = 3.88\epsilon_0 \right)$ and an isotropic dielectric material with permittivity $\epsilon^B = 4.5\epsilon_0$, when the optic axis of calomel is aligned parallel to the x axis.

   i. Dyakonov-wave propagation is possible only if the inequalities (4.14) are satisfied.
  ii. Generally, Dyakonov waves are restricted to very small AEDs.
 iii. The AED depends upon both the permittivity scalar of the isotropic partnering material and the permittivity dyadic of the uniaxial partnering material.
 iv. The AED increases as the difference $\epsilon_t^A - \epsilon_s^A$ increases.
  v. Typically, Dyakonov waves are neither tightly confined to the vicinity of the interface (unlike the simple SPP waves in Chapter 2) nor do the maximums of their field magnitudes occur precisely at the interface.
 vi. In the half-space occupied by the uniaxial dielectric material, the Dyakonov wave is a linear combination of ordinary and extraordinary components. In the half-space occupied by the isotropic dielectric material, the Dyakonov wave is a linear combination of s- and p-polarized components.

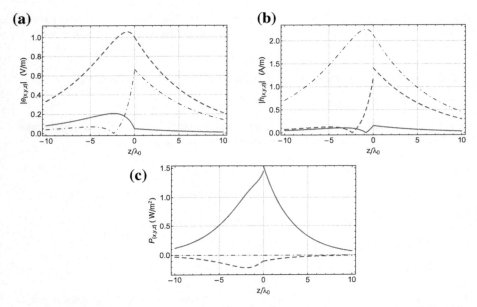

Figure 4.5 Variations with z/λ_0 of the magnitudes of the Cartesian components of (a) $\underline{e}(z)$, (b) $\underline{h}(z)$, and (c) $\underline{P}(x, y, z)$ of a Dyakonov wave guided by the interface of calomel ($\epsilon_t^{\mathcal{A}} = 7.02\epsilon_0$ and $\epsilon_s^{\mathcal{A}} = 3.88\epsilon_0$) and an isotropic dielectric material with permittivity $\epsilon^{\mathcal{B}} = 4.5\epsilon_0$, when $\psi = (\psi_{max} + \psi_{min})/2 = 33.97°$. The components parallel to $\underline{u}_x$, $\underline{u}_y$, and $\underline{u}_z$ are represented by red solid, blue dashed, and black chain-dashed lines, respectively. The data were computed by setting $C_s^{\mathcal{B}} = 1$ V m^{-1} in Eq. (4.9). (For interpretation of the references to color in this figure legend, the reader is referred to the web version of this book.)

Analogous characteristics characterize Dyakonov waves in more complex cases, described in the remainder of this Chapter.

4.2.1.2   Optic Axis not in Interface Plane

Let us now suppose that the optic axis specified in Eq. (4.1) is no longer parallel to the interface plane. Instead, the unit vector $\underline{u}_c$ is inclined at angle $\phi \in (0°, 90°]$ with respect to the interface plane. In this case the inequalities (4.14) must still be satisfied for Dyakonov-wave propagation but closed-form expressions for the limits on ψ describing the AED are not available.

Tipping of the optical axis away from the interface plane has several notable consequences for Dyakonov-wave propagation [162]. First, $\alpha_2^{\mathcal{A}}$ acquires a non-zero real part. Accordingly, the field phasors of the Dyakonov wave oscillate as well as decay as distance from the interface increases in the $+z$ direction; however, the z-directed component of the time-averaged Poynting vector remains null-valued [165]. Second, the width of the AED decreases continuously as ϕ increases, becoming null-valued in the limit $\phi \to 90°$. Third, the Dyakonov wave becomes steadily less localized to the vicinity of the interface as ϕ increases, becoming completely delocalized—and hence ceasing to exist as a surface wave—as $\phi \to 90°$. Thus, when the optic axis lies parallel to the interface plane (i.e. $\phi = 0°$), the AED is the widest and the Dyakonov wave is most strongly bound to the interface.

4.2.2   Interface of a Biaxial Material and an Isotropic Material

4.2.2.1   Optic Ray Axes in Interface Plane

Generalizing further, let us suppose next that the material $\mathcal{A}$ occupying the half-space $z > 0$ is a biaxial dielectric material with permittivity dyadic

$$\underline{\underline{\epsilon}}^{\mathcal{A}} = \epsilon_1^{\mathcal{A}} \underline{u}_1 \underline{u}_1 + \epsilon_2^{\mathcal{A}} \underline{u}_2 \underline{u}_2 + \epsilon_3^{\mathcal{A}} \underline{u}_3 \underline{u}_3, \tag{4.21}$$

where the unit vectors $\underline{u}_1$, $\underline{u}_2$, and $\underline{u}_3$ are arbitrary but form a right-handed triad:

$$\underline{u}_1 \times \underline{u}_2 = \underline{u}_3, \quad \underline{u}_3 \times \underline{u}_1 = \underline{u}_2, \quad \underline{u}_2 \times \underline{u}_3 = \underline{u}_1. \tag{4.22}$$

Without loss of generality, let us assume that

$$\epsilon_1^{\mathcal{A}} > \epsilon_2^{\mathcal{A}} > \epsilon_3^{\mathcal{A}}. \tag{4.23}$$

The material $\mathcal{B}$ occupying the half-space $z < 0$ remains an isotropic dielectric material described by Eq. (4.2).

It is helpful to recast $\underline{\underline{\epsilon}}^{\mathcal{A}}$ in terms of the optic ray axes aligned with the unit vectors

$$\left. \begin{aligned} \underline{a}_+^{\mathcal{A}} &= \underline{u}_1 \sqrt{\frac{\epsilon_1^{\mathcal{A}} - \epsilon_2^{\mathcal{A}}}{\epsilon_1^{\mathcal{A}} - \epsilon_3^{\mathcal{A}}}} + \underline{u}_3 \sqrt{\frac{\epsilon_2^{\mathcal{A}} - \epsilon_3^{\mathcal{A}}}{\epsilon_1^{\mathcal{A}} - \epsilon_3^{\mathcal{A}}}} \\ \underline{a}_-^{\mathcal{A}} &= \underline{u}_1 \sqrt{\frac{\epsilon_1^{\mathcal{A}} - \epsilon_2^{\mathcal{A}}}{\epsilon_1^{\mathcal{A}} - \epsilon_3^{\mathcal{A}}}} - \underline{u}_3 \sqrt{\frac{\epsilon_2^{\mathcal{A}} - \epsilon_3^{\mathcal{A}}}{\epsilon_1^{\mathcal{A}} - \epsilon_3^{\mathcal{A}}}} \end{aligned} \right\}, \tag{4.24}$$

as

$$\underline{\underline{\epsilon}}^{\mathcal{A}} = \epsilon_2^{\mathcal{A}} \underline{\underline{I}} + \frac{\epsilon_1^{\mathcal{A}} - \epsilon_3^{\mathcal{A}}}{2} \left(\underline{a}_+^{\mathcal{A}} \underline{a}_-^{\mathcal{A}} + \underline{a}_-^{\mathcal{A}} \underline{a}_+^{\mathcal{A}} \right). \tag{4.25}$$

Now let us restrict attention to the case where the optic ray axes lie parallel to the interface plane, i.e. the xy plane. A surface wave may be guided provided the inequalities

$$\epsilon_2^{\mathcal{A}} < \epsilon^{\mathcal{B}} < \epsilon_1^{\mathcal{A}} \tag{4.26}$$

are satisfied [162].

If the coordinate axes are oriented such that $\underline{u}_1 = \underline{u}_x$ and $\underline{u}_3 = \underline{u}_y$, then the optic ray axes $\underline{a}_\pm^{\mathcal{A}}$ are directed symmetrically about the coordinate axes in the xy plane and the direction of propagation relative to the x axis is restricted to the angular range $\psi_{min} < \psi < \psi_{max}$, where the limits are [163]

$$\psi_{min} = \sin^{-1} \left\{ \frac{\nu}{\nu_1} \left[1 + \frac{\nu(\nu_1 - \nu)^2}{(1 + \nu_1)(\nu_2 - \nu) + \nu(\nu_1 - \nu)} \right]^{-1} \right\}^{1/2} \tag{4.27}$$

and

$$
\psi_{max} = \sin^{-1} \left[\frac{\nu}{\nu_1} + \frac{1}{2\nu_1(1+\nu)^2} \left((1+\nu_1)(\nu_2 - \nu) + 2\nu(1+\nu)(\nu_1 - \nu) \right) \right.
$$

$$
\left. - \{ (1+\nu_1)(\nu_2 - \nu)[(1+\nu_1)(\nu_2 - \nu) + 4\nu(1+\nu)(\nu_1 - \nu)] \}^{1/2} \right)^{1/2} \right],
$$

(4.28)

with the scalar parameters

$$
\nu = \frac{\epsilon_3^A}{\epsilon^B} - 1, \qquad \nu_1 = \frac{\epsilon_3^A}{\epsilon_1^A} - 1, \qquad \nu_2 = \frac{\epsilon_3^A}{\epsilon_2^A} - 1. \qquad (4.29)
$$

The biaxial/isotropic interface can provide slightly larger AEDs as well as a slightly greater degree of localization to the vicinity of the interface, than a uniaxial/isotropic interface, provided that the degrees of anisotropy of material $\mathcal{A}$ are approximately the same for both interfaces—i.e. the ratio $\epsilon_1^A/\epsilon_3^A$ is comparable in magnitude to the ratio $\epsilon_t^A/\epsilon_s^A$ of Section 4.2.1.1 [162]. Furthermore, Dyakonov-wave propagation is supported by both positive and negative *biaxial* dielectric materials [305]. Given the inequalities (4.23), the material $\mathcal{A}$ is positive biaxial if

$$
\frac{\epsilon_1^A - \epsilon_2^A}{\epsilon_1^A - \epsilon_3^A} \in \left(\frac{1}{2}, 1 \right), \qquad (4.30)
$$

but it is negative biaxial if

$$
\frac{\epsilon_1^A - \epsilon_2^A}{\epsilon_1^A - \epsilon_3^A} \in \left(0, \frac{1}{2} \right). \qquad (4.31)
$$

In contrast, if material $\mathcal{A}$ is uniaxial, then it will only support Dyakonov-wave propagation provided that $\epsilon_1^A > \epsilon_2^A = \epsilon_3^A$.

Finally, if Dyakonov waves can propagate in the angular range $\psi_{min} < \psi < \psi_{max}$ with ψ_{min} and ψ_{max} specified in Eqs. (4.27) and (4.28), respectively, then they can also propagate in the angular ranges $(180° - \psi_{max}) < \psi < (180° - \psi_{min})$, $(180° + \psi_{min}) < \psi < (180° + \psi_{max})$, and $-\psi_{max} < \psi < -\psi_{min}$.

4.2.2.2   Optic Ray Axes not in Interface Plane

Let us next suppose that at least one of the optic ray axes of the biaxial dielectric material $\mathcal{A}$ is not parallel to the interface plane. The consequences for Dyakonov waves of tipping the optical axes away from orientations parallel to the interface are similar to those described in Section 4.2.1.2 for the case when material $\mathcal{A}$ is uniaxial. That is, the inequalities (4.26) must still be satisfied for Dyakonov-wave propagation but closed-form expressions for the limits on ψ describing the AED are not available. The width of the AED decreases continuously and the Dyakonov wave becomes steadily less localized to the vicinity of the interface, as at least one of the two optic ray axes becomes more obliquely inclined to the interface plane [162]. Thus, when both optic ray axes of material $\mathcal{A}$ are oriented parallel to the interface, both the width of the AED and the degree of localization of the Dyakonov wave to the vicinity of the interface are the greatest.

4.3 Interface of Two Anisotropic Materials

4.3.1 Interface of Two Uniaxial Materials

4.3.1.1 Optic Axes in Interface Plane

The simplest anisotropic/anisotropic interface for Dyakonov-wave propagation may be envisaged as follows [306]: Take material $\mathcal{A}$ to be a uniaxial dielectric material characterized by the permittivity dyadic

$$\underline{\underline{\epsilon}}^{\mathcal{A}} = \epsilon_s \underline{\underline{I}} + (\epsilon_t - \epsilon_s)\underline{u}_c^{\mathcal{A}}\underline{u}_c^{\mathcal{A}}. \tag{4.32}$$

Material $\mathcal{B}$ is the same uniaxial dielectric material as $\mathcal{A}$ but oriented differently. Thus, its permittivity dyadic has the form

$$\underline{\underline{\epsilon}}^{\mathcal{B}} = \epsilon_s \underline{\underline{I}} + (\epsilon_t - \epsilon_s)\underline{u}_c^{\mathcal{B}}\underline{u}_c^{\mathcal{B}}. \tag{4.33}$$

Furthermore, the optic axes of both materials are taken to lie parallel to the interface plane. For convenience, let us orient our Cartesian coordinate axes such that they bisect the angles between the unit vectors $\underline{u}_c^{\mathcal{A}}$ and $\underline{u}_c^{\mathcal{B}}$; i.e.

$$\left.\begin{array}{l} \underline{u}_c^{\mathcal{A}} = \underline{u}_x \cos\Psi + \underline{u}_y \sin\Psi \\ \underline{u}_c^{\mathcal{B}} = \underline{u}_x \cos\Psi - \underline{u}_y \sin\Psi \end{array}\right\} \tag{4.34}$$

with 2Ψ being the angle by which the orientation of material $\mathcal{B}$ is twisted relative to the orientation of material $\mathcal{A}$.

The field phasors in the half-space $z > 0$ are given by

$$\left.\begin{array}{l} \underline{e}(z) = \sum\limits_{m=1}^{2} C_m^{\mathcal{A}} \left\{ v_{1m}^{\mathcal{A}}\underline{u}_x + v_{2m}^{\mathcal{A}}\underline{u}_y \right. \\ \qquad\qquad \left. + \left[v_{zx}^{eh,\mathcal{A}} v_{3m}^{\mathcal{A}} + v_{zy}^{eh,\mathcal{A}} v_{4m}^{\mathcal{A}} \right] \underline{u}_z \right\} \exp\left(i\alpha_m^{\mathcal{A}}z\right) \\ \underline{h}(z) = \sum\limits_{m=1}^{2} C_m^{\mathcal{A}} \left\{ v_{3m}^{\mathcal{A}}\underline{u}_x + v_{4m}^{\mathcal{A}}\underline{u}_y \right. \\ \qquad\qquad \left. + \left[v_{zx}^{he,\mathcal{A}} v_{1m}^{\mathcal{A}} + v_{zy}^{he,\mathcal{A}} v_{2m}^{\mathcal{A}} \right] \underline{u}_z \right\} \exp\left(i\alpha_m^{\mathcal{A}}z\right) \end{array}\right\}, \quad z > 0, \tag{4.35}$$

wherein the wavenumbers

$$\alpha_1^{\mathcal{A}} = \sqrt{\omega^2 \mu_0 \epsilon_s - q^2} \tag{4.36}$$

and

$$\alpha_2^A = \sqrt{\omega^2 \mu_0 \epsilon_t - q^2 \left(\sin^2 \psi + \frac{\epsilon_t}{\epsilon_s} \cos^2 \psi \right)}, \tag{4.37}$$

obey the restrictions

$$\text{Im} \left\{ \alpha_m^A \right\} > 0, \quad m \in [1, 2]. \tag{4.38}$$

The field phasors in the half-space $z < 0$ are given by

$$\left.\begin{aligned}
\underline{e}(z) &= \sum_{m=3}^{4} C_m^B \left\{ v_{1m}^B \underline{u}_x + v_{2m}^B \underline{u}_y \right. \\
&\quad + \left. \left[v_{zx}^{eh,B} v_{3m}^B + v_{zy}^{eh,B} v_{4m}^B \right] \underline{u}_z \right\} \exp \left(i \alpha_m^B z \right) \\
\underline{h}(z) &= \sum_{m=3}^{4} C_m^B \left\{ v_{3m}^B \underline{u}_x + v_{4m}^B \underline{u}_y \right. \\
&\quad + \left. \left[v_{zx}^{he,B} v_{1m}^B + v_{zy}^{he,B} v_{2m}^B \right] \underline{u}_z \right\} \exp \left(i \alpha_m^B z \right)
\end{aligned}\right\}, \quad z < 0, \tag{4.39}$$

wherein

$$\left.\begin{aligned}
\alpha_3^B &= -\alpha_1^A \\
\alpha_4^B &= -\alpha_2^A
\end{aligned}\right\}. \tag{4.40}$$

Explicit expressions for v_{1m}^A, etc., $v_{zx}^{eh,A}$, etc., v_{1m}^B, etc., and $v_{zx}^{eh,B}$, etc., can be found by following the methodology of Chapter 3. The coefficients C_1^A, C_2^A, C_3^B, and C_4^B are not known. Enforcement of the standard boundary conditions across the interface $z = 0$ yields the dispersion equation for Dyakonov-wave propagation.

As in Sections 4.2.1.1 and 4.2.1.2, propagation of Dyakonov waves is only possible here for positive uniaxial materials ($\epsilon_t > \epsilon_s$). Furthermore, this propagation is restricted to certain directional ranges $\psi_{min} < \psi < \psi_{max}$ which are centered on the bisectors of the optic axes. Thus, in general, there are four allowed directional ranges of propagation as follows:

- Two angular ranges are centered about the bisectors of the acute angles between the unit vectors $\underline{u}_c^A$ and $\underline{u}_c^B$. These two are identical in width and the directions allowed therein are diametrically opposite.
- Two angular ranges are centered about the bisectors of the obtuse angles between the unit vectors $\underline{u}_c^A$ and $\underline{u}_c^B$. Both ranges have the same width and are diametrically opposite in directions.

The widths of all four angular ranges depend upon the anisotropy parameter

$$\Lambda = \frac{\epsilon_t}{\epsilon_s} - 1 > 0. \tag{4.41}$$

Explicit expressions for the limits of the AED are not available for general values of Λ, but they can be derived for $\Lambda \ll 1$ and $\Lambda \gg 1$.

Naturally occurring positive uniaxial crystals are often characterized by $\Lambda \ll 1$. In this case, the common width of the two AEDs centered about the bisector of the acute angle between the optic axes is [306]

$$\psi_{max} - \psi_{min} = 2\Lambda^2 \cos^2 \Psi \sin \Psi, \tag{4.42}$$

and the common width of the two AEDs centered about the bisector of the obtuse angle between the optic axes is

$$\psi_{max} - \psi_{min} = 2\Lambda^2 \sin^2 \Psi \cos \Psi. \tag{4.43}$$

Equations (4.42) and (4.43) indicate that Dyakonov-wave propagation is possible for $\Psi \in (0°, 90°)$ provided that $\Lambda > 0$.

Rutile is an example of a mineral characterized by $\Lambda \ll 1$, with $\epsilon_s = 7.02\epsilon_0$ and $\epsilon_t = 8.70\epsilon_0$; an illustration of the four angular ranges of propagation for rutile with $\Psi = 60°$ is provided in Figure 4.6. The width of the AEDs centered about the bisector of the acute angle between the optic axes is substantially larger than the width of the AEDs centered about the bisector of the obtuse angle between the same axes; i.e. 2.46° as opposed to 1.42°. Furthermore, the AEDs in the anisotropic/anisotropic case are substantially larger than for the anisotropic/isotropic interface case illustrated in Figure 4.2.

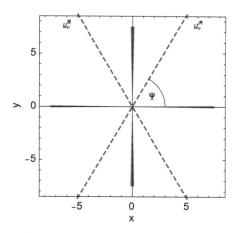

Figure 4.6 Ranges (shaded red) of angle ψ for which a Dyakonov wave is guided by the planar interface of two uniaxial dielectric materials. Both materials are rutile, characterized by the permittivity parameters $\epsilon_s = 7.02\epsilon_0$ and $\epsilon_t = 8.70\epsilon_0$. The optic axes for both materials are parallel to the interface plane: $\underline{u}_c^A = \underline{u}_x \cos 60° + \underline{u}_y \sin 60°$ and $\underline{u}_c^B = \underline{u}_x \cos 60° - \underline{u}_y \sin 60°$. The optic axes are represented by dashed lines. The width of the AEDs centered about the bisector of the acute angle between the optic axes is 2.46°, while that of the AEDs centered about the bisector of the obtuse angle between the optic axes is 1.42°. (For interpretation of the references to color in this figure legend, the reader is referred to the web version of this book.)

The case of $\Lambda \gg 1$ appears achievable using engineered materials, at least in principle. The common width of the two AEDs centered about the bisector of the acute angles between the optic axes is [306]

$$\psi_{max} - \psi_{min} = 180° - 2 \left(\Psi + \frac{1}{\sqrt{\Lambda}} \right), \tag{4.44}$$

and the common width of the two AEDs centered about the bisector of the obtuse angles between the optic axes is

$$\psi_{max} - \psi_{min} = 2 \left(\Psi - \frac{1}{\sqrt{\Lambda}} \right). \tag{4.45}$$

Dyakonov-wave propagation is forbidden in the remaining angular space which consists of four separate ranges, each of width $2/\sqrt{\Lambda}$ and centered about a line perpendicular to an optic axis. An illustration of the four angular ranges of propagation for $\epsilon_s = 1.2\epsilon_0$, $\epsilon_t = 15\epsilon_0$, and $\Psi = 60°$ is provided in Figure 4.7. As in Figure 4.6, the width of the AEDs centered about the bisector of the acute angle between the optic axes is substantially larger than the width of the AEDs centered about the bisector of the obtuse angle between the optic axes; i.e. 86.21° as opposed to 26.21°.

By restricting attention to propagation along directions parallel to the bisectors of the angles between the optic axes of the materials $\mathcal{A}$ and $\mathcal{B}$, further analytical progress can be made. For this specialization, the phase speed of the Dyakonov wave is lower than the phase speed of the extraordinary plane wave traveling in the same direction in the bulk of either of the two partnering materials [164]. In addition, the phase speed decreases as the degree of uniaxiality of the two partnering materials increases.

Figure 4.8 shows the solutions q in relation to the half-twist angle Ψ for Dyakonov waves guided by a planar calomel/calomel interface. Among naturally occurring

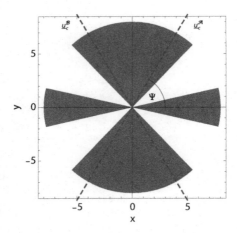

Figure 4.7 Same as Figure 4.6 except that $\epsilon_s = 1.2\epsilon_0$ and $\epsilon_t = 15\epsilon_0$. The width of the AEDs corresponding to the bisector of the acute angle between the optic axes is 86.21°, while that of the AEDs corresponding to the bisector of the obtuse angle between the optic axes is 26.21°.

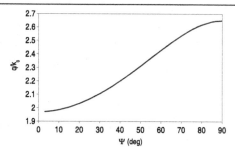

Figure 4.8 Solutions q in relation to the half-twist angle Ψ for Dyakonov waves guided by a planar calomel/calomel interface, when the optic axes on both sides of the interface are parallel to the xy plane, and the direction of propagation is along the x axis. No solution exists at $\Psi = 0°$ and in its neighborhood.

uniaxial crystals, calomel is distinguished by a high degree of positive uniaxiality ($\Lambda = 0.81$). No solution exists for $\Psi = 0°$, because the physical interface of the partnering materials then vanishes. No solution exists for $\Psi \in (0°, \Psi_{min}]$ either, where $\Psi_{min} > 0°$. As Ψ increases over the range $\Psi_{min} < \Psi \leqslant 90°$, the phase speed of the Dyakonov wave decreases to its minimum value at $\Psi = 90°$.

Magnitudes of the Cartesian components of $\underline{e}(z)$ are plotted in Figure 4.9 as functions of z/λ_0 for $\Psi \in \{10°, 80°\}$. The z-directed component of the electric field phasor is the strongest component for both values of Ψ. The x- and the z-directed components are more tightly bound to the interface than the y-directed component, with the 3-dB widths of their peaks at $z = 0$ along the z axis on the order of λ_0 for $\Psi = 10°$. The widths of the peaks of these components decrease by a factor on the order of $1/4$ as Ψ increases from $10°$ to $80°$. In addition, the peak of the x-directed component grows in amplitude relative to that of the z-directed component. On the other hand, the peak of

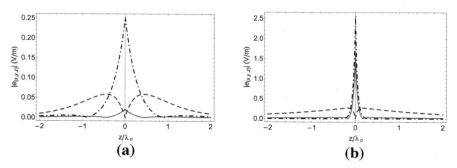

Figure 4.9 Variations with z/λ_0 of the magnitudes of the Cartesian components of $\underline{e}(z)$ for the calomel/calomel interface of Figure 4.8. The components parallel to $\underline{u}_x, \underline{u}_y$, and $\underline{u}_z$ are represented by red solid, blue dashed, and black chain-dashed lines, respectively. The data were computed by setting $C_1^A = 1$ V m^{-1} in Eq. (4.35). (a) $\Psi = 10°$ and (b) $\Psi = 80°$. (For interpretation of the references to color in this figure legend, the reader is referred to the web version of this book.)

the y-directed component is much broader ($\sim\lambda_0$ for $\Psi = 10°$) and increases in width as Ψ increases. Also, the amplitude of the peak of the y-directed component decreases relative to those of the other two components as Ψ increases.

4.3.1.2   Optic Axes not in Interface Plane

Next, let us turn to the case wherein materials $\mathcal{A}$ and $\mathcal{B}$ continue to be the uniaxial dielectric materials characterized by the permittivity dyadics (4.32) and (4.33), but now their optic axes are tilted by an angle χ relative to the xy plane as follows:

$$\left.\begin{aligned}\underline{u}_c^{\mathcal{A}} &= \left(\underline{u}_x \cos\Psi + \underline{u}_y \sin\Psi\right)\cos\chi + \underline{u}_z \sin\chi \\ \underline{u}_c^{\mathcal{B}} &= \left(\underline{u}_x \cos\Psi - \underline{u}_y \sin\Psi\right)\cos\chi + \underline{u}_z \sin\chi\end{aligned}\right\}. \tag{4.46}$$

Again focusing on the phase speed for propagation along the x axis in the xy plane, numerical studies reveal the effect of varying χ [165]. In Figure 4.10, solutions q are plotted against the half-twist angle Ψ for several values of χ for a calomel/calomel interface. No solution exists for $\Psi \in [0°, \Psi_{min}]$. As the tilt angle χ increases from $0°$, Ψ_{min} increases, so that range of the half-twist angle Ψ allowing Dyakonov-wave propagation shrinks. For the particular example illustrated in the figure, Dyakonov waves cannot exist for χ slightly greater than $20°$. Additionally, the phase speed of the Dyakonov wave decreases as χ increases.

Figure 4.11 shows the magnitudes of the Cartesian components of $\underline{e}(z)$ as functions of z/λ_0 for $\Psi \in \{25°, 55°\}$, when $\chi = 15°$ in Figure 4.10. At this value of χ, Figure 4.10 indicates that Dyakonov waves can exist only over the approximate range $\Psi \in [15°, 64°]$. The peaks in the y-directed component are much stronger for $\chi = 15°$ than for $\chi = 0°$ in Figure 4.9. The z-directed component of $\underline{e}(z)$ weakens, while its x-directed component strengthens slightly, as Ψ increases from $25°$ to $55°$. The widths

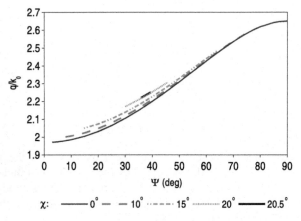

Figure 4.10 Solutions q against the half-twist angle Ψ for Dyakonov waves guided by a planar calomel/calomel interface, when the optic axes on both sides are inclined at an angle χ with respect to the xy plane.

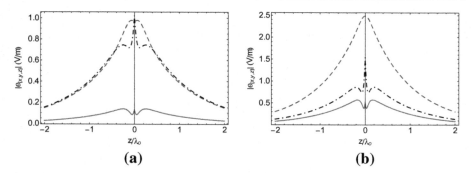

Figure 4.11 Variations with z/λ_0 of the magnitudes of the Cartesian components of $\underline{e}(z)$ for the calomel/calomel interface of Figure 4.10. The optic axes on both sides are inclined at an angle $\chi = 15°$ with respect to the xy plane. The components parallel to $\underline{u}_x$, $\underline{u}_y$, and $\underline{u}_z$ are represented by red solid, blue dashed, and black chain-dashed lines, respectively. The data were computed by setting $C_1^A = 1$ V m^{-1} in Eq. (4.35). (a) $\Psi = 25°$ and (b) $\Psi = 55°$.

of the peaks of all three Cartesian components are of comparable magnitude and similar to the width of the peaks of the y-directed component for $\chi = 0°$ in Figure 4.9.

4.3.2   Interface of Two Biaxial Materials

4.3.2.1   Optic Ray Axes in Interface Plane

The simplest biaxial/biaxial interface for Dyakonov-wave propagation may be envisaged as follows. Suppose that the half-space $z > 0$ is occupied by a biaxial dielectric material. Both of its optic ray axes lie in the xy plane, separated by an angle 2δ. Thus, the permittivity dyadic of material $\mathcal{A}$ may be expressed as

$$\underline{\underline{\epsilon}}^{\mathcal{A}} = \epsilon_\alpha \underline{\underline{I}} + \epsilon_\beta \left(\underline{a}_+^{\mathcal{A}} \underline{a}_+^{\mathcal{A}} + \underline{a}_-^{\mathcal{A}} \underline{a}_-^{\mathcal{A}} \right), \tag{4.47}$$

where

$$\underline{a}_-^{\mathcal{A}} = \underline{\underline{R}}(2\delta) \cdot \underline{a}_+^{\mathcal{A}}, \tag{4.48}$$

the dyadic

$$\underline{\underline{R}}(\theta) = \cos\theta (\underline{u}_x \underline{u}_x + \underline{u}_y \underline{u}_y) + \sin\theta (\underline{u}_y \underline{u}_x - \underline{u}_x \underline{u}_y) \tag{4.49}$$

represents a counterclockwise rotation about the z axis by θ in the xy plane, and $\underline{a}_+^{\mathcal{A}} \cdot \underline{u}_z = 0$. The half-space $z < 0$ is occupied by the same biaxial dielectric material but oriented differently. The optic ray axes of material $\mathcal{B}$ are twisted about the z axis by an angle 2Ψ in the xy plane relative to those of material $\mathcal{A}$. Thus, the permittivity dyadic of material $\mathcal{B}$ may be expressed as

$$\underline{\underline{\epsilon}}^{\mathcal{B}} = \epsilon_\alpha \underline{\underline{I}} + \epsilon_\beta \left(\underline{a}_+^{\mathcal{B}} \underline{a}_+^{\mathcal{B}} + \underline{a}_-^{\mathcal{B}} \underline{a}_-^{\mathcal{B}} \right), \tag{4.50}$$

with

$$
\left.
\begin{aligned}
\underline{a}_+^\mathcal{B} &= \underline{\underline{R}}(2\Psi) \cdot \underline{a}_+^\mathcal{A} \\
\underline{a}_-^\mathcal{B} &= \underline{\underline{R}}(2\Psi) \cdot \underline{a}_-^\mathcal{A}
\end{aligned}
\right\}.
\tag{4.51}
$$

The plethora of parameters in Eqs. (4.47) and (4.50) has restricted investigation so far only to the special case of propagation along the direction parallel to the bisector of the twist angle, i.e. $\underline{u}_{\mathrm{prop}} = \underline{\underline{R}}(\Psi + \delta) \cdot \underline{a}_+^\mathcal{A}$ [166,167]. The changes that arise for Dyakonov-wave propagation as one moves from the uniaxial/uniaxial interface to the biaxial/biaxial interface mirror those associated with the transition from the uniaxial/isotropic interface to the biaxial/isotropic interface described in Section 4.2. Thus, Dyakonov waves are supported by both positively and negatively biaxial materials and the phase speed of Dyakonov waves tends to be lower than it is for the case of the uniaxial/uniaxial interface ($\delta = 0°$). In addition, the range of half-twist angles for which Dyakonov waves can exist is here restricted to $\Psi \in (0°, \Psi_{\max})$, where $\Psi_{\max} \leqslant 90° - \delta$ [167]. This Ψ-range is larger for materials which exhibit lower birefringence, as defined in Appendix B, and have a smaller angle 2δ between their optic ray axes.

Figure 4.12 presents the relative wavenumbers q/k_0 of Dyakonov waves against the half-twist angle Ψ for a variety of mineral/mineral interfaces [166,167]. Hemimorphite has mid-range positive biaxiality ($\delta = 21.6°$) but a small birefringence ($\Delta n = 0.022$, as defined in Appendix B), and Dyakonov-wave propagation is possible for $\Psi \in [0°, 67.5°]$. Crocoite has comparable positive biaxiality ($\delta = 23.7°$) to hemimorphite, but its birefringence is more than an order of magnitude greater at 0.35. The Ψ-range for crocoite is smaller by about $10°$ compared to that of hemimorphite. Tellurite is barely positively biaxial ($\delta = 44.7°$), and its Ψ-range is even smaller than crocoite's. Witherite is highly negatively biaxial ($\delta = 85.2°$) and moderately birefringent ($\Delta n = 0.148$), but allows Dyakonov-wave propagation in the chosen configuration of just $\Psi \in [0°, 3.4°]$. Finally, cerussite has a similar negative biaxiality ($\delta = 84.9°$) to that of witherite, but as its birefringence ($\Delta n = 0.273$) is twice as much, the Ψ-range is marginally smaller.

Clearly therefore, the AED is larger for positive biaxiality than for negative biaxiality and reduces as $\delta \to 90°$. Furthermore, the AED decreases if the birefringence increases without change of biaxiality.

4.3.2.2 Optic Ray Axes not in Interface Plane

Some general relations are available for Dyakonov waves guided by biaxial/biaxial interfaces for certain orientations of the optic ray axes which do not lie in the interface plane [38]. These emerge from analyses wherein bianisotropic partnering materials are considered [37,39]. However, the practicality of such bianisotropic schemes for exciting Dyakonov waves has yet to be demonstrated.

4.4 Nanostructured Materials

Nature provides us with a wealth of anisotropic dielectric crystals which can be used to guide Dyakonov waves. However, from the viewpoint of potential applications, it

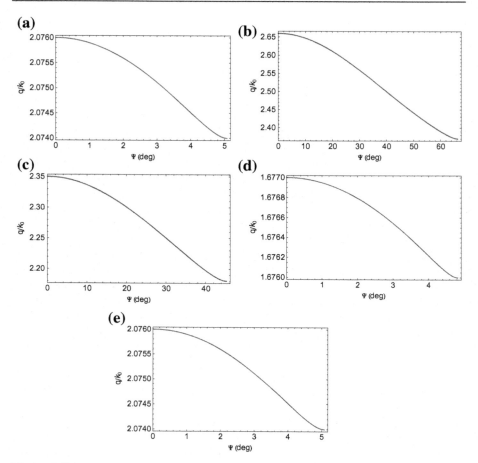

Figure 4.12 Solutions q against the half-twist angle Ψ for Dyakonov waves guided by a planar interface of two materials described by Eqs. (4.47) and (4.50). Both partnering materials are (a) hemimorphite ($\epsilon_\alpha = 2.6147\epsilon_0$, $\epsilon_\beta = 0.0358\epsilon_0$, and $\delta = 21.6°$), (b) crocoite ($\epsilon_\alpha = 5.6169\epsilon_0$, $\epsilon_\beta = 0.8697\epsilon_0$, and $\delta = 23.7°$), (c) tellurite ($\epsilon_\alpha = 4.7524\epsilon_0$, $\epsilon_\beta = 0.7612\epsilon_0$, and $\delta = 44.7°$), (d) witherite ($\epsilon_\alpha = 2.809\epsilon_0$, $\epsilon_\beta = 0.2372\epsilon_0$, and $\delta = 85.2°$), or (e) cerussite ($\epsilon_\alpha = 4.3015\epsilon_0$, $\epsilon_\beta = 0.5295\epsilon_0$, and $\delta = 84.9°$).

may be more useful to use instead engineered materials which possess structures of nanometer dimensions. By exploiting such nanostructured materials, the degree and nature of the anisotropy may be tailored to order, to a considerable extent. Three type of such nanostructured materials—namely, liquid crystals, columnar thin films, and photonic crystals—are highly relevant.

4.4.1   Liquid Crystals

As described in Section 1.4.4, nematic and smectic liquid crystals function optically as uniaxial and biaxial dielectric materials, respectively. Typically, liquid crystals are strongly anisotropic with respect to their optical properties and can therefore support relatively large AEDs for Dyakonov-wave propagation. For example, the E7 liquid crystal mixture, a commercially available mixture of 4-pentyl-4′-cyanobiphenyl and triphenyl, is nematic at room temperature [307]; i.e. it is a uniaxial dielectric material at room temperature. For the planar interface of an isotropic dielectric material and the E7 liquid crystal mixture, the AED can be as high as 0.94°, as illustrated in Figure 4.2.

The use of liquid crystals offers the potential for controlling the AED for Dyakonov-wave propagation, as well as the location and confinement of the Dyakonov wave relative to the interface, by means of temperature, pressure, and the application of quasistatic electric fields. These possibilities are ripe for both theoretical and experimental research, now that Dyakonov waves have been observed in a laboratory [41].

Our attention here is focused on liquid crystals which are not structurally chiral. As chiral liquid crystals are nonhomogeneous, they support Dyakonov-Tamm waves rather than Dyakonov waves. Dyakonov-Tamm waves are discussed in Chapter 7.

4.4.2   Columnar Thin Films

The vapor deposition of a material which is an isotropic dielectric material in bulk form can yield a columnar thin film that may be viewed as a biaxial dielectric continuum from the macroscopic standpoint, as outlined in Section 1.4.3.1. CTFs are attractive for Dyakonov-wave propagation because not only can their anisotropy be tailored during the manufacturing stage, but they can also be modified post-manufacturing by infiltrating the void regions in between the constituent nanowires of a CTF with an appropriate fluid [97,308]. Indeed, the prospect of a fluid infiltrating a CTF and thereby influencing Dyakonov-wave propagation opens up the possibility of harnessing this mechanism for optical sensors. However, the relatively small AEDs for Dyakonov waves are likely to be a major obstacle for such applications.

For the planar interface of a CTF and an isotropic dielectric material, numerical studies [43] have revealed that the width of the AED decreases as the vapor flux angle χ_v, shown in Figure 1.7, increases toward 90°; the tilt angle χ of the nanowires with respect to the substrate plane in Eq. (1.29) also increases toward 90° in this limit. This is not surprising, since the three principal permittivity scalars of the CTF come closer to each other in magnitude as χ_v approaches 90° [94]. Furthermore, the permittivity of the isotropic dielectric material required for Dyakonov-wave propagation in a particular direction increases as χ_v (and χ) increases [43].

4.4.3   Photonic Crystals

Photonic crystals, as described in Section 1.4.3.3, present another method of artificially achieving dielectric anisotropy. Provided that the dimensions of the unit cell are much smaller than the electromagnetic wavelengths involved, these nanostructured materials may be regarded as being effectively homogeneous and characterized accordingly in terms of homogeneous permittivity dyadics. By carefully selecting the unit cell's shape and dimensions, anisotropy may be tailored to order.

The planar interface of a 2D photonic crystal (which functions as a biaxial dielectric continuum at sufficiently long wavelengths) and an isotropic dielectric material has been demonstrated theoretically to be capable of supporting Dyakonov-wave propagation [168]. The tunability and the potentially high degree of positive uniaxiality of such a photonic crystal are attractive attributes for practical applications.

4.5 Electro-Optic Materials

Along with the potential of liquid crystals and CTFs to control the propagation of Dyakonov waves, an alternative method of control is offered by electro-optic materials [188,189]. These are materials whose optical properties are sensitive to an applied DC (or quasistatic) electric field [271].

As an example, let us consider the planar interface of an anisotropic dielectric material $\mathcal{A}$ which displays the Pockels effect and an isotropic dielectric material $\mathcal{B}$. The Pockels effect is a linear electro-optic effect, whereby the application of a DC electric field $\underline{E}^{DC}$ modifies the *inverse* permittivity dyadic of the material in question. The modification is described by means of as many as 18 electro-optic coefficients. These coefficients—written traditionally as r_{JK} ($J \in [1, 6]$ and $K \in [1, 3]$), following the abbreviated notation for representing symmetric second-order tensors [286]—may or may not be independent of each other, depending upon the point group symmetry of the material they characterize [271].

While the permittivity dyadic of material $\mathcal{B}$ is simply given by Eq. (4.2), the inverse permittivity dyadic of material $\mathcal{A}$ has the general form

$$
\left(\underline{\underline{\epsilon}}^{\mathcal{A}}\right)^{-1} = \sum_{J=1}^{3} \left(\epsilon_J^{\mathcal{A}0}\right)^{-1} \underline{u}_J \underline{u}_J + \sum_{J=1}^{3} \sum_{K=1}^{3} \delta_{JK} r_{JK}^{\mathcal{A}} \left(\underline{u}_K \cdot \underline{E}^{DC}\right) \underline{u}_J \underline{u}_K
$$

$$
+ \left(\sum_{K=1}^{3} r_{4K}^{\mathcal{A}} \underline{u}_K\right) \cdot \underline{E}^{DC} \left(\underline{u}_2 \underline{u}_3 + \underline{u}_3 \underline{u}_2\right)
$$

$$
+ \left(\sum_{K=1}^{3} r_{5K}^{\mathcal{A}} \underline{u}_K\right) \cdot \underline{E}^{DC} \left(\underline{u}_1 \underline{u}_3 + \underline{u}_3 \underline{u}_1\right)
$$

$$
+ \left(\sum_{K=1}^{3} r_{6K}^{\mathcal{A}} \underline{u}_K\right) \cdot \underline{E}^{DC} \left(\underline{u}_1 \underline{u}_2 + \underline{u}_2 \underline{u}_1\right), \tag{4.52}
$$

where $\epsilon_1^{\mathcal{A}0}$, $\epsilon_2^{\mathcal{A}0}$, and $\epsilon_3^{\mathcal{A}0}$ are the eigenvalues of the permittivity dyadic of material $\mathcal{A}$ when the DC electric field is absent; the unit vectors

$$
\left.\begin{aligned}
\underline{u}_1 &= -\left(\underline{u}_x \cos \phi^{\mathcal{A}} + \underline{u}_y \sin \phi^{\mathcal{A}}\right) \cos \theta^{\mathcal{A}} + \underline{u}_z \sin \theta^{\mathcal{A}} \\
\underline{u}_2 &= \underline{u}_x \sin \phi^{\mathcal{A}} - \underline{u}_y \cos \phi^{\mathcal{A}} \\
\underline{u}_3 &= \left(\underline{u}_x \cos \phi^{\mathcal{A}} + \underline{u}_y \sin \phi^{\mathcal{A}}\right) \sin \theta^{\mathcal{A}} + \underline{u}_z \cos \theta^{\mathcal{A}}
\end{aligned}\right\},
$$

$$
\theta^{\mathcal{A}} \in [0°, 180°], \quad \phi^{\mathcal{A}} \in [0°, 360°) \tag{4.53}
$$

are the eigenvectors of the permittivity dyadic of material $\mathcal{A}$ when the DC electric field is absent; $r_{JK}^{\mathcal{A}}$ are the electro-optic coefficients; and δ_{JK} is the Kronecker delta. Valid up to first order in the components of the DC electric field (a common approximation in electro-optics [309]), an approximation of the permittivity dyadic may be expressed as [310]

$$
\underline{\underline{\epsilon}}^A \approx \sum_{J=1}^{3} \epsilon_J^{A0} \underline{u}_J \underline{u}_J - \epsilon_0^{-1} \sum_{J=1}^{3} \sum_{K=1}^{3} \delta_{JK} \epsilon_J^{A0} \epsilon_K^{A0} r_{JK}^{A} \left(\underline{u}_K \cdot \underline{E}^{DC} \right) \underline{u}_J \underline{u}_K
$$

$$
- \epsilon_0^{-1} \epsilon_2^{A0} \epsilon_3^{A0} \sum_{K=1}^{3} r_{4K} \underline{u}_K \cdot \underline{E}^{DC} (\underline{u}_2 \underline{u}_3 + \underline{u}_3 \underline{u}_2)
$$

$$
- \epsilon_0^{-1} \epsilon_1^{A0} \epsilon_3^{A0} \sum_{K=1}^{3} r_{5K} \underline{u}_K \cdot \underline{E}^{DC} (\underline{u}_1 \underline{u}_3 + \underline{u}_3 \underline{u}_1)
$$

$$
- \epsilon_0^{-1} \epsilon_1^{A0} \epsilon_2^{A0} \sum_{K=1}^{3} r_{6K} \underline{u}_K \cdot \underline{E}^{DC} (\underline{u}_1 \underline{u}_2 + \underline{u}_2 \underline{u}_1), \tag{4.54}
$$

under the assumption that

$$
\epsilon_0^{-1} \left\{ \max_{K \in [1,3]} \left| \epsilon_K^{A0} \right| \right\} \left\{ \max_{J \in [1,6]} \left| \sum_{K=1}^{3} r_{JK} \underline{u}_K \cdot \underline{E}^{DC} \right| \right\} \ll 1. \tag{4.55}
$$

The material $\mathcal{A}$ is clearly anisotropic in the absence of an applied DC electric field, provided that ϵ_1^{A0}, ϵ_2^{A0}, and ϵ_3^{A0} are not all the same. However, the material can be isotropic in the absence of an applied DC electric field but become anisotropic when the DC field is switched on [289].

For the specific case where material $\mathcal{A}$ is potassium niobate, numerical studies have demonstrated that the application of a DC electric field can significantly affect the directions along which Dyakonov waves may propagate [188]. Furthermore, an exhaustive search of the parameter space revealed that the Pockels effect could be harnessed to expand the AED to approximately 1.4° in this case [189].

Higher-order electro-optic effects, such as the Kerr effect which is a quadratic electro-optic effect, may also be exploited to control the propagation of Dyakonov waves [187].

4.6 Magnetic Analogs

The description of Dyakonov waves so far in this chapter has focused on two partnering materials which have different dielectric symmetries but are not magnetic. Since the Maxwell postulates are form-invariant under the duality transformation which interchanges the electric and magnetic fields [17,292], it follows that the propagation of Dyakonov waves may also be supported by two partnering materials which have different magnetic symmetries. While materials with anisotropic dielectric properties are commonly encountered at optical frequencies, materials with anisotropic magnetic properties are more commonly found at microwave and radio frequencies. However, recent research has shown that artificial magnetism can be realized at optical frequencies by judicious dispersals of metallic nanoparticles in a dielectric host material [311–313].

For example, suppose that material $\mathcal{A}$ is a magnetically anisotropic material specified by the permeability dyadic [172]

$$\underline{\underline{\mu}}^{\mathcal{A}} = \mu_\alpha \underline{\underline{I}} + (\mu_\beta - \mu_\alpha)\underline{u}_z\underline{u}_z + i\mu_g(\underline{u}_y\underline{u}_x - \underline{u}_x\underline{u}_y), \tag{4.56}$$

where the gyrotropic permeability parameter μ_g owes its existence to the presence of an applied quasistatic magnetic field [17,314]. Let material $\mathcal{B}$ be the same as material $\mathcal{A}$ but with an orientational change which reverses the direction of the gyration vector, i.e.

$$\underline{\underline{\mu}}^{\mathcal{B}} = \mu_\alpha \underline{\underline{I}} + (\mu_\beta - \mu_\alpha)\underline{u}_z\underline{u}_z - i\mu_g(\underline{u}_y\underline{u}_x - \underline{u}_x\underline{u}_y). \tag{4.57}$$

Furthermore, it is assumed that the dissipation is negligibly small (i.e. μ_α, μ_β, and μ_g are purely real), $\mu_\alpha > 0$, $\mu_\beta > 0$, and the permittivity of both partnering materials equals ϵ_0. Then, Dyakonov-wave propagation guided by the planar interface of materials $\mathcal{A}$ and $\mathcal{B}$ may arise provided that the conditions [172]

$$\mu_\alpha > \mu_\beta, \quad \mu_g \neq 0 \tag{4.58}$$

are satisfied.

In principle, magneto-optic effects could be exploited to control Dyakonov waves, in a similar manner to that described for the electro-optic effect in Section 4.5. But, in practice, magneto-optic effects may not be as useful as electro-optic effects, because the change in optical properties induced by an applied quasistatic magnetic field for a magneto-optic material is commonly much smaller than that induced by an applied quasistatic electric field for an electro-optic material, for typically encountered field strengths. Theory, however, indicates that the influence of an applied quasistatic magnetic field on Dyakonov-wave propagation may be intensified if one of the two partnering materials were a nonlinear dielectric material [186].

The magnetic properties of the two partnering materials can have a significant bearing on the propagation of a Dyakonov wave even when the permeabilities of both materials are isotropic. For example, suppose that material $\mathcal{A}$ is a uniaxial dielectric material characterized by the permittivity dyadic (4.1) with its optic axis lying parallel to the $z = 0$ plane (i.e. $\underline{u}_c \cdot \underline{u}_z = 0$), while material $\mathcal{B}$ is an isotropic dielectric-magnetic material with permittivity and permeability dyadics

$$\underline{\underline{\epsilon}}^{\mathcal{B}} = \epsilon^{\mathcal{B}}\underline{\underline{I}}, \quad \underline{\underline{\mu}}^{\mathcal{B}} = \mu^{\mathcal{B}}\underline{\underline{I}} \tag{4.59}$$

such that $\epsilon^{\mathcal{B}} \neq \epsilon_0$ and $\mu^{\mathcal{B}} \neq \mu_0$. Furthermore, let $\epsilon^{\mathcal{B}} > 0$ and $\mu^{\mathcal{B}} > 0$. Numerical studies have revealed that the corresponding AEDs for Dyakonov-wave propagation guided by the planar interface of these two partnering materials may be greater for $\mu^{\mathcal{B}} \neq \mu_0$ than for $\mu^{\mathcal{B}} = \mu_0$ [169].

4.7 More Exotic Materials

The introduction of magnetic properties can lead to larger AEDs for Dyakonov waves. These AEDs may be further enlarged by the use of engineered composite materials which support the propagation of plane waves with negative phase velocity. A succinct introduction to these materials is available in Section 1.5, whereas Ref. 68 is recommended for an in-depth treatment.

For example, as in Section 4.6, consider the interface of a uniaxial dielectric material $\mathcal{A}$ specified by the permittivity dyadic (4.1) with its optic axis lying parallel to the $z = 0$ plane (i.e. $\underline{u}_c \cdot \underline{u}_z = 0$) and an isotropic dielectric-magnetic material $\mathcal{B}$ specified by the permittivity and permeability dyadics of the form given in Eq. (4.59). Here, however, material $\mathcal{B}$ is taken to be a lossless material which supports NPV propagation; i.e. $\epsilon^{\mathcal{B}} < 0$ and $\mu^{\mathcal{B}} < 0$. A numerical study has revealed that the NPV characteristic of material $\mathcal{B}$ results in considerably larger AEDs for Dyakonov waves, as well as a greater degree of localization to the planar interface of materials $\mathcal{A}$ and $\mathcal{B}$, than would be the case if material $\mathcal{B}$ did not support NPV propagation [69]. Furthermore, Dyakonov-wave propagation arises when material $\mathcal{A}$ is negatively uniaxial—as opposed to the cases described in Section 4.2 which involved positive uniaxial materials—and multiple Dyakonov-wave modes may be excited at a given frequency along a specified direction of propagation. However, the neglect of dissipation in this numerical study [69]—as well as in related numerical studies involving NPV materials partnered with biaxial dielectric materials [70] and gyrotropic materials [170]—is a major limitation, since engineered composite materials that support NPV propagation are notoriously dissipative in practice [71,72].

Parenthetically, let us note that because an isotropic dielectric-magnetic material that supports NPV propagation typically possesses either (i) a negative-valued permittivity in the lossless case or (ii) a permittivity whose real part is negative-valued in the dissipative case, a semantic issue arises here as to whether the surface waves discussed here are classified as SPP waves or Dyakonov waves. We have opted for the latter classification but a convincing case can be put forth even for the former classification.

Further possibilities for Dyakonov-wave propagation are presented if one (or both) of the partnering materials is an indefinite dielectric-magnetic material. Such a material is specified by a permittivity dyadic and/or a permeability dyadic whose eigenvalues are not all of the same sign, for the lossless case. Clearly, indefinite materials are anisotropic. They may be realized as engineered composite materials and they may support the propagation of plane waves with NPV [315]. Conditions for the excitation of surface waves guided by the interface of an indefinite dielectric-magnetic material and an isotropic dielectric-magnetic material have been established for various cases, including that where the isotropic dielectric-magnetic material supports NPV propagation [316].

Dyakonov waves are often thought of as experiencing little attenuation, since they are typically guided by the planar interface of dielectric materials which exhibit negligible dissipation. However, a notable exception exists when a partnering material is an NPV material.

A further exception is furnished by piezoelectric materials. Let us consider the Dyakonov wave that is guided by the planar interface of two materials, at least one of which possesses piezoelectric properties [286]. If the chosen optical frequency is less than typical phonon frequencies (say, 10^{12}–10^{13} Hz), then attenuation can arise due to coupling with elastodynamic waves in the bulk piezoelectric material(s). This attenuation is directly proportional to the maximum phase speed of all elastodynamic waves allowed to propagate in the bulk material and inversely proportional to the phase speed of the Dyakonov wave [317].

4.8 Experimental Observation

The first experimental observation of Dyakonov waves was reported by Takayama *et al.* [41] in 2009, some 21 years after the seminal theoretical description of these surface waves [35]. The 21-year delay is indicative of the challenging nature of the practicalities that experimentalists must wrestle with in order to observe these elusive waves.

4.8.1 *Prism-Coupled Configuration*

The prism-coupled configuration of Figure 3.3 was used by Takayama *et al.* [41] for their experimental study. Monochromatic light of known polarization state and intensity is incident on one slanted face of a prism of refractive index n_{prism}. The intensity of light exiting the other slanted face of the prism is measured. Attached to the base of the prism is a layer of the partnering isotropic material followed by a layer of the partnering anisotropic material. As the angle of incidence θ_{inc} increases from $0°$, no transmission occurs across the two partnering materials. A sharp dip in the measured intensity of light exiting the second slanted face of the prism with respect to θ_{inc} is indicative of the launch of a Dyakonov wave.

    The underlying boundary-value problem is shown in Figure 3.4d. The half-space $z < 0$ is occupied by an isotropic dielectric material of refractive index n_{prism} and the half-space $z > d_A + d_B$ by air ($n_{air} = 1$). The isotropic partnering material in the region $0 < z < d_A$ has a permittivity ϵ^A, while the anisotropic partnering material in the region $d_A < z < d_A + d_B$ has a permittivity dyadic $\underline{\underline{\epsilon}}^B$.

    Suppose that $\epsilon^A = 4.5\epsilon_0$ and $\epsilon^B = (3.97\underline{\underline{I}} + 0.94\underline{u}_x\underline{u}_x)\epsilon_0$. Then, according to the curves in Figure 4.2 for YVO$_4$, Dyakonov-wave propagation is possible for $\psi \in [51.4501°, 51.8848°]$. For $\psi = 51.6674°$, the solution of the dispersion equation (4.12) (with superscripts A and B interchanged) of the canonical boundary-value problem is $q/k_0 = 3.212521$.

    In the prism-coupled configuration, let $n_{prism} = 3.58$. Furthermore, a tiny imaginary part is added to ϵ^A in order to obtain non-zero absorptances; thus, $\epsilon^A = 4.5 \left(1 + 2 \times 10^{-4}i\right)\epsilon_0$. Plots of absorptances A_p and A_s of incident s- and p-polarized light, respectively, against θ_{inc} in Figure 4.13 show ultranarrow peaks at $\theta_{inc} \simeq 63.8126°$, and $n_{prism} \sin 63.8126° = 3.21253$ in close conformity with the solution predicted by Eq. (4.12).

4.8.2 *First Observation of Dyakonov Waves*

Figure 4.13 underscores the difficulty of observing Dyakonov waves, putting the experimental accomplishment of Takayama *et al.* [41] in the proper perspective.

    The partnering anisotropic material chosen for their experiment was a biaxial dielectric material—namely, potassium titanyl phosphate (KTP) specified by a permittivity

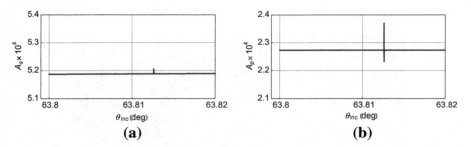

Figure 4.13 Calculated absorptances (a) A_s and (b) A_p as functions of the angle of incidence θ_{inc} for the boundary-value problem of Figure 3.4d underlying the excitation of a Dyakonov wave guided by the interface of layers of an isotropic material and an anisotropic material, both homogeneous. The refractive index of the material of incidence and reflection is 3.58, the 50-nm thick layer of the isotropic partnering material labeled $\mathcal{A}$ has a permittivity $4.5(1+2\times10^{-4}i)\epsilon_0$, and the 100-nm thick layer of the anisotropic partnering material labeled $\mathcal{B}$ is YVO$_4$ with permittivity $(3.97\underline{I}+0.94\underline{u}_x\underline{u}_x)\epsilon_0$. The material of transmission is air.

dyadic with eigenvalues $3.104\epsilon_0$, $3.1371\epsilon_0$, and $3.4775\epsilon_0$ at a free-space wavelength of 632.8 nm. The partnering isotropic material chosen was an index-matching liquid whose permittivity is intermediate in value between the two largest eigenvalues of the permittivity dyadic of the partnering biaxial dielectric material, in conformity with the condition (4.26) to be satisfied for Dyakonov-wave propagation. The orientation of the biaxial dielectric material was such that its optic ray axes were parallel to the interface plane. The calculated width of the AED was 0.13°.

Shown in Figure 4.14, the experimental setup allowed the angle of incidence to be controlled with a resolution of ±0.0001°, the orientation angle of the biaxial dielectric material relative to the incident beam of light to be controlled with a resolution of ±0.01°, and the angular beam width was approximately 0.05°. Although the intensity of a leaky waveguide mode dominated that of the Dyakonov wave, as shown in Figure 4.15, the observed direction of propagation of the Dyakonov wave agreed well with the theoretical estimate. The width of the observed dip in the reflectance as a function of the angle of incidence, which provided the identifying signature of the Dyakonov wave, was approximately 0.0001°. This width is three orders of magnitude smaller than the corresponding width for typical SPP waves; see Figure 2.7.

In order to observe Dyakonov waves, tolerances in the alignments of various optical components, including the symmetry axes of the anisotropic material(s), are very stringent. While this may pose an impediment to certain practical applications of Dyakonov waves, it may also present opportunities for others. For example, in the experimental study of Takayama *et al.* [41], a change in the refractive index of the isotropic dielectric partnering material of 0.005 resulted in a shift in the propagation angle of the Dyakonov wave by 3°. This sensitivity of 600°/RIU bodes very well for optical sensing applications. Indeed, this figure compares very favorably to optical sensors that exploit SPP waves [4]. For example, with an SPP-wave-based sensor using a nonporous aluminum thin film, a sensitivity of approximately 79°/RIU has been reported at a wavelength of 653 nm [318].

Figure 4.14 Experimental setup used by Takayama *et al.* [41] to observe a Dyakonov wave. Courtesy: Osamu Takayama, Institut de Ciències Fotòniques, Barcelona, Spain.

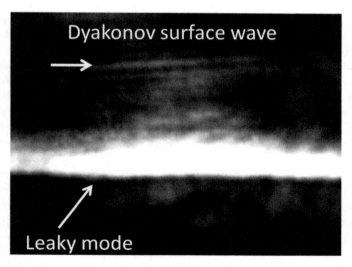

Figure 4.15 Experimental observation on a CCD camera of a Dyakonov wave by Takayama *et al.* [41]. Courtesy: Osamu Takayama, Institut de Ciències Fotòniques, Barcelona, Spain.

4.9 Outlook

As a consequence of the small widths of the AEDs for Dyakonov waves, together
with the attendant difficulties associated with their observation, Dyakonov waves have
received relatively little attention in research circles and beyond, as compared with
the much more conspicuous SPP waves. However, there are methods by which the
AEDs may be enlarged—by means of electro-optic, magnetic, magneto-optic, and
nonlinear effects, and by the use of exotic materials which support NPV propagation,
for examples.

The very fact that Dyakonov waves are highly sensitive to the optical environment
needed for their existence offers attractive opportunities for optical sensing applications. As dissipation in both partnering materials is negligibly small, Dyakonov waves
are not expected to play a significant role in the present-day push for the harvesting of
solar energy.

5 Tamm Waves

5.1 Introduction

Chapters 2 and 4 provide descriptions of surface-wave propagation when both partnering materials are homogeneous. This is the first chapter in this book wherein at least one of the two partnering materials is periodically nonhomogeneous normal to the interface guiding surface waves. Both partnering materials are isotropic dielectric, and the specific type of surface waves are called Tamm waves.

Igor Yevgenyevich Tamm did not discover or predict this class of surface waves, but these waves are named after him because he pointed out that electron states at the surface of a material of finite extent are different from the electron states inside the same material [47]. The optical analog of a surface electron state is a surface wave guided by the interface of two dissimilar materials, at least one of which is periodically nonhomogeneous in the direction normal to the interface. The simplest case is when both partnering materials are isotropic dielectric.

Investigations of Tamm waves have been carried out now for several decades [173], and the existence of s-polarized Tamm waves has been experimentally validated [174–176]. Optical biosensing with two s-polarized Tamm-wave modes has been experimentally demonstrated in the prism-coupled configuration [177].

Most often, the periodically nonhomogeneous dielectric material has been taken as piecewise homogeneous [173,175,176,178,179], i.e. it is a periodically multilayered material. But the periodic nonhomogeneity can be continuous as well, such as exhibited by a rugate filter [62,180]. Since both partnering materials are isotropic, the total angular existence domain is 360°, i.e. Tamm waves can propagate along any direction in the interface plane. With proper selection of the two partnering materials, numerous Tamm-wave modes can be obtained for a given interface, some s polarized and the others p polarized [62,180].

5.2 Canonical Boundary-Value Problem

The canonical boundary-value problem of Tamm-wave propagation can be formulated in the same way as in Section 3.5 to determine the relevant dispersion equation. Let a periodically nonhomogeneous material labeled $\mathcal{A}$ occupy the half-space $z > 0$, while a homogeneous material labeled $\mathcal{B}$ occupies the half-space $z < 0$, as shown in Figure 5.1.

Electromagnetic Surface Waves. http://dx.doi.org/10.1016/B978-0-12-397024-4.00005-0

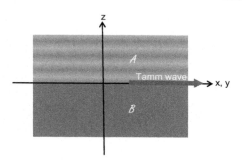

Figure 5.1 Schematic of the canonical boundary-value problem for Tamm-wave propagation guided by the planar interface of two isotropic dielectric materials $\mathcal{A}$ and $\mathcal{B}$. Whereas material $\mathcal{A}$ is periodically nonhomogeneous along the z axis, material $\mathcal{B}$ is homogeneous.

Both materials are isotropic, dielectric, non-magnetic, and achiral [17]. The permittivity of material $\mathcal{A}$ varies periodically so that

$$\epsilon^{\mathcal{A}}(z + 2\Omega^{\mathcal{A}}) = \epsilon^{\mathcal{A}}(z), \quad z > 0, \tag{5.1}$$

with $2\Omega^{\mathcal{A}}$ as the period, whereas the permittivity $\epsilon^{\mathcal{B}}$ of material $\mathcal{B}$ does not depend on z.

The canonical boundary-value problem can be formulated exactly as done in Section 3.5 leading to the dispersion equation (3.87) involving the determinant of a 4×4 matrix $\left[\underline{\underline{Y}}(q)\right]$. However, as both partnering materials are isotropic, that 4×4 matrix can be partitioned into two 2×2 null matrixes, and two 2×2 non-null matrixes. The singularity of one of the 2×2 non-null matrixes indicates the propagation of an s-polarized Tamm wave, whereas the singularity of the other 2×2 non-null matrix indicates the propagation of a p-polarized Tamm wave. Accordingly, the two types of Tamm waves can be analyzed separately [62].

5.2.1 s-Polarized Tamm Wave

Without any loss of generality, we can choose $\underline{u}_{\mathrm{prop}} = \underline{u}_x$. Then, for all $z \in (-\infty, \infty)$, the field phasors of an s-polarized Tamm wave may be written as

$$\left.\begin{array}{l} \underline{E}(\underline{r}) = e_y(z)\,\underline{u}_y \exp\left(iqx\right) \\[2mm] \underline{H}(\underline{r}) = \left[h_x(z)\,\underline{u}_x + \frac{q}{\omega\mu_0}e_y(z)\,\underline{u}_z\right]\exp\left(iqx\right) \end{array}\right\}. \tag{5.2}$$

The column 2-vector

$$[\underline{f}_s(z)] = \begin{bmatrix} e_y(z) \\ h_x(z) \end{bmatrix} \tag{5.3}$$

obeys the 2×2 matrix ordinary differential equation

$$\frac{d}{dz}[\underline{f}_s(z)] = \begin{cases} i\left[\underline{\underline{P}}_s^{\mathcal{A}}(z)\right] \cdot [\underline{f}_s(z)], & z > 0, \\[2mm] i\left[\underline{\underline{P}}_s^{\mathcal{B}}\right] \cdot [\underline{f}_s(z)], & z < 0, \end{cases} \tag{5.4}$$

where the matrixes

$$\left[\underline{\underline{P}}_s^A(z)\right] = \begin{bmatrix} 0 & -\omega\mu_0 \\ \frac{q^2}{\omega\mu_0} - \omega\epsilon^A(z) & 0 \end{bmatrix} \tag{5.5}$$

and

$$\left[\underline{\underline{P}}_s^B\right] = \begin{bmatrix} 0 & -\omega\mu_0 \\ \frac{q^2}{\omega\mu_0} - \omega\epsilon^B & 0 \end{bmatrix}. \tag{5.6}$$

As in Section 3.4.1, application of the piecewise-uniform-approximation method to Eq. (5.4) for $z > 0$ determines the matrix $\left[\underline{\underline{Q}}_s^A\right]$ that appears in the relation

$$\left[\underline{f}_s(2\Omega^A)\right] = \left[\underline{\underline{Q}}_s^A\right] \cdot [\underline{f}_s(0+)] \equiv \begin{bmatrix} Q_{s11}^A & Q_{s12}^A \\ Q_{s21}^A & Q_{s22}^A \end{bmatrix} \cdot [\underline{f}_s(0+)] \tag{5.7}$$

to characterize the optical response of one period of the periodically nonhomogeneous partnering material. The determinant of $\left[\underline{\underline{Q}}_s^A\right]$ equals unity, which means that either of its two eigenvalues is the reciprocal of the other eigenvalue. Let the eigenvalue denoted by σ_s^A be the one that satisfies the inequality

$$\text{Re}\left\{\ln\sigma_s^A\right\} < 0. \tag{5.8}$$

This eigenvalue represents decay as $z \to \infty$. In contrast, the eigenvalue $1/\sigma_s^A$ represents decay as $z \to -\infty$. Accordingly, with the proviso that the electromagnetic field must decay as $z \to \infty$, we get

$$[\underline{f}_s(0+)] = C_s^A \begin{bmatrix} 1 \\ \frac{\sigma_s^A - Q_{s11}^A}{Q_{s12}^A} \end{bmatrix}, \tag{5.9}$$

where C_s^A is an unknown coefficient.

Equation (5.4) can be solved analytically for $z < 0$ because the matrix $\left[\underline{\underline{P}}_s^B\right]$ does not depend on z. Thus,[1]

$$[\underline{f}_s(z)] = C_s^B \begin{bmatrix} 1 \\ -\frac{\alpha^B}{\omega\mu_0} \end{bmatrix} \exp\left(i\alpha^B z\right), \quad z \leqslant 0, \tag{5.10}$$

[1] In Chapters 1, 2, and 6, the half-space $z < 0$ in the canonical boundary-value problem for SPP-wave propagation is occupied by a metal, and the spatial dependence of the field phasors is taken to be $\exp[i(q\underline{u}_{\text{prop}} - \alpha_{\text{met}}\underline{u}_z) \cdot \underline{r}]$ with $\text{Im}\{\alpha_{\text{met}}\} > 0$. In Chapters 4, 5, and 7, sometimes the half-space $z < 0$ in the canonical boundary-value problem for surface-wave propagation is occupied by an isotropic homogeneous dielectric material. In consonance with Chapter 3, we then adopt the spatial dependence $\exp[i(q\underline{u}_{\text{prop}} + \alpha^B\underline{u}_z) \cdot \underline{r}]$ with $\text{Im}\{\alpha^B\} < 0$. Both representations are identical, as can be ascertained by setting $\alpha_{\text{met}} = -\alpha^B$.

where the coefficient C_s^B is unknown and the wavenumber

$$\alpha^B = \sqrt{\omega^2 \mu_0 \epsilon^B - q^2} \tag{5.11}$$

obeys the inequality

$$\text{Im}\{\alpha^B\} < 0. \tag{5.12}$$

Enforcement of the boundary condition $[\underline{f}_s(0+)] = [\underline{f}_s(0-)]$ yields the matrix equation

$$\begin{bmatrix} 1 & -1 \\ \dfrac{\sigma_s^A - Q_{s11}^A}{Q_{s12}^A} & \dfrac{\alpha^B}{\omega \mu_0} \end{bmatrix} \cdot \begin{bmatrix} C_s^A \\ C_s^B \end{bmatrix} = \begin{bmatrix} 0 \\ 0 \end{bmatrix}, \tag{5.13}$$

which is analogous to Eq. (3.85). A non-trivial solution exists provided the 2×2 matrix on the left side of Eq. (5.13) is singular. Therefore, the dispersion equation for s-polarized Tamm waves is

$$Q_{s12}^A \alpha^B + \omega \mu_0 \left(\sigma_s^A - Q_{s11}^A \right) = 0. \tag{5.14}$$

Concurrently, the unknown coefficients obey the relationship

$$C_s^A = C_s^B \tag{5.15}$$

for an s-polarized Tamm wave.

The right sides of both Eqs. (5.5) and (5.6) depend quadratically on q. Therefore, if q is a solution of the dispersion equation (5.14), then $-q$ also is. The two solutions indicate s-polarized Tamm waves propagating in opposite directions along the x axis.

5.2.2   p-Polarized Tamm Wave

The procedure to obtain the dispersion equation for p-polarized Tamm waves is similar to the one provided in Section 5.2.1. For all $z \in (-\infty, \infty)$, the field phasors of a p-polarized Tamm wave may be written as

$$\left. \begin{array}{l} \underline{E}(\underline{r}) = [e_x(z)\, \underline{u}_x + e_z(z)\, \underline{u}_z] \exp{(iqx)} \\ \underline{H}(\underline{r}) = h_y(z)\, \underline{u}_y \exp{(iqx)} \end{array} \right\}, \tag{5.16}$$

where

$$e_z(z) = \begin{cases} -\dfrac{q}{\omega \epsilon^A(z)} h_y(z), & z > 0, \\ -\dfrac{q}{\omega \epsilon^B} h_y(z), & z < 0, \end{cases} \tag{5.17}$$

The column 2-vector

$$[\underline{f}_p(z)] = \begin{bmatrix} e_x(z) \\ h_y(z) \end{bmatrix} \tag{5.18}$$

obeys the 2×2 matrix ordinary differential equation

$$
\frac{d}{dz}[\underline{f}_p(z)] = \begin{cases} i \left[\underline{\underline{P}}_p^{\mathcal{A}}(z)\right] \bullet [\underline{f}_p(z)], & z > 0, \\ i \left[\underline{\underline{P}}_p^{\mathcal{B}}\right] \bullet [\underline{f}_p(z)], & z < 0, \end{cases}
\tag{5.19}
$$

The matrixes involved in this equation for the two half-spaces are given by

$$
\left[\underline{\underline{P}}_p^{\mathcal{A}}(z)\right] = \begin{bmatrix} 0 & \omega\mu_0 - \frac{q^2}{\omega\epsilon^{\mathcal{A}}(z)} \\ \omega\epsilon^{\mathcal{A}}(z) & 0 \end{bmatrix}
\tag{5.20}
$$

and

$$
\left[\underline{\underline{P}}_s^{\mathcal{B}}\right] = \begin{bmatrix} 0 & \omega\mu_0 - \frac{q^2}{\omega\epsilon^{\mathcal{B}}} \\ \omega\epsilon^{\mathcal{B}} & 0 \end{bmatrix}.
\tag{5.21}
$$

As in Section 5.2.1, the matrix $\left[\underline{\underline{Q}}_p^{\mathcal{A}}\right]$ appearing in the relation

$$
\left[\underline{f}_p(2\Omega^{\mathcal{A}})\right] = \left[\underline{\underline{Q}}_p^{\mathcal{A}}\right] \bullet [\underline{f}_p(0+)] \equiv \begin{bmatrix} Q_{p11}^{\mathcal{A}} & Q_{p12}^{\mathcal{A}} \\ Q_{p21}^{\mathcal{A}} & Q_{p22}^{\mathcal{A}} \end{bmatrix} \bullet [\underline{f}_p(0+)]
\tag{5.22}
$$

can be obtained by the application of a numerical procedure. Just like $\left[\underline{\underline{Q}}_s^{\mathcal{A}}\right]$, $\left[\underline{\underline{Q}}_p^{\mathcal{A}}\right]$ also has a determinant equal to unity. Provided the eigenvalue $\sigma_p^{\mathcal{A}}$ of $\left[\underline{\underline{Q}}_p^{\mathcal{A}}\right]$ obeys the inequality

$$
\text{Re}\left\{\ln \sigma_p^{\mathcal{A}}\right\} < 0,
\tag{5.23}
$$

it represents decay as $z \to \infty$; the other eigenvalue $1/\sigma_p^{\mathcal{A}}$ represents decay as $z \to -\infty$. Thus, with $\sigma_p^{\mathcal{A}}$ being that solution of the quadratic equation

$$
\left(Q_{p11}^{\mathcal{A}} - \sigma_p^{\mathcal{A}}\right)\left(Q_{p22}^{\mathcal{A}} - \sigma_p^{\mathcal{A}}\right) - Q_{p12}^{\mathcal{A}} Q_{p21}^{\mathcal{A}} = 0,
\tag{5.24}
$$

which satisfies the inequality (5.23), we set

$$
[\underline{f}_p(0+)] = C_p^{\mathcal{A}} \begin{bmatrix} 1 \\ \frac{\sigma_p^{\mathcal{A}} - Q_{p11}^{\mathcal{A}}}{Q_{p12}^{\mathcal{A}}} \end{bmatrix},
\tag{5.25}
$$

where $C_p^{\mathcal{A}}$ is an unknown coefficient.

The solution of Eq. (5.19) for $z < 0$ is given by

$$
[\underline{f}_p(z)] = -C_p^B \begin{bmatrix} \frac{\alpha^B}{k_0 n^B} \\ \frac{n^B}{\eta_0} \end{bmatrix} \exp(i\alpha^B z), \quad z \leqslant 0,
\tag{5.26}
$$

where the coefficient C_p^B is unknown and the wavenumber α^B is specified by Eqs. (5.11) and (5.12).

The continuity of $[\underline{f}(z)]$ across the interface plane $z = 0$ yields the matrix equation

$$
\begin{bmatrix} 1 & \frac{\alpha^B}{k_0 n^B} \\ \frac{\sigma_p^A - Q_{p11}^A}{Q_{p12}^A} & \frac{n^B}{\eta_0} \end{bmatrix} \cdot \begin{bmatrix} C_p^A \\ C_p^B \end{bmatrix} = \begin{bmatrix} 0 \\ 0 \end{bmatrix},
\tag{5.27}
$$

whence emerges the dispersion equation

$$
\omega \epsilon^B Q_{p12}^A - \alpha^B \left(\sigma_p^A - Q_{p11}^A \right) = 0
\tag{5.28}
$$

for p-polarized Tamm waves. For a p-polarized Tamm wave, the unknown coefficients obey the relation

$$
C_p^A = -\frac{\alpha^B}{k_0 n^B} C_p^B.
\tag{5.29}
$$

If q is a solution of the dispersion equation (5.28), then $-q$ also is. The two solutions indicate p-polarized Tamm waves propagating in opposite directions along the x axis.

5.2.3   Illustrative Numerical Results

5.2.3.1   Interface of a Homogeneous Dielectric Material and a Periodic Multilayer

Suppose that the periodically nonhomogeneous partnering dielectric material is periodically multilayered, the unit cell comprising a layer of thickness d_1^A and refractive index n_1^A and another layer of thickness $d_2^A = 2\Omega^A - d_1^A$ and refractive index n_2^A; i.e.

$$
\epsilon^A(z) = \begin{cases} \left(n_1^A\right)^2 \epsilon_0, & z \in (0, d_1^A), \\ \left(n_2^A\right)^2 \epsilon_0, & z \in (d_1^A, 2\Omega^A), \end{cases}
\tag{5.30}
$$

in the unit cell nearest to the interface plane $z = 0$. The refractive index of the homogeneous partnering dielectric material is denoted by

$$
n^B = \sqrt{\epsilon^B / \epsilon_0}.
\tag{5.31}
$$

Tamm waves guided by this configuration have been theoretically studied by many researchers—see, e.g. Refs. 173, 178 and 319.

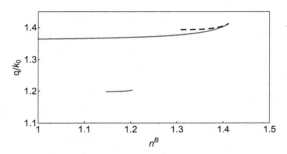

Figure 5.2 Solutions q of the dispersion equations (5.14) and (5.28) plotted against $n^{\mathcal{B}}$ for Tamm waves localized to the interface of (i) a periodically multilayered material described by Eq. (5.30) with $n_1^{\mathcal{A}} = 1.45, d_1^{\mathcal{A}} = 487\,\text{nm}, n_2^{\mathcal{A}} = 2.32$, and $d_2^{\mathcal{A}} = 779\,\text{nm}$, and (ii) a homogeneous dielectric material of refractive index $n^{\mathcal{B}} \in [1, 1.5]$, when the free-space wavelength $\lambda_0 = 633\,\text{nm}$. If q is a solution, then $-q$ also is. The red solid lines indicate s-polarized Tamm waves and the black dashed lines are for p-polarized Tamm waves. (For interpretation of the references to color in this figure legend, the reader is referred to the web version of this book.)

Both complex and purely real solutions q of the dispersion equations (5.14) and (5.28) *may* exist, even when dissipation is negligible in both partnering materials. A solution with a positive imaginary part represents a leaky surface wave, which may be called a *leaky Tamm wave*. Unless Im$\{q\}$ is minute, a leaky Tamm wave is of little value for applications [174], and is therefore ignored in the remainder of this chapter.

Figure 5.2 shows the solutions q of the dispersion equations (5.14) and (5.28) for $\lambda_0 = 633\,\text{nm}$, when $n_1^{\mathcal{A}} = 1.45, d_1^{\mathcal{A}} = 487\,\text{nm}, n_2^{\mathcal{A}} = 2.32$, and $d_2^{\mathcal{A}} = 779\,\text{nm}$, whereas $n^{\mathcal{B}} \in [1, 1.5]$. The selected refractive indexes are commonplace in the optical regime, and the selected thicknesses are on the order of the free-space wavelength. Only real-valued solutions are presented—which means that the Tamm waves thereby indicated can propagate indefinitely. That, of course, cannot be strictly true because dissipation can be very small but not absent in any passive material; nevertheless, the propagation length Δ_prop of a Tamm wave can be very large, extending into the range of several millimeters [174].

According to Figure 5.2, a Tamm wave can be guided by the interface of a periodically multilayered, isotropic, dielectric material, and free space. This is an essential difference between Tamm waves and SPP waves, with the latter requiring that one partnering material be a metal. Another difference, as indicated in Figure 5.2, is that Tamm waves can be either s or p polarized, while the simple SPP waves in Chapter 2 are exclusively p polarized. The preponderance of s-polarized Tamm waves over p-polarized Tamm waves indicated by the figure helps explain why only s-polarized Tamm waves have been experimentally observed thus far [174–177].

Yet another difference is that, while only one simple SPP wave can propagate at a specified frequency for a particular choice of partnering materials, it is possible for a single interface to guide multiple Tamm-wave modes at a fixed frequency. The propagation of two s-polarized Tamm-wave modes has been experimentally confirmed and even exploited for biosensing [177]. The different Tamm-wave modes certainly differ in phase speed and spatial profiles of the associated electromagnetic field phasors.

The multiplicity has to be a consequence of the periodic nonhomogeneity of one of the partnering materials, and can become more pronounced when the unit cell of the partnering material $\mathcal{A}$ is made of more than two layers which are, of course, thinner. A unit cell comprising numerous ultrathin layers of dissimilar materials is, in effect, continuously nonhomogeneous.

Two solutions exist in Figure 5.2 for limited ranges of $n^{\mathcal{B}}$. Therefore, once a periodic multilayer has been chosen as one of the two partnering materials, the homogeneous partner must be carefully selected for experimental work on Tamm-wave propagation.

Parenthetically, were $\epsilon^{\mathcal{B}} < 0$, Tamm waves would transmute into Fano waves. The periodic nonhomogeneity of $\epsilon^{\mathcal{A}}(z)$ then would enable the emergence of multiple Fano-wave modes [61]. If, in addition, $\text{Im}\{\epsilon^{\mathcal{B}}\} > 0$, multiple SPP-wave modes would arise, as discussed in Chapter 6.

5.2.3.2   Interface of a Homogeneous Dielectric Material and a Rugate Filter

Solutions of the dispersion equations (5.14) and (5.28) for Tamm waves guided by the interface of a semi-infinite rugate filter and a homogeneous dielectric material are plotted against $n^{\mathcal{B}}$ in Figure 5.3. The rugate filter is described by

$$\epsilon^{\mathcal{A}}(z) = \epsilon_0 \left[n^{\mathcal{A}}_{\text{avg}} + \frac{\Delta n^{\mathcal{A}}}{2} \sin \left(\frac{\pi z}{\Omega^{\mathcal{A}}} \right) \right]^2, \tag{5.32}$$

and thus has a continuous variation of the refractive index along the z axis. The number of solutions decreases as $n^{\mathcal{B}}$ increases toward the maximum value $n^{\mathcal{A}}_{\text{avg}} + \Delta n^{\mathcal{A}}/2$ of the refractive index $n^{\mathcal{A}}(z) = \sqrt{\epsilon^{\mathcal{A}}(z)/\epsilon_0}$ in the rugate filter. Thus, only two solutions—one p polarized and the other s polarized—exist for the highest value of $n^{\mathcal{B}}$ in Figure 5.3, although eight solutions exist for the lowest value of $n^{\mathcal{B}}$. The multiplicity of solutions is just like that of SPP waves guided by the interface of a metal and a rugate filter in

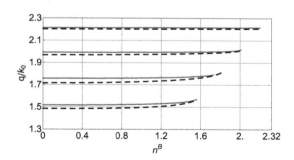

Figure 5.3 Solutions q of the dispersion equations (5.14) and (5.28) plotted against $n^{\mathcal{B}}$ for Tamm waves localized to the planar interface of (i) a rugate filter described by Eq. (5.32) with $n^{\mathcal{A}}_{\text{avg}} = 1.885$, $\Delta n^{\mathcal{A}} = 0.87$, and $\Omega^{\mathcal{A}} = \lambda_0$, and (ii) a homogeneous dielectric material of refractive index $n^{\mathcal{B}} \in (0, 2.32]$, when the free-space wavelength $\lambda_0 = 633$ nm. No solution was found for $n^{\mathcal{B}} > 2.207$. If q is a solution, then $-q$ also is. The red solid lines indicate s-polarized Tamm waves and the black dashed lines are for p-polarized Tamm waves [61,62]. (For interpretation of the references to color in this figure legend, the reader is referred to the web version of this book.)

Table 5.1 Normalized wavenumbers q/k_0 of Tamm waves obtained from Figure 5.3 for $n^B = 1.515$. If q is a solution of the dispersion equation (5.14) or (5.28), then $-q$ also is.

Polarization state	q/k_0	q/k_0	q/k_0	q/k_0
s-pol	1.5560	1.7735	1.9966	2.2144
p-pol	1.5496	1.7553	1.9822	2.2034

Section 6.4.1, the periodic nonhomogeneity of the rugate filter being responsible for multiplicity in both cases. However, whereas those SPP waves are much more likely to be p polarized than s polarized [63], the polarization states of Tamm waves here are almost equally—but not always equally—split between s and p.

The phase speed $v_p = \omega/q$ of a Tamm wave is smaller than the phase speed in the homogeneous partnering material, because $q > k_0 n^B$. Furthermore, as $k_0 \left(n_{\text{avg}}^A - \Delta n^A/2 \right) < q < k_0 \left(n_{\text{avg}}^A + \Delta n^A/2 \right)$, the phase speed of the Tamm wave lies between the minimum and the maximum values of the phase speed $1/\sqrt{\mu_0 \epsilon^A(z)}$ in the rugate filter. Exploration of the q-range beyond $k_0 \left(n_{\text{avg}}^A + \Delta n^A/2 \right)$ failed to turn up solutions [62].

For a fixed value of n^B, the different Tamm-wave modes possible have distinct phase speeds. They also have different spatial variations of the components of the electromagnetic field. For instance, eight different Tamm-wave modes can be identified for $n^B = 1.515$ from Figure 5.3. All eight wavenumbers q of Tamm waves propagating along the $+x$ axis are cataloged in Table 5.1, whereas the negatives of these wavenumbers hold valid for Tamm waves propagating along the $-x$ axis. Figure 5.4 shows the spatial profiles of the fields of three of the eight Tamm-wave modes propagating along the $+x$ axis. Inside the homogeneous partnering material B, the non-zero field components decay exponentially as $z \to -\infty$, the variation being of the type

$$\exp \left[z \sqrt{q^2 - (k_0 n^B)^2} \right]$$ with $q > k_0 n^B$. Thus, the higher the magnitude of q (i.e. the lower the phase speed) at fixed n^B, the more tightly bound is the Tamm wave to the plane $z = 0-$.

In contrast, the spatial variations have to be far more complicated inside the partnering periodically nonhomogeneous material A ($z > 0$). According to the Floquet theory presented in Appendix D, the z-dependence of each non-zero component of a field phasor is the product of two factors. One of these factors decays exponentially as $z \to \infty$, but the other factor is faithfully reproduced unit cell by unit cell. Hence, in Figure 5.4, the average of a non-zero component over the ℓth unit cell $z \in [2(\ell - 1)\Omega^A, 2\ell\Omega^A]$ is higher than the average over the $(\ell + 1)$th unit cell $z \in [2\ell\Omega^A, 2(\ell + 1)\Omega^A]$, $\ell \in \{1, 2, ...\}$, but the variations inside two adjacent unit cells look quite similar in form. Moreover, the maximum field strength of a Tamm wave does not necessarily lie at the plane $z = 0+$. Thus, in Figure 5.4, the maximum magnitudes inside the rugate filter lie at $z \simeq 1.4\Omega^A$ in the plots in the top and middle rows, and at $z = 0.5\Omega^A$ in the bottom row.

Whereas the decay rate $-\text{Im}\{\alpha^B\}$ is meaningful because material B is homogeneous, the somewhat analogous decay rates $-\text{Re}\{\ln \sigma_s^A\}/2\Omega^A$ and $-\text{Re}\{\ln \sigma_p^A\}/2\Omega^A$ are

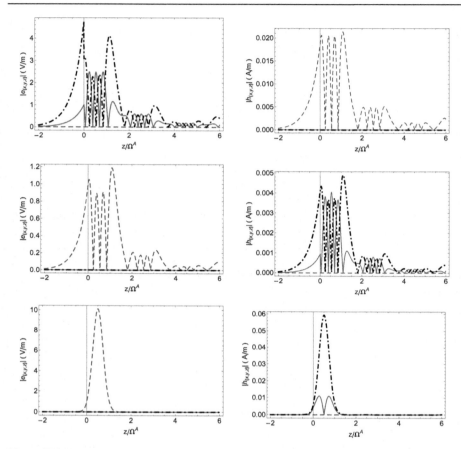

Figure 5.4 Variations with z of the magnitudes of the non-zero Cartesian components of (left) $\underline{e}(z)$ and (right) $\underline{h}(z)$, for three of the eight Tamm-wave modes possible for propagation along the $+x$ axis, when $n^{\mathcal{B}} = 1.515$ and all other parameters are the same as for Figure 5.3 (i.e. $\lambda_0 = 633$ nm, $n_{\mathrm{avg}}^{\mathcal{A}} = 1.885$, $\Delta n^{\mathcal{A}} = 0.87$, and $\Omega^{\mathcal{A}} = \lambda_0$). The components parallel to $\underline{u}_x$, $\underline{u}_y$, and $\underline{u}_z$ are represented by red solid, blue dashed, and black chain-dashed lines, respectively. The data were computed by setting $C_p^{\mathcal{A}} = -(\alpha^{\mathcal{B}}/k_0 n^{\mathcal{B}})C_p^{\mathcal{B}} = 1$ V m^{-1} for p-polarized Tamm-wave modes and $C_s^{\mathcal{A}} = C_s^{\mathcal{B}} = 1$ V m^{-1} for s-polarized Tamm-wave modes. (top) p-polarization state and $q/k_0 = 1.5496$, (middle) s-polarization state and $q/k_0 = 1.5560$, and (bottom) s-polarization state and $q/k_0 = 2.2144$ [62]. (For interpretation of the references to color in this figure legend, the reader is referred to the web version of this book.)

of little use because material $\mathcal{A}$ is periodically nonhomogeneous. Instead, the decay constants [50]

$$\left.\begin{array}{l} \beta_s^{\mathcal{A}} = \exp\left[\mathrm{Re}\left\{\ln \sigma_s^{\mathcal{A}}\right\}\right] \\ \beta_p^{\mathcal{A}} = \exp\left[\mathrm{Re}\left\{\ln \sigma_p^{\mathcal{A}}\right\}\right] \end{array}\right\}, \tag{5.33}$$

indicate decay over one period of material $\mathcal{A}$ and therefore appropriately quantify the degree of localization of Tamm-wave modes to the plane $z = 0+$ inside that material. The restrictions (5.8) and (5.23) imply that

$$\beta_s^{\mathcal{A}} \in (0, 1), \quad \beta_p^{\mathcal{A}} \in (0, 1). \tag{5.34}$$

The smaller the decay constant is, the more tightly is the Tamm wave bound to the interface. For the Tamm-wave modes in Figure 5.4, $\beta_p^{\mathcal{A}} = 0.2362$ (top row), whereas $\beta_s^{\mathcal{A}} = 0.1956$ (middle row) and $\beta_s^{\mathcal{A}} = 0.0006$ (bottom row). These decay constants quantitate the weaker localization of the p-polarized Tamm-wave mode in comparison to the stronger localization of the two s-polarized Tamm-wave modes in the figure.

The abundance of solutions in Figure 5.3 in relation to their paucity in Figure 5.2 is remarkable, all the more so because the period, the maximum refractive index, and the minimum refractive index of the nonhomogeneous partnering material are identical in both cases. This observation suggests that experimental observation and exploitation of Tamm waves would be facilitated if $\epsilon^{\mathcal{A}}(z)$ varies somewhat rapidly and significantly in the neighborhood of the interface $z = 0$.

The period $2\Omega^{\mathcal{A}}$ must not be so large in comparison to λ_0 that the rugate filter becomes virtually homogeneous close to the interface; otherwise, Tamm-wave propagation would not be supported. If $2\Omega^{\mathcal{A}}/\lambda_0$ is very small, the number of Tamm-wave modes can be very small, even zero [62].

According to Section 2.2, neither Eq. (5.14) nor Eq. (5.28) can have a solution if $\Delta n^{\mathcal{A}} = 0$—because then both partnering materials are homogeneous with permittivities that are positive and real valued. A minimum value of $\Delta n^{\mathcal{A}}$ may be necessary for Tamm-wave propagation. As $\Delta n^{\mathcal{A}}$ increases, the number of solutions increases, the decay constants become smaller to indicate greater localization to the interface, solutions with higher phase speed $v_p = \omega/q$ appear, and the phase-speed distribution generally widens [62].

5.3   Practical Configurations

5.3.1   Prism-Coupled Configuration

Being quite straightforward to implement, the prism-coupled configuration has been employed for experiments with Tamm waves [174–177]. Materials $\mathcal{A}$ and $\mathcal{B}$ are taken to be of finite thicknesses $d_{\mathcal{A}}$ and $d_{\mathcal{B}}$, respectively. The layer of material $\mathcal{B}$ is in contact with a prism of refractive index $n_{\text{prism}} > 1$, whereas the layer of material $\mathcal{A}$ is in contact with air ($n_{\text{air}} = 1$). The experimental situation is similar to the one depicted in Figure 3.3 for Dyakonov waves. The situation wherein the roles of materials $\mathcal{A}$ and $\mathcal{B}$ are reversed has also been experimentally implemented [176].

For analytical purposes, a linearly polarized plane wave is taken to be incident on the interface of the prism and material $\mathcal{B}$ at an angle θ_{inc} with respect to the z axis, as shown in Figure 5.5. The plane wave transmitted into air has a wave vector inclined to the z axis at an angle θ_{tr}, which may be complex valued. The mathematical technique described in Section 3.7 is straightforward to apply for analysis and prediction. The

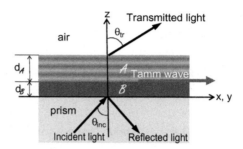

Figure 5.5 Schematic of the prism-coupled configuration to excite Tamm waves.

angle of incidence is varied from $0°$ toward $90°$, and the reflectance into the prism is computed as a function of that angle. Even though both partnering materials for Tamm-wave propagation are supposed to have negligible dissipation, $n^\mathcal{B}$ must be endowed with a small positive imaginary part [180] so that some reflectance dips for $\theta_{\mathrm{inc}} >$ $\sin^{-1}(1/n_{\mathrm{prism}})$ indicate the excitation of Tamm waves, as is illustrated for simple SPP waves in Section 2.3.1.4. The reflectance dips for $\theta_{\mathrm{inc}} > \sin^{-1}(1/n_{\mathrm{prism}})$ also manifest themselves as absorptance peaks.

5.3.1.1 Interface of a Homogeneous Dielectric Material and a Periodic Multilayer

Figure 5.6 presents the absorptances A_s and A_p calculated as functions of θ_{inc} when material $\mathcal{A}$ is the periodic multilayer used for Figure 5.2, $n^\mathcal{B} = 1.35(1 + 10^{-4}i)$, $n_{\mathrm{prism}} = 2.6$, $d_\mathcal{A} = 6\Omega^\mathcal{A}$, $d_\mathcal{B} = 200$ nm, and $\lambda_0 = 633$ nm. According to Figure 5.2,

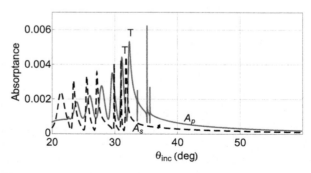

Figure 5.6 Calculated linear absorptances A_s and A_p as functions of the angle of incidence θ_{inc} in the prism-coupled configuration of Figure 5.5, when material $\mathcal{A}$ is the periodic multilayer described by Eq. (5.30) with $n_1^\mathcal{A} = 1.45$, $d_1^\mathcal{A} = 487$ nm, $n_2^\mathcal{A} = 2.32$, and $d_2^\mathcal{A} = 2\Omega^\mathcal{A} - d_1^\mathcal{A} = 779$ nm; the refractive index of the homogeneous partnering material $\mathcal{B}$ is $n^\mathcal{B} = 1.35(1 + 10^{-4}i)$; while the thicknesses are $d_\mathcal{A} = 6\Omega^\mathcal{A}$ and $d_\mathcal{B} = 200$ nm. The free-space wavelength is $\lambda_0 = 633$ nm. The prism material is rutile ($n_{\mathrm{prism}} = 2.6$). The letter T identifies the absorptance peaks that represent the excitation of Tamm waves.

when $n^B = 1.35$, an s-polarized Tamm wave should exist with wavenumber $q = 1.3844k_0$. Therefore, in the prism-coupled configuration, an s-polarized Tamm wave should be excited at $\theta_{inc} \approx \sin^{-1}(1.3844/n_{prism}) = 32.17°$. An A_s-peak at $\theta_{inc} = 31.80°$ in Figure 5.6 is close enough to that prediction that it can be associated with the excitation of an s-polarized Tamm wave. Likewise, a p-polarized Tamm wave with wavenumber $q = 1.3944k_0$ is indicated by Figure 5.2, and the consequent prediction $\theta_{inc} \approx 32.43°$ is very close to $\theta_{inc} = 32.47°$ for an A_p-peak in Figure 5.6.

In the canonical boundary-value problem, a wave that is evanescent as $z \to \infty$ is present in the partnering material occupying the half-space $z > 0$, while a wave that evanesces as $z \to -\infty$ is present in the partnering material occupying the half-space $z < 0$. In contrast, due to the finite thickness of each partnering material in the prism-coupled configuration, both types of waves are present in each partnering material. Their interaction is responsible for the difference between the angle of incidence predicted by the canonical boundary-value problem for a specific Tamm-wave mode and the one gleaned for it from an absorptance plot for the prism-coupled configuration.

Although the absorptance peaks are prominent in Figure 5.6, the peak absorptances are minuscule because dissipation is either assumed to be absent or very small in the partnering dielectric materials. Significantly larger imaginary parts of $\epsilon^A(z)$ and/or ϵ^B would enhance the peak absorptances, as experiments suggest [175,176]. But the propagation length Δ_{prop} would be decreased thereby, reducing the potential of Tamm waves for long-range communication. A comparison with the magnitudes of absorptance peaks in Section 6.6 is however sufficient to confirm that, unlike SPP waves, Tamm waves do not have direct relevance for the harvesting of solar energy.

5.3.1.2 Interface of a Homogeneous Dielectric Material and a Rugate Filter

When material A is the rugate filter described by Eq. (5.32) with $n_{avg}^A = 1.885$, $\Delta n^A = 0.87$, and $\Omega^A = \lambda_0$, and material B has a refractive index $n^B = 1.515$, whereas $n_{prism} = 2.6$ and $\lambda_0 = 633$ nm remain unchanged from Section 5.3.1.1, the angles of incidence in the prism-coupled configuration predicted for the excitation of Tamm-wave modes using the solutions q of the canonical boundary-value problem are provided in Table 5.2. Figure 5.7 contains plots of the absorptances A_s and A_p with respect to θ_{inc} when $d_A = 6\Omega^A$ and $d_B = 150$ nm in the prism-coupled configuration. Peaks of A_p and A_s occur in Figure 5.7 at values of θ_{inc} that are very close to the predictions in Table 5.2.

Table 5.2 Predictions of the angles of incidence to excite s- and p-polarized Tamm waves in the prism-coupled configuration with $n_{prism} = 2.6$. These predictions follow from the normalized wavenumbers q/k_0 provided in Table 5.1 for Tamm waves propagating along the $+x$ axis in the canonical boundary-value problem.

Polarization state	θ_{inc} (°)	θ_{inc} (°)	θ_{inc} (°)	θ_{inc} (°)
s-pol	36.76	43.01	50.17	58.40
p-pol	36.58	42.46	49.67	57.94

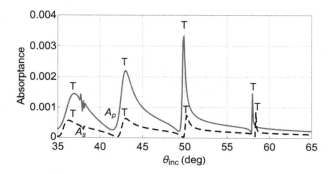

Figure 5.7 Calculated linear absorptances A_s and A_p as functions of the angle of incidence θ_{inc} in the prism-coupled configuration of Figure 5.5, when material $\mathcal{A}$ is the rugate filter described by Eq. (5.32) with $n_{avg}^{\mathcal{A}} = 1.885$, $\Delta n^{\mathcal{A}} = 0.87$, and $\Omega^{\mathcal{A}} = \lambda_0$; the refractive index of the homogeneous partnering material $\mathcal{B}$ is $n^{\mathcal{B}} = 1.515(1 + 10^{-4}i)$; while the thicknesses are $d_{\mathcal{A}} = 8\Omega^{\mathcal{A}}$ and $d_{\mathcal{B}} = 50$ nm. The free-space wavelength is $\lambda_0 = 633$ nm. The prism material is rutile ($n_{prism} = 2.6$). The letter T identifies the absorptance peaks that represent the excitation of Tamm waves.

5.3.2 Grating-Coupled Configuration

Although not yet experimentally confirmed, theory shows that Tamm waves can be excited in the grating-coupled configuration [180]. The general framework provided in Section 3.8 can be used for calculations; however, the following description of the periodically corrugated boundary between materials $\mathcal{A}$ and $\mathcal{B}$ is clearer than the one in that section for the illustrative numerical results provided here.

As shown in Figure 5.8, the half-spaces $z < 0$ and $z > d_3$ are occupied by air ($n_{air} = 1$), the region $0 \leqslant z \leqslant d_1$ is occupied by material $\mathcal{A}$ of permittivity $\epsilon^{\mathcal{A}}(z)$, and the region $d_2 \leqslant z < d_3$ by material $\mathcal{B}$ of permittivity $\epsilon^{\mathcal{B}}$. The region $d_1 < z < d_2$ contains a periodically undulating interface with period L_x along the x axis. The permittivity

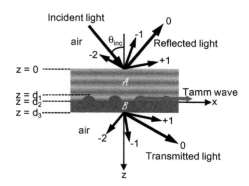

Figure 5.8 Schematic of the grating-coupled configuration to excite Tamm waves. The incident plane wave propagates in the xz plane, so that the method of Section 3.8 can be applied with $\psi = 0°$ and $n \in \{0\}$.

$\epsilon_g(x, z) = \epsilon_g(x \pm L_x, z)$ is expressed in the reference unit cell $0 < x < L_x$ of this region as

$$\epsilon_g(x, z) = \begin{cases} \epsilon^B - [\epsilon^B - \epsilon^A(z)] \, \mathcal{U} \, [d_2 - z - w(x)], & x \in (0, L_1), \\ \epsilon^A(z), & x \in (L_1, L_x), \end{cases} \quad (5.35)$$

with $\mathcal{U}(\sigma)$ as the unit step function defined by Eq. (2.102) and

$$w(x) = (d_2 - d_1) \sin\left(\frac{\pi x}{L_1}\right), \quad L_1 \in (0, L_x). \tag{5.36}$$

Thus the grating is singly periodic and the grating plane is the xz plane. Although the function $w(x)$ is a sinusoid in Eq. (5.36), other suitable functions can also be used as long as they are subject to the restrictions mentioned in Section 3.8.

A linearly polarized plane wave, propagating in the half-space $z < 0$ at an angle $\theta_{inc} \in [0°, 90°)$ to the z axis and at an angle $\psi = 0°$ to the x axis in the xy plane, is incident on the plane $z = 0$. The field phasors of the incident plane wave are described by Eqs. (3.141) with $n \in \{0\}$ and $n_1 = n_{air}$. Linear absorptances A_s and A_p defined in Eqs. (3.151) can be calculated after using the RCWA described in Section 3.8.5.

5.3.2.1 Interface of a Homogeneous Dielectric Material and a Periodic Multilayer

Let us begin with material $\mathcal{A}$ being the periodic multilayer described by Eq. (5.30) with $n_1^A = 1.45$, $d_1^A = 487$ nm, $n_2^A = 2.32$, and $d_2^A = 2\Omega^A - d_1^A = 779$ nm. Material $\mathcal{B}$ has a refractive index $n^B = 1.35(1 + 10^{-4}i)$. Figure 5.9 contains plots of A_s and A_p as functions of the angle of incidence θ_{inc} at $\lambda_0 = 633$ nm. These plots contain several peaks. The angular locations of several peaks shift significantly as d_2 and/or $d_3 - d_2$ increase, and the spatial field profiles change significantly as well. Therefore, these absorptance peaks cannot represent surface waves.

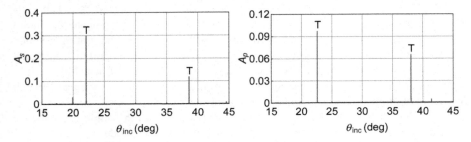

Figure 5.9 Calculated linear absorptances A_s and A_p as functions of the angle of incidence θ_{inc} at $\lambda_0 = 633$ nm, for the grating-coupled configuration of Figure 5.8. Material $\mathcal{A}$ is the periodic multilayer described by Eq. (5.30) with $n_1^A = 1.45$, $d_1^A = 487$ nm, $n_2^A = 2.32$, and $d_2^A = 779$ nm, so that $\Omega^A = 633$ nm. Material $\mathcal{B}$ has a refractive index $n^B = 1.35(1 + 10^{-4}i)$. The grating is described via Eqs. (5.35) and (5.36), with $d_3 - d_2 = 100$ nm, $d_2 - d_1 = 50$ nm, $d_2 = 6\Omega^A$, $L_x = \Omega^A$, and $L_1 = 0.5L_x$. The letter T identifies the absorptance peaks that represent the excitation of Tamm waves.

Table 5.3 Relative wavenumbers $k_{xy}^{(m,n)}/k_0$ of linear Floquet harmonics of order (m, n), $m \in$ [−2, 2] and $n = 0$, at the A_s-peaks marked by T in Figure 5.9. Boldface entries signify Tamm waves.

θ_{inc} (°)	$k_{xy}^{(m,0)}/k_0$				
	$m = -2$	$m = -1$	$m = 0$	$m = 1$	$m = 2$
22.58	−1.6160	−0.6160	0.3840	**1.3840**	2.3840
38.02	**−1.3841**	−0.3841	0.6159	1.6159	2.6159

Table 5.4 Relative wavenumbers $k_{xy}^{(m,n)}/k_0$ of linear Floquet harmonics of order (m, n), $m \in$ [−2, 2] and $n = 0$, at the A_p-peaks marked by T in Figure 5.9. Boldface entries signify Tamm waves.

θ_{inc} (°)	$k_{xy}^{(m,0)}/k_0$				
	$m = -2$	$m = -1$	$m = 0$	$m = 1$	$m = 2$
22.11	−1.6236	−0.6236	0.3764	**1.3764**	2.3764
38.58	**−1.3764**	−0.3764	0.6236	1.6236	2.6236

However, the angular locations of two A_s-peaks do not shift significantly as d_2 and/or $d_3 - d_2$ increase beyond certain thresholds.[2] The field phasors are also bound tightly to the interface of materials $\mathcal{A}$ and $\mathcal{B}$. For these two peaks—located at $\theta_{\text{inc}} = 22.58°$ and $\theta_{\text{inc}} = 38.02°$—the wavenumbers $k_{xy}^{(m,n)}$ defined in Eq. (3.1.43)$_4$ are provided in Table 5.3 for $m \in [-2, 2]$ and $n = 0$. The relative wavenumber $q/k_0 = 1.3844$ of the sole s-polarized Tamm wave from Figure 5.2 for $n^{\mathcal{B}} = 1.35$ is close enough to $k_{xy}^{(1,0)}/k_0 = 1.3840$ for $\theta_{\text{inc}} = 22.58°$ that we can conclude that the Tamm wave is excited as a linear Floquet harmonic of order $(1, 0)$. Likewise, $q/k_0 = -1.3944$ is also close enough to $k_{xy}^{(-2,0)}/k_0 = -1.3841$ for $\theta_{\text{inc}} = 38.02°$ to indicate the Tamm wave is excited as a linear Floquet harmonic of order $(-2, 0)$ in the grating-coupled configuration. Thus, as the angle of incidence θ_{inc} rises from $0°$ toward $90°$, the same s-polarized Tamm-wave mode is excited twice, the first time as a linear Floquet harmonic of order $(1, 0)$ propagating along the $+x$ axis, and the next as a linear Floquet harmonic of order $(-2, 0)$ propagating along the $-x$ axis.

With the assistance of Table 5.4, we can also see that, as the angle of incidence θ_{inc} increases from $0°$ toward $90°$, a p-polarized Tamm-wave mode is excited twice, first at $\theta_{\text{inc}} = 22.11°$ as a linear Floquet harmonic of order $(1, 0)$ propagating along the $+x$ axis, and then at $\theta_{\text{inc}} = 38.58°$ as a linear Floquet harmonic of order $(-2, 0)$ propagating along the $-x$ axis. The phase speed v_p is the same in both instances.

---

[2]The ratio $d_2/2\Omega^{\mathcal{A}}$ should be increased in increments of unity, for the identification of Tamm waves. It would then be convenient to always choose d_2 as an integral multiple of $2\Omega^{\mathcal{A}}$.

5.3.2.2   Interface of a Homogeneous Dielectric Material and a Rugate Filter

Next, let material $\mathcal{A}$ be changed to the rugate filter described by

$$\epsilon^{\mathcal{A}}(z) = \epsilon_0 \left\{ n^{\mathcal{A}}_{\text{avg}} + \frac{\Delta n^{\mathcal{A}}}{2} \sin \left[\frac{\pi [d_2 - z)]}{\Omega^{\mathcal{A}}} \right] \right\}^2 \tag{5.37}$$

with $n^{\mathcal{A}}_{\text{avg}} = 1.885$, $\Delta n^{\mathcal{A}} = 0.87$, and $\Omega^{\mathcal{A}} = 633$ nm, and the refractive index of material $\mathcal{B}$ be set equal to $n^{\mathcal{B}} = 1.515(1 + 10^{-4}i)$. Figure 5.10 contains plots of A_s and A_p as functions of the angle of incidence θ_{inc} at $\lambda_0 = 633$ nm. These plots contain several peaks, of which four A_s-peaks and four A_p-peaks indicate the excitation of Tamm-wave modes—because (i) their angular locations do not change significantly after the thickness $d_3 - d_2$ and/or the integer $d_2/2\Omega^{\mathcal{A}}$ increase beyond certain thresholds, and (ii) the field phasors are also bound well to the interface of materials $\mathcal{A}$ and $\mathcal{B}$ [180].

The wavenumbers $k_{xy}^{(m,n)}$, $m \in [-2, 2]$ and $n = 0$, defined in Eq. (3.143)$_4$ are provided in Table 5.5 for the four A_s-peaks and in Table 5.6 for the four A_p-peaks identified by T in Figure 5.10. Comparison with the solutions q of the dispersion

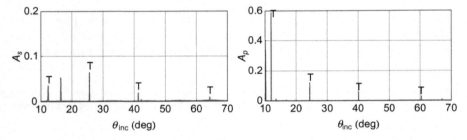

Figure 5.10 Calculated linear absorptances A_s and A_p as functions of the angle of incidence θ_{inc} at $\lambda_0 = 633$ nm, for the grating-coupled configuration of Figure 5.8. Material $\mathcal{A}$ is the rugate filter described by Eq. (5.37) with $n^{\mathcal{A}}_{\text{avg}} = 1.885$, $\Delta n^{\mathcal{A}} = 0.87$, and $\Omega^{\mathcal{A}} = \lambda_0$. Material $\mathcal{B}$ has a refractive index $n^{\mathcal{B}} = 1.515(1 + 10^{-4}i)$. The grating is described via Eqs. (5.35) and (5.36), with $d_3 - d_2 = \Omega^{\mathcal{A}}$, $d_2 - d_1 = 50$ nm, $d_1 = 8\Omega^{\mathcal{A}}$, $L_x = 0.75\Omega^{\mathcal{A}}$, and $L_1 = 0.5L_x$. The letter T identifies the absorptance peaks that represent the excitation of Tamm waves [180].

Table 5.5 Relative wavenumbers $k_{xy}^{(m,n)}/k_0$ of linear Floquet harmonics of order (m, n), $m \in [-2, 2]$ and $n = 0$, at the A_s-peaks marked by T in Figure 5.10. Boldface entries signify Tamm waves.

θ_{inc} (°)	$k_{xy}^{(m,0)}/k_0$				
	$m = -2$	$m = -1$	$m = 0$	$m = 1$	$m = 2$
12.26	−2.45432	−1.12099	0.21235	**1.54568**	2.87901
25.59	−2.23474	−0.90141	0.43193	**1.76526**	3.09860
41.33	−2.00627	−0.67294	0.66040	**1.99373**	3.32706
64.33	−1.76536	−0.43203	0.90130	**2.23464**	3.56797

equations (5.14) and (5.28) provided in Table 5.5 lets us conclude that all eight Tamm-wave modes are excited as linear Floquet harmonics of order $(1, 0)$ and propagate along the $+x$ axis in the grating-coupled configuration.

5.3.3 Prospects for Optical Sensing

The reflectance dips in Figures 2.7 and 2.8 indicative of the excitation of a simple SPP wave are a few degrees wide on the θ_{inc} axis. The absorptance peaks in Figures 5.6 and 5.7 indicating the excitation of a Tamm wave in the prism-coupled configuration are less wide and should therefore lead to a finer sensing modality. In the grating-coupled configuration, the absorptance peaks for Tamm waves are less than a tenth of a degree in width, as exemplified by the magnified plots of absorptance in Figure 5.11, suggesting possibilities for even finer sensing.

Moreover, while $\mathrm{Im}\{q\}$ is negligibly small for Tamm waves, it is significant for simple SPP waves. The much longer propagation length of a Tamm wave should be an advantage over a simple SPP wave for optical sensing because the surface wave can then be affected by a larger amount of the analyte. The possibility of simultaneously employing two or more Tamm-wave modes of the same frequency should also be advantageous toward error-free sensing [177].

But, as the analyte must be put in the half-space $z < 0$ close to the plane $z = 0$ in Figure 5.8 for the grating-coupled configuration, a big disadvantage is noise that comes about from the traversal of light through the analyte before and after interaction

Table 5.6 Relative wavenumbers $k_{xy}^{(m,n)}/k_0$ of linear Floquet harmonics of order (m, n), $m \in [-2, 2]$ and $n = 0$, at the A_p-peaks marked by T in Figure 5.10. Boldface entries signify Tamm waves.

θ_{inc} (°)	$k_{xy}^{(m,0)}/k_0$				
	$m = -2$	$m = -1$	$m = 0$	$m = 1$	$m = 2$
11.98	−2.45910	−1.12576	0.20757	**1.54090**	2.87424
24.37	−2.25404	−0.92071	0.41263	**1.74596**	3.07929
40.13	−2.02214	−0.68881	0.64452	**1.97886**	3.31119
60.39	−1.79726	−0.46393	0.86941	**2.20274**	3.53608

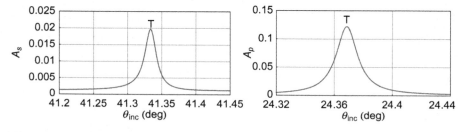

Figure 5.11 Two magnified portions of the absorptance plots of Figure 5.10. The letter T identifies the absorptance peaks that represent the excitation of Tamm waves.

with the two partnering materials. However that noise does not show up in the prism-coupled configuration, and a very minute amount of an analyte should be detectable by exploiting Tamm waves [177].

5.4 Interface of Two Periodically Nonhomogeneous Dielectric Materials

Tamm-wave propagation can also be guided by the interface of two periodically non-homogeneous materials, as can be seen by solving the underlying canonical boundary-value problem. For example, let us consider the interface of two rugate filters jointly specified by the refractive-index profile [62]

$$n(z) = \begin{cases} n^A(z) = n^A_{avg} + \frac{\Delta n^A}{2} \sin\left(\tau^A + \frac{\pi z}{\Omega^A}\right), & z > 0, \\ n^B(z) = n^B_{avg} + \frac{\Delta n^B}{2} \sin\left(\tau^B + \frac{\pi z}{\Omega^B}\right), & z < 0, \end{cases} \tag{5.38}$$

for all $z \in (-\infty, \infty)$, where n^A_{avg} and n^B_{avg} are the mean refractive indexes, whereas Δn^A and Δn^B are the amplitudes, $2\Omega^A$ and $2\Omega^B$ are the periods, and τ^A and τ^B are the phases of the refractive-index modulation in the two half-spaces $z > 0$ and $z < 0$.

The boundary-value problem can be solved either by following the methodology of Section 3.5 or by suitably adapting Sections 5.2.1 and 5.2.2. The latter is shown next.

5.4.1 s-Polarized Tamm Wave

All equations pertaining solely to material A in Section 5.2.1 still apply. However, the column 2-vector $[\underline{f}_s(z)]$ in material B now obeys the 2×2 matrix ordinary differential equation

$$\frac{d}{dz}[\underline{f}_s(z)] = i\left[\underline{\underline{P}}^B_s(z)\right] \cdot [\underline{f}_s(z)], \quad z < 0, \tag{5.39}$$

where the matrix

$$\left[\underline{\underline{P}}^B_s(z)\right] = \begin{bmatrix} 0 & -\omega\mu_0 \\ \frac{q^2}{\omega\mu_0} - \omega\epsilon^B(z) & 0 \end{bmatrix} \tag{5.40}$$

employs $\epsilon^B(z) = \epsilon_0[n^B(z)]^2$.

Application of the piecewise-uniform-approximation method to Eq. (5.40) for $z < 0$ determines the matrix $\left[\underline{\underline{Q}}^B_s\right]$ that appears in the relation

$$\left[\underline{f}_s(-2\Omega^B)\right] = \left[\underline{\underline{Q}}^B_s\right] \cdot [\underline{f}_s(0-)] \equiv \begin{bmatrix} Q^B_{s11} & Q^B_{s12} \\ Q^B_{s21} & Q^B_{s22} \end{bmatrix} \cdot [\underline{f}_s(0-)]. \tag{5.41}$$

As the electromagnetic field must decay as $z \to -\infty$, we get

$$[\underline{f}_s(0-)] = C^B_s \begin{bmatrix} 1 \\ \frac{\sigma^B_s - Q^B_{s11}}{Q^B_{s12}} \end{bmatrix}, \tag{5.42}$$

where σ_s^B is that solution of the quadratic equation

$$\left(Q_{s11}^B - \sigma_s^B \right) \left(Q_{s22}^B - \sigma_s^B \right) - Q_{s12}^B Q_{s21}^B = 0 \tag{5.43}$$

which satisfies the inequality

$$\text{Re}\left\{ \ln \sigma_s^B \right\} < 0. \tag{5.44}$$

A decay constant $\beta_s^B = \exp \left[\text{Re} \left\{ \ln \sigma_s^B \right\} \right] \in (0, 1)$ is appropriate to denote the decay of fields over one period of material B.

Enforcement of the boundary condition $[\underline{f}_s(0+)] = [\underline{f}_s(0-)]$ yields the matrix equation

$$\begin{bmatrix} 1 & -1 \\ \dfrac{\sigma_s^A - Q_{s11}^A}{Q_{s12}^A} & -\dfrac{\sigma_s^B - Q_{s11}^B}{Q_{s12}^B} \end{bmatrix} \cdot \begin{bmatrix} C_s^A \\ C_s^B \end{bmatrix} = \begin{bmatrix} 0 \\ 0 \end{bmatrix}, \tag{5.45}$$

leading to the dispersion equation

$$\left(\sigma_s^A - Q_{s11}^A \right) Q_{s12}^B = \left(\sigma_s^B - Q_{s11}^B \right) Q_{s12}^A \tag{5.46}$$

for s-polarized Tamm waves. Equation (5.15) is still satisfied by C_s^A and C_s^B for an s-polarized Tamm wave.

5.4.2   p-Polarized Tamm Wave

The 2×2 matrix ordinary differential equation

$$\frac{d}{dz}[\underline{f}_p(z)] = i \left[\underline{\underline{P}}_p^B(z) \right] \cdot [\underline{f}_p(z)], \quad z < 0, \tag{5.47}$$

with

$$\left[\underline{\underline{P}}_p^B(z) \right] = \begin{bmatrix} 0 & \omega\mu_0 - \dfrac{q^2}{\omega\epsilon^B(z)} \\ \omega\epsilon^B(z) & 0 \end{bmatrix} \tag{5.48}$$

has to be solved numerically in order to determine the matrix $\left[\underline{\underline{Q}}_p^B \right]$ that appears in the relation

$$\left[\underline{f}_p(-2\Omega^B) \right] = \left[\underline{\underline{Q}}_p^B \right] \cdot [\underline{f}_p(0-)] \equiv \begin{bmatrix} Q_{p11}^B & Q_{p12}^B \\ Q_{p21}^B & Q_{p22}^B \end{bmatrix} \cdot [\underline{f}_p(0-)]. \tag{5.49}$$

Next, the two eigenvalues σ_p^B and $1/\sigma_p^B$ of $\left[\underline{\underline{Q}}_p^B \right]$ are obtained by solving a quadratic equation. The eigenvalue σ_p^B is subject to the restriction

$$\text{Re}\left\{ \ln \sigma_p^B \right\} < 0 \tag{5.50}$$

and a decay constant $\beta_p^B = \exp \left[\text{Re} \left\{ \ln \sigma_p^B \right\} \right] \in (0, 1)$ is defined.

Then, the equality of

$$[\underline{f}_p(0-)] = C_p^B \begin{bmatrix} 1 \\ \frac{\sigma_p^B - Q_{p11}^B}{Q_{p12}^B} \end{bmatrix} \tag{5.51}$$

and $[\underline{f}_p(0+)]$ given by Eq. (5.25) yields the dispersion equation

$$\left(\sigma_p^A - Q_{p11}^A\right) Q_{p12}^B = \left(\sigma_p^B - Q_{p11}^B\right) Q_{p12}^A \tag{5.52}$$

for p-polarized Tamm waves. The coefficients C_p^A and C_p^B of a p-polarized Tamm wave still satisfy Eq. (5.29).

5.4.3 Illustrative Numerical Results

Table 5.7 presents the wavenumbers q of all seven Tamm waves obtained for the interface of two rugate filters $\mathcal{A}$ $\left(n_{\mathrm{avg}}^A = 1.885, \Delta n^A = 0.87, \tau^A = 0°, \text{ and } \Omega^A = \lambda_0\right)$ and $\mathcal{B}$ $\left(n_{\mathrm{avg}}^B = 1.6, \Delta n^B = 0.6, \tau^B = 90°, \text{ and } \Omega^B = 0.5\lambda_0\right)$ that differ in all parameters of the refractive-index profile, when $\lambda_0 = 633$ nm. Four of these Tamm waves are s polarized, while the remaining three are p polarized.

The spatial profiles of the electromagnetic field of one of the p-polarized Tamm waves are given in Figure 5.12. The decay constants for this Tamm wave are $\beta_p^A = 0.5906$ and $\beta_p^B = 0.1001$, thereby indicating stronger localization to the interface $z = 0$ in the half-space $z < 0$ than in the half-space $z > 0$.

The interface created by an abrupt phase defect in a rugate filter can also guide Tamm waves. Suppose that $n_{\mathrm{avg}}^A = n_{\mathrm{avg}}^B$, $\Delta n^A = \Delta n^B$, and $\Omega^A = \Omega^B$, but $\tau^A \neq \tau^B$. Then, the entire space can be considered to be occupied by a single rugate filter which has a phase defect $\tau^A - \tau^B$ at the plane $z = 0$, provided that $|\tau^A - \tau^B| \notin \{0°, 360°\}$.

The solutions of the dispersion equations for Tamm waves bound to the phase-defect plane $z = 0$ are shown in Figure 5.13, when $n_{\mathrm{avg}}^A = n_{\mathrm{avg}}^B = 1.885$, $\Delta n^A = \Delta n^B = 0.87$, $\Omega^A = \Omega^B = \lambda_0 = 633$ nm, and $\tau^A = 0°$ but $\tau^B \in (0°, 360°)$ is variable. Although a phase defect does exist despite the continuity of $n(z)$ across the plane $z = 0$ when $\tau^B = 180°$, no solutions of the dispersion equations signifying Tamm-wave propagation exist.

Table 5.7 Normalized wavenumbers q/k_0 of Tamm waves guided by the interface of two rugate filters $\mathcal{A}$ $\left(n_{\mathrm{avg}}^A = 1.885, \Delta n^A = 0.87, \tau^A = 0°, \text{ and } \Omega^A = \lambda_0\right)$ and $\mathcal{B}$ $\left(n_{\mathrm{avg}}^B = 1.6, \Delta n^B = 0.6, \tau^B = 90°, \text{ and } \Omega^B = 0.5\lambda_0\right)$ that differ in all parameters of the refractive-index profile (5.38), when $\lambda_0 = 633$ nm [62]. If q is a solution of the dispersion equation (5.46) or (5.52), then $-q$ also is.

Polarization state	q/k_0	q/k_0	q/k_0	q/k_0
s-pol	1.6178	1.8013	2.0021	2.2147
p-pol	1.6101	–	1.9927	2.2045

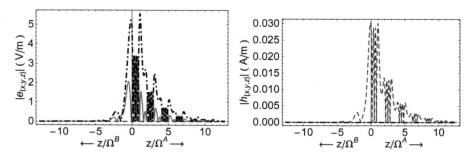

Figure 5.12 Variations with z of the magnitudes of the non-zero Cartesian components of (left) $\underline{e}(z)$ and (right) $\underline{h}(z)$, for a p-polarized Tamm wave propagating along the $+x$ axis, guided by the interface of two rugate filters described in Table 5.7. The components parallel to $\underline{u}_x$, $\underline{u}_y$, and $\underline{u}_z$ are represented by red solid, blue dashed, and black chain-dashed lines, respectively. The distance z is differently scaled for the two half-spaces. The data were computed by setting $C_p^{\mathcal{A}} = C_p^{\mathcal{B}} = 1$ V m^{-1} and $q/k_0 = 1.6101$ [62]. (For interpretation of the references to color in this figure legend, the reader is referred to the web version of this book.)

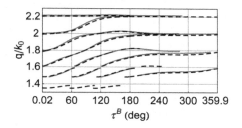

Figure 5.13 Solutions q of the dispersion equations (5.46) and (5.52) in relation to $\tau^{\mathcal{B}}$ for Tamm waves localized to the phase-defect interface $z = 0$ in a rugate filter described by Eq. (5.38) with $n_{\text{avg}}^{\mathcal{A}} = n_{\text{avg}}^{\mathcal{B}} = 1.885$, $\Delta n^{\mathcal{A}} = \Delta n^{\mathcal{B}} = 0.87$, $\Omega^{\mathcal{A}} = \Omega^{\mathcal{B}} = \lambda_0 = 633$ nm, and $\tau^{\mathcal{A}} = 0°$. The red solid lines indicate s-polarized Tamm waves and the black dashed lines are for p-polarized Tamm waves [62]. (For interpretation of the references to color in this figure legend, the reader is referred to the web version of this book.)

Clearly, multiple Tamm-wave modes can be guided by a phase defect in a rugate filter. For $\tau^{\mathcal{B}} \notin \{0°, 180°\}$, up to 13 solutions of Eqs. (5.46) and (5.52) are possible with $q > k_0(n_{\text{avg}}^{\mathcal{A}} - \Delta n^{\mathcal{A}}/2)$, almost always equally divided between the s- and p-polarization states. As $\tau^{\mathcal{B}}$ increases from $0°$, the number of solutions increases episodically. All branches of s-polarized solutions are paired with branches of p-polarized solutions.

Four unpaired branches of p-polarized solutions also exist in Figure 5.13. One branch satisfies the inequality $q > k_0(n_{\text{avg}}^{\mathcal{A}} - \Delta n^{\mathcal{A}}/2)$, but the other three unpaired branches do not. Solutions on these three branches are special because q/k_0 is less than the minimum value of the refractive index of the rugate filter on either side of the phase-defect interface, and the corresponding surface waves may be called high-phase-speed Tamm waves. Such solutions cannot exist if one of the regions on either side of the phase defect is replaced by a homogeneous dielectric material.

Despite the fact that no solution of Eqs. (5.46) and (5.52) exists for $\tau^{\mathcal{B}} = 0°$ (or $360°$), some solution branches, but not all, appear to wrap about $\tau^{\mathcal{B}} = 360°$ in

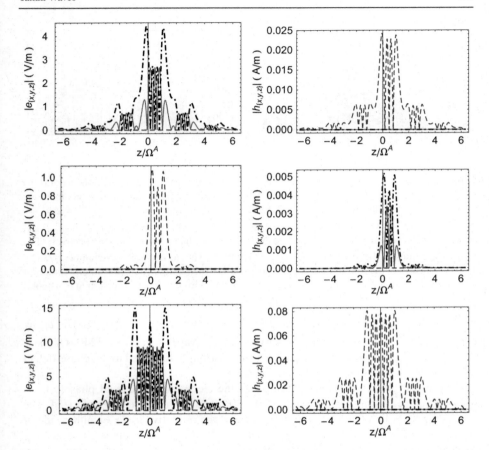

Figure 5.14 Variations with z of the magnitudes of the non-zero Cartesian components of (left) $\underline{e}(z)$ and (right) $\underline{h}(z)$, for Tamm waves localized to the phase-defect interface $z = 0$ in a rugate filter described by Eq. (5.38) with $n_{avg}^{\mathcal{A}} = n_{avg}^{\mathcal{B}} = 1.885$, $\Delta n^{\mathcal{A}} = \Delta n^{\mathcal{B}} = 0.87$, $\Omega^{\mathcal{A}} = \Omega^{\mathcal{B}} = \lambda_0 = 633$ nm, and $\tau^{\mathcal{A}} = 0°$. The components parallel to $\underline{u}_x$, $\underline{u}_y$, and $\underline{u}_z$ are represented by red solid, blue dashed, and black chain-dashed lines, respectively. The data were computed by setting $C_p^{\mathcal{A}} = C_p^{\mathcal{B}} = 1$ V m^{-1} for p-polarized Tamm waves and $C_s^{\mathcal{A}} = C_s^{\mathcal{B}} = 1$ V m^{-1} for s-polarized Tamm waves. (top) p-polarization state, $\tau^{\mathcal{B}} = 8°$, and $q/k_0 = 1.6155$; (middle) s-polarization state, $\tau^{\mathcal{B}} = 8°$, and $q/k_0 = 1.8007$; and (bottom) p-polarization state, $\tau^{\mathcal{B}} = 174°$, and $q/k_0 = 1.5718$ [62]. (For interpretation of the references to color in this figure legend, the reader is referred to the web version of this book.)

Figure 5.13. The branches that appear to wrap around have solutions with decay constant $\beta_{p,s}^{\mathcal{A}} = \beta_{p,s}^{\mathcal{B}}$ much smaller than unity for $\tau^{\mathcal{B}}$ in the neighborhood of 0° and 360°. Branches of solutions with decay constant close to unity for $\tau^{\mathcal{B}}$ in the neighborhood of 0° do not show the wrapping characteristic.

Spatial profiles of the electromagnetic field phasors are shown in Figure 5.14 for two p-polarized Tamm waves at $\tau^{\mathcal{B}} = 8°$ and 174°, and one s-polarized Tamm wave at $\tau^{\mathcal{B}} = 8°$. The decay constant at $\tau^{\mathcal{B}} = 8°$ for the p-polarized Tamm wave is

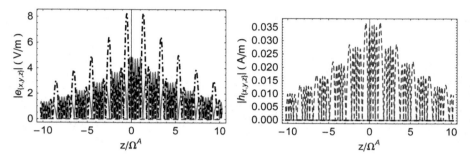

Figure 5.15 Variations with z of the magnitudes of the non-zero Cartesian components of (left) $\underline{e}(z)$ and (right) $\underline{h}(z)$, for a high-phase-speed Tamm wave localized to the phase-defect interface $z = 0$ in a rugate filter described by Eq. (5.38) with $n^{\mathcal{A}}_{avg} = n^{\mathcal{B}}_{avg} = 1.885$, $\Delta n^{\mathcal{A}} = \Delta n^{\mathcal{B}} = 0.87$, $\Omega^{\mathcal{A}} = \Omega^{\mathcal{B}} = \lambda_0 = 633$ nm, $\tau^{\mathcal{A}} = 0°$, and $\tau^{\mathcal{B}} = 30°$. The components parallel to $\underline{u}_x$, $\underline{u}_y$, and $\underline{u}_z$ are represented by red solid, blue dashed, and black chain-dashed lines, respectively. The data were computed by setting $C^{\mathcal{A}}_p = C^{\mathcal{B}}_p = 1$ V m^{-1}. The normalized wavenumber $q/k_0 = 1.3611$ is less than the minimum refractive index in the rugate filter [62]. (For interpretation of the references to color in this figure legend, the reader is referred to the web version of this book.)

$\beta^{\mathcal{A}}_p = 0.2821$, and that for the s-polarized Tamm wave is $\beta^{\mathcal{A}}_s = 0.0221$. The decay constant at $\tau^{\mathcal{B}} = 174°$ for the p-polarized Tamm wave is $\beta^{\mathcal{A}}_p = 0.3240$. Of these three Tamm waves, the s-polarized wave is most strongly localized to the phase-defect interface because its decay constant is the least.

Figure 5.15 shows the spatial profiles of the electromagnetic field phasors for a high-phase-speed Tamm wave, when $\tau^{\mathcal{B}} = 30°$. The Tamm wave is p polarized. The value of q/k_0 is less than the minimum value of the refractive index on either side of the phase-defect interface. The decay constant $\beta^{\mathcal{A}}_p = 0.7765$ is quite large, indicating weak localization to the phase-defect interface.

An abrupt change of either the mean refractive index or the amplitude of the sinusoidal variation of the refractive index also creates an interface that can guide multiple Tamm-wave modes with different phase speed, polarization state, and degree of localization [62].

5.5 Outlook

In contrast to the very narrow AEDs of Dyakonov waves, Tamm waves have the maximum possible total AED: a full 360°. Therefore, Tamm waves were experimentally observed much earlier and more easily [174–176] than Dyakonov waves. Easier observation also led to application to optical biosensing with two s-polarized Tamm-wave modes in the prism-coupled configuration [177].

However, the full potential of Tamm waves for optical sensing is very far from being realized and the best is still to come. Appropriate nonhomogeneous partnering materials can be made of porous materials which can be infiltrated by a fluid containing the analyte(s) to be sensed [124]. Both p- and s-polarized Tamm-wave modes can be excited. Tamm-wave modes of both linear polarization states appear to be more numerous when the periodically nonhomogeneous partnering material is continuously nonhomogeneous as opposed to comprising a few homogeneous layers in its unit cell. Therefore, numerous channels become available for more reliable sensing of a single analyte, as well as for simultaneous sensing of multiple analytes.

6 Surface-Plasmon-Polariton Waves II

6.1 Introduction

The practical application of simple SPP waves guided by the interface of a metal and a homogeneous isotropic dielectric material in chemical sensors has flourished during the past two decades [3,4]. To a lesser extent, simple SPP waves have been pursued for microscopy. Recently, the excitation of SPP waves, guided by an interface of a metallic surface-relief grating and a periodically nonhomogeneous semiconductor, has been examined to enhance the absorption of light by thin-film solar cells [84]. As other unidirectionally nonhomogeneous dielectric materials are investigated for deployment as a partnering material for a suitable metal for the propagation of SPP waves, other applications may be identified and developed.

This chapter is focused on SPP-wave propagation guided by a planar interface of a metal and a dielectric material that may be anisotropic and/or periodically nonhomogeneous. The chapter commences with SPP waves guided by the interface of a metal and a multilayered isotropic dielectric material. Next, the use of homogeneous anisotropic dielectric materials of various types is considered, with emphasis on columnar thin films. Then, multiple SPP-wave modes guided by a single interface of a metal and a periodically nonhomogeneous dielectric material are discussed. With sculptured thin films and liquid crystals, nanotechnology offers a large palette of possibilities. The final section of this chapter examines the employment of chiral STFs for SPR sensors.

6.2 Interface of a Metal and an Isotropic Dielectric Multilayer

The investigation of SPP waves guided by the interface of a metal and multilayered dielectric materials arose out of both a necessity in applications wherein the partnering dielectric material is inherently multilayered and a desire to improve the propagation length for communications as well as chemical sensing. An early study dealt with the application of a voltage for the alignment of a liquid crystal on the surface of an aligning layer atop a thick metallic layer [28,29]. In this study, the partnering dielectric material was layered in two ways. The liquid crystal and the aligning layer are two

Electromagnetic Surface Waves. http://dx.doi.org/10.1016/B978-0-12-397024-4.00006-2

dielectric materials of different thickness. Additionally, the orientation of the liquid crystal changes with distance from the aligning layer, necessitating the use of the piecewise-uniform-approximation method introduced in Section 2.3.4.4 for analysis.

A limitation of SPP waves in some applications is the small propagation length due to the dissipative nature of the metal. The Sarid configuration of Figure 2.11, commonly adopted to launch LRSPP waves, can be modified to yield even greater propagation lengths by inserting thin intermediate layers of dielectric materials between the metal film and the substrate and/or between the metal film and the cover. The arrangement of materials and layer thicknesses on the two sides of the metal film can be either asymmetric [320] or symmetric [242, 243]. The insertion of these dielectric layers engenders multiple SPP-wave modes, the number of which depends on the thickness of the intermediate layers. SPP-wave modes of even symmetry about the the mid-plane of the metal film are particularly long ranged, with extremely long propagation lengths as the intermediate layers reach a critical thickness [239]. Intermediate layers may facilitate the excitation of LRSPP waves even with a thick metallic film. In addition to alleviating manufacturing problems when employing thin films of a metal to obtain longer propagation lengths, the use of thicker metal films may allow the propagation of both electrical and optical signals with the same structure. Normalized admittance diagrams [244] are useful for the design of dielectric multilayers for SPP-wave propagation.

The quite recent exploitation of SPR [321–323] for characterizing lipid membranes has provided impetus to research on SPP waves guided by metal/multilayer interfaces. Beyond improving the sensitivity of SPR sensors of biomolecules, these SPP waves can be excited with both s- and p-polarized light. Thus, the SPP waves can be either s polarized or p polarized. Because the electric field of a p-polarized SPP wave is oriented perpendicular to the electric field of an s-polarized SPP wave, it becomes possible to determine the anisotropic optical parameters of the biomolecules.

When many layers of dissimilar dielectric materials are arranged in a periodic fashion, the resultant structure is a periodic multilayer that is often called a one-dimensional photonic crystal. Light within a certain range of frequencies called the band gap cannot propagate as a freely traveling wave inside the photonic crystal in certain ranges of directions. When propagation is disallowed in all directions, the band gap is said to be a complete band gap.

Suppose that a one-dimensional photonic crystal occupies a half-space. Inside this photonic crystal, light may propagate as an evanescent wave which decays with distance away from the boundary. Several researchers have investigated, both theoretically [154, 155, 157] and experimentally [160], the properties of SPP waves, guided by the interface of a metal and a periodic multilayer, which propagate with frequencies within the band gap and are even excited by normally incident light without the use of either a prism or a grating. Such SPP-wave modes have been called surface plasmon-Bragg modes and Tamm plasmon polariton (waves). The phase speeds of these SPP-wave modes can be very high, much larger than c_0 [157]. Furthermore, some of these SPP-wave modes may be excited by s-polarized light and others by p-polarized light. In addition, SPP-wave modes with lower phase speeds are also guided by the same interface, the phase speed of such a surface wave being lower than of any plane wave in any material within the periodic multilayer [154]. When the periodic multilayer is replaced by a rugate filter, theory indicates that SPP-wave modes with phase speeds ranging from very small to very high are still possible [159].

6.3 Interface of a Metal and a Homogeneous Anisotropic Dielectric Material

The nonhomogeneity of an isotropic dielectric multilayer imparts interesting characteristics to SPP waves—including very high phase speeds and excitation by s-polarized light. Propagation occurs in any direction in the interface plane and without any dependence on that direction. Thus, the total AED is the maximum it can be: 360°.

Dependence on the direction of propagation shows up when a homogeneous anisotropic dielectric material is partnered with a metal. Early investigations [324,325] of SPP waves guided by the interface of a metal and an anisotropic dielectric material elucidated basic characteristics. Present-day researchers continue to show an interest in the use of anisotropic dielectric materials, because ongoing developments in nanotechnology provide various means of engineering the anisotropy and later manipulating it by the application of an external stimulus.

6.3.1 General Considerations

In order to solve the canonical boundary-value problem, the starting part of the framework of Section 2.2 is still useful. The field phasors are given by

$$\underline{E}(\underline{r}) = \begin{cases} \underline{\mathcal{E}}_{met} \exp{(i\underline{k}_{met} \cdot \underline{r})}, & z < 0, \\ \underline{\mathcal{E}}_{diel} \exp{(i\underline{k}_{diel} \cdot \underline{r})}, & z > 0, \end{cases} \tag{6.1}$$

and

$$\underline{H}(\underline{r}) = \begin{cases} \underline{\mathcal{H}}_{met} \exp{(i\underline{k}_{met} \cdot \underline{r})}, & z < 0, \\ \underline{\mathcal{H}}_{diel} \exp{(i\underline{k}_{diel} \cdot \underline{r})}, & z > 0, \end{cases} \tag{6.2}$$

Equations (2.8) still apply to $\underline{\mathcal{E}}_{met}$ and $\underline{\mathcal{H}}_{met}$ in the half-space $z \leqslant 0$ occupied by the metal, but Eqs. (2.9) for the half-space $z \geqslant 0$ require modification to

$$\left. \begin{aligned} \underline{k}_{diel} \times \underline{\mathcal{E}}_{diel} &= \omega \mu_0 \underline{\mathcal{H}}_{diel} \\ \underline{k}_{diel} \times \underline{\mathcal{H}}_{diel} &= -\omega \underline{\underline{\epsilon}}_{diel} \cdot \underline{\mathcal{E}}_{diel} \\ \underline{k}_{diel} \cdot \underline{\underline{\epsilon}}_{diel} \cdot \underline{\mathcal{E}}_{diel} &= 0 \\ \underline{k}_{diel} \cdot \underline{\mathcal{H}}_{diel} &= 0 \end{aligned} \right\}, \quad z > 0, \tag{6.3}$$

in which the permittivity of the partnering dielectric material is not a scalar but the dyadic $\underline{\underline{\epsilon}}_{diel}$ [17]. In the most general case, $\underline{\underline{\epsilon}}_{diel}$ has nine complex-valued components as described in Eqs. (1.28), all functions of the angular frequency ω. Anisotropic dielectric materials may or may not be gyrotropic. If an anisotropic dielectric material is not gyrotropic, it can belong to either a uniaxial or a biaxial class, as discussed in

Section 1.4.1. In almost all research reported on SPP waves guided by a metal/dielectric interface with a partnering biaxial dielectric material, the biaxiality is of the orthorhombic system defined through Eq. (1.25).

The amplitude vectors of the field phasors in the metal are written as

$$
\left.
\begin{aligned}
\underline{\mathcal{E}}_{\text{met}} &= b_s \underline{u}_s + b_p \left(\frac{\alpha_{\text{met}} \underline{u}_{\text{prop}} + q \underline{u}_z}{k_0 n_{\text{met}}} \right) \\
\underline{\mathcal{H}}_{\text{met}} &= \frac{n_{\text{met}}}{\eta_0} \left[b_s \left(\frac{\alpha_{\text{met}} \underline{u}_{\text{prop}} + q \underline{u}_z}{k_0 n_{\text{met}}} \right) - b_p \underline{u}_s \right]
\end{aligned}
\right\},
\tag{6.4}
$$

with unknown coefficients b_p and b_s referring to the p- and s-polarization states, respectively. Completion of the analysis requires a representation for the amplitude vectors $\underline{\mathcal{E}}_{\text{diel}}$ and $\underline{\mathcal{H}}_{\text{diel}}$ stemming from Eqs. (6.3), followed by the imposition of boundary conditions (2.11). Alternatively, the 4×4-matrix approach of Section 3.5 can be adopted.

The incorporation of an anisotropic dielectric material to partner a metal vastly expands the complexity of SPP waves. Even so, with the results of several studies [31, 149, 326] carried out during the last two decades, a few general comments can be made before delving into specific situations. Unlike the simple SPP waves in Chapter 2, the SPP wave on the metal side of the interface is, in general, composed of both s- and p-polarized components when the partnering dielectric material is anisotropic. As a result, the SPP wave may be excited, in general, with either s- or p-polarized incident light. Both the relative proportions of s- and p-polarized components in the metal and the efficiency with which the SPP wave can be excited by s- and p-polarized incident light depend on the permittivity dyadic $\underline{\underline{\epsilon}}_{\text{diel}}$.

The SPP wave on the dielectric side of the interface also has two components. In the case of a uniaxial dielectric material, we can speak of one component being an ordinary and the other component being an extraordinary wave; more generally for any biaxial dielectric material, both components are distinct extraordinary waves [17, 78]. As these components depend on $\underline{u}_{\text{prop}}$, so too do the characteristics of the SPP wave. Unlike the interface of a metal and an isotropic dielectric material, the interface of a metal and an anisotropic dielectric material may not guide SPP waves in all directions parallel to the interface plane and the total AED is often less than 360°. Reorientation of the anisotropic dielectric material—which can be described mathematically through a transformation of the type $\underline{\underline{\epsilon}}_{\text{diel}} \rightarrow \underline{\underline{S}} \cdot \underline{\underline{\epsilon}}_{\text{diel}} \cdot \underline{\underline{S}}^{-1}$ with the dyadic $\underline{\underline{S}}$ representing a rotation about a fixed axis—can drastically alter the characteristics of SPP-wave propagation.

6.3.2 Columnar Thin Films

Ongoing developments in thin-film technology over the last century [42, 327] have made significant control possible over the permittivity dyadic of biaxial dielectric materials with large birefringence. The permittivity dyadic of a columnar thin film, as described in Section 1.4.3.1, is influenced by the selection of the material to be evaporated and the vapor flux angle χ_v during fabrication. Once the evaporant material has been chosen, the choice of χ_v affects the orientations of the optic ray axes $\underline{a}_+$ and $\underline{a}_-$, the biaxiality δ, and the birefringence Δn of the CTF. These nanoengineered materials, thus, make it possible to engineer interfaces supporting SPP waves with a considerable diversity of attributes.

Although the permittivity dyadics of CTFs have not been widely related to the vapor flux angle, experimental data are available on CTFs grown by the evaporation of a few materials. In particular, CTFs fabricated of oxides of tantalum, titanium, and zirconium have been characterized [94, 328]. Thus, with the dyadic $\underline{\underline{S}}_y(\chi)$ defined in Eq. (1.30), the eigenvalues of the permittivity dyadic

$$\underline{\underline{\epsilon}}_{CTF} = \underline{\underline{S}}_y(\chi) \cdot (\epsilon_a \underline{u}_z \underline{u}_z + \epsilon_b \underline{u}_x \underline{u}_x + \epsilon_c \underline{u}_y \underline{u}_y) \cdot \underline{\underline{S}}_y^{-1}(\chi) \tag{6.5}$$

may be conveniently characterized in terms of simple quartic functions of the normalized vapor flux angle $\chi_v^{norm} = 2\chi_v/\pi$ as

$$\left.\begin{aligned} \epsilon_a &= \epsilon_0 \left[n_{a0} + n_{a1}\chi_v^{norm} + n_{a2} \left(\chi_v^{norm} \right)^2 \right]^2 \\ \epsilon_b &= \epsilon_0 \left[n_{b0} + n_{b1}\chi_v^{norm} + n_{b2} \left(\chi_v^{norm} \right)^2 \right]^2 \\ \epsilon_c &= \epsilon_0 \left[n_{c0} + n_{c1}\chi_v^{norm} + n_{c2} \left(\chi_v^{norm} \right)^2 \right]^2 \end{aligned}\right\}, \tag{6.6}$$

and the tilt angle χ of the CTF as

$$\chi = \tan^{-1} (a \tan \chi_v). \tag{6.7}$$

The 10 constants $n_{j\ell}$ and a for CTFs of each of the three evaporated materials, as reported by Hodgkinson et al. [94], are provided in Table 6.1. A word of caution: The data in this table are applicable to CTFs produced by one particular experimental

Table 6.1 Empirical constants to be used in Eqs. (6.6) and (6.7) for CTFs of three evaporated materials [94]. The data hold at $\lambda_0 = 633$ nm.

	Tantalum oxide	Titanium oxide	Zirconium oxide
n_{a0}	1.1961	1.0443	1.2394
n_{a1}	1.5439	2.7394	1.2912
n_{a2}	−0.7719	−1.3697	−0.6456
n_{b0}	1.4600	1.6765	1.4676
n_{b1}	1.0400	1.5649	0.9428
n_{b2}	−0.5200	−0.7825	−0.4714
n_{c0}	1.3532	1.3586	1.3861
n_{c1}	1.2296	2.1109	0.9979
n_{c2}	−0.6148	−1.0554	−0.4990
a	3.1056	2.8818	3.5587

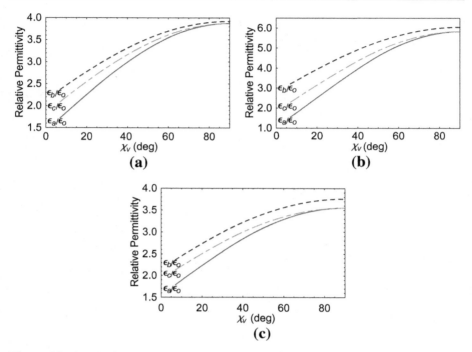

Figure 6.1 The principal relative permittivity scalars ϵ_a, ϵ_b, and ϵ_c of CTFs fabricated by evaporating three different materials. The data were calculated using Eqs. (6.6) and Table 6.1 for $\chi_v \in [7°, 90°]$. (a) Tantalum oxide, (b) titanium oxide, and (c) zirconium oxide.

apparatus, but may have to be modified for CTFs produced by other researchers on different apparatuses.

In order to illustrate the variation of the permittivity dyadic of a CTF with χ_v, Figure 6.1 shows the values of the principal permittivity scalars ϵ_a, ϵ_b, and ϵ_c determined using Eqs. (6.6) and Table 6.1 as functions of $\chi_v \in [7°, 90°]$. This range of χ_v is roughly the widest obtainable with current technology and corresponds to the range $[20°, 90°]$ of the tilt angle χ.

Instead of the quartic equations (6.6), higher-order polynomial equations can be devised from experimental data. For instance, the principal permittivity scalars ϵ_a, ϵ_b, and ϵ_c of silicon-oxide CTFs have been represented as sextic functions of χ_v^{norm} [57]. More complicated functionalities have also been devised for silicon-oxide CTFs and magnesium-fluoride CTFs [329]. All such functionalities strongly depend on the apparatus being used for fabrication.

6.3.3   Metal/CTF Interface

Calculations for the interface of bulk aluminum and a titanium-oxide CTF indicate that SPP-wave propagation is possible, in general, in four limited ranges of directions along the interface plane [44,45]. Recalling that $\underline{u}_{\text{prop}} = \underline{u}_x \cos \psi + \underline{u}_y \sin \psi$ and that the morphologically significant plane of the CTF described by Eq. (6.5) is the xz plane, we can see that propagation characteristics must not change if ψ were to be replaced by $-\psi$. For each angle $\psi \in [0, \psi_{\text{max}}]$, $\psi_{\text{max}} \leqslant 90°$, the SPP wavenumber q does not change greatly if ψ were to be replaced by $180° - \psi$. Thus, there is a twofold symmetry [45]. The value of ψ_{max} depends on χ_v. When $\chi_v = 90°$, the generally biaxial CTF becomes uniaxial with its sole optic ray axis aligned parallel to the z axis. SPP-wave propagation then occurs in all directions (i.e. $\psi_{\text{max}} = 90°$), as symmetry demands. At some smaller value of χ_v, the AED begins to shrink, and ψ_{max} generally decreases as χ_v decreases.

When $\psi = 0°$, the dispersion equation for SPP-wave propagation is [44]

$$q^2 \epsilon_{\text{met}} + q \alpha_{\text{met}} (\epsilon_a - \epsilon_b) \sin \chi \cos \chi$$
$$-(\alpha_{\text{met}} \alpha_{\text{diel}} + \omega^2 \mu_0 \epsilon_{\text{met}})(\epsilon_a \cos^2 \chi + \epsilon_b \sin^2 \chi) = 0. \qquad (6.8)$$

The sole SPP wave has to be p polarized and is entirely unaffected by ϵ_c.

Figure 6.2 shows the variation of $\text{Re}\{q\}/k_0$ with ψ of SPP-wave modes guided by the planar interface of bulk aluminum ($n_{\text{met}} = 1.38 + 7.61i$) and a titanium-oxide CTF described by Eqs. (6.6) and Table 6.1 for $\chi_v \in \{7.2°, 20°, 60°, 90°\}$ and $\lambda_0 = 633$ nm [45]. In general, the phase speed $v_p = \omega/\text{Re}\{q\}$ increases as χ_v decreases, but the maximum phase speed does not exceed c_0.

As the phase speed v_p increases, so does the propagation length $\Delta_{\text{prop}} = 1/\text{Im}\{q\}$ of the SPP wave; furthermore, Δ_{prop} can increase by as much as two orders of magnitude as $\psi \to \psi_{\text{max}}$. These characteristics are evident in Figure 6.3. The increase in Δ_{prop} is consistent with a concurrent weakening of the localization of the SPP wave to the interface on the CTF side. With a large fraction of the volume over which the wave has significant amplitude occurring in the partnering dielectric material wherein dissipation is negligible, increased propagation lengths are to be expected. In contrast, the strong localization on the metal side of the interface is completely insensitive to ψ.

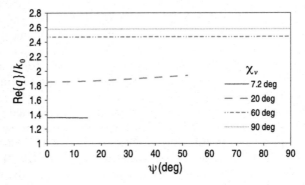

Figure 6.2 Dependence of $\text{Re}\{q\}/k_0$ on the propagation angle ψ in the xy plane of SPP-wave modes guided by the planar interface of bulk aluminum ($n_{\text{met}} = 1.38 + 7.61i$) and a titanium-oxide CTF described by Eqs. (6.6) and Table 6.1 for $\chi_v \in \{7.2°, 20°, 60°, 90°\}$ and $\lambda_0 = 633$ nm [45].

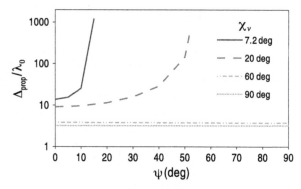

Figure 6.3 Dependence of Δ_{prop} on the propagation angle ψ in the xy plane of SPP-wave modes guided by the planar interface of bulk aluminum ($n_{\text{met}} = 1.38 + 7.61i$) and a titanium-oxide CTF described by Eqs. (6.6) and Table 6.1 for $\chi_v \in \{7.2°, 20°, 60°, 90°\}$ and $\lambda_0 = 633$ nm [45].

6.4 Interface of a Metal and a Continuously and Periodically Nonhomogeneous Dielectric Material

Nanotechnology can quite easily produce dielectric materials whose constitutive properties change continuously along one direction [111] for specific optical applications. Some, such as rugate filters with sinusoidally varying refractive index [60], have been around for several decades. Others, such as STFs [42], have been conceptualized and developed in the last two decades or so. In either case, investigations of SPP-wave propagation guided by the interface of a metal and a dielectric material whose permittivity varies continuously and periodically in the direction normal to the interface plane constitute a recent development [332]. The most remarkable result of these investigations is the discovery [54,56] of multiple SPP-wave modes guided by a single interface at a single frequency in a specific direction in the interface plane, with ramifications for improved error-free sensing, harvesting of solar energy, and communications. Each of the multiple SPP-wave modes has a distinct phase speed, propagation length, localization characterstics, and field profile. If the partnering dielectric material is also anisotropic, the SPP wave on the metal side of the interface has, in general, both s- and p-polarized components.

Detailed investigations on SPP-wave propagation have been reported for three types of continuously nonhomogeneous periodic dielectric materials as partnering dielectric materials: rugate filters [63,153,159,333], SNTFs [54,55,259,334], and chiral STFs [56,161,228,335,336]. Multiple SPP-wave modes guided by metal/SNTF interfaces and metal/chiral-STF interfaces were discovered theoretically at nearly the same time. Since both SNTFs and chiral STFs are anisotropic as well as periodically nonhomogeneous materials, it was uncertain whether anisotropy or periodicity was responsible for the existence of multiple SPP-wave modes. Perhaps, anisotropy and periodicity were together responsible.

Shortly after the experimental validations of multiple SPP-wave modes guided by metal/SNTF interfaces [57] and metal/chiral-STF interfaces [58,59], calculations for metal/rugate-filter interfaces—in the canonical [63], Turbadar-Kretschmann-Raether [333], and grating-coupled configurations [86,153]—showed that they too support multiple SPP-wave modes. The question was settled: periodicity of the partnering dielectric material in the direction normal to the interface alone engenders multiplicity of SPP-wave modes.

As experimental investigations have verified the existence of multiple SPP-wave modes at a single frequency [57–59], development of practical applications is likely to occur in the not-too-distant future. Moreover, as some liquid crystals have continuously varying permittivity dyadics of the same form as chiral STFs [337], metal/liquid-crystal interfaces could also support the propagation of multiple SPP-wave modes. While STFs afford much greater flexibility in the spatial variation of the permittivity dyadic, liquid crystals make sensitivity to an external stimulus such as voltage or pressure an enticing possibility for application.

6.4.1   Metal/Rugate-Filter Interface

A rugate filter [60,119] is commonly conceptualized as being made of an isotropic dielectric material with a refractive index that varies periodically in one direction. The periodic variation in the permittivity of a rugate filter may take on a variety of forms, but is commonly taken to be sinusoidal.

6.4.1.1   Canonical Boundary-Value Problem

The canonical boundary-value problem of SPP-wave propagation can be formulated in the same way as in Section 3.5 to determine the relevant dispersion equation. Let a periodically nonhomogeneous dielectric material labeled $\mathcal{A}$ occupy the half-space $z > 0$, while a homogeneous metal labeled $\mathcal{B}$ occupies the half-space $z < 0$, as shown in Figure 6.4. The permittivity of material $\mathcal{A}$ varies periodically so that

$$\epsilon^{\mathcal{A}}(z + 2\Omega^{\mathcal{A}}) = \epsilon^{\mathcal{A}}(z), \quad z > 0, \tag{6.9}$$

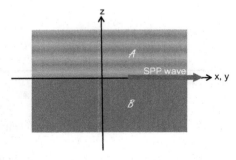

Figure 6.4 Schematic of the canonical boundary-value problem for SPP-wave propagation guided by the planar interface of a periodically nonhomogeneous dielectric material $\mathcal{A}$ and a metal $\mathcal{B}$. Material $\mathcal{A}$ can be isotropic or anisotropic and is periodically nonhomogeneous along the z axis.

with $2\Omega^{\mathcal{A}}$ as the period, whereas the permittivity $\epsilon^{\mathcal{B}}$ of the metal $\mathcal{B}$ does not depend on z.

As material $\mathcal{A}$ throughout Section 6.4.1 is a rugate filter with a sinusoidally varying refractive index, its permittivity is given by

$$\epsilon^{\mathcal{A}}(z) = \epsilon_0 \left[n_{\text{avg}}^{\mathcal{A}} + \frac{\Delta n^{\mathcal{A}}}{2} \sin\left(\frac{\pi z}{\Omega^{\mathcal{A}}}\right) \right]^2, \quad z > 0, \quad (6.10)$$

whereas the refractive index of the metal $\mathcal{B}$ is denoted by $n^{\mathcal{B}} = \sqrt{\epsilon^{\mathcal{B}}/\epsilon_0}$. Field analysis can be approached as described in Section 3.5 for propagation parallel to $\underline{u}_{\text{prop}}$. Much more easily, because both the rugate filter and the metal are isotropic, the SPP wave can be taken to propagate along the x axis without loss of generality, and the procedures of Section 5.2 can be adapted. Either way, some SPP-wave modes will turn out to be p polarized and the remainder will be s polarized.

With $n_{\text{avg}}^{\mathcal{A}}$, $\Delta n^{\mathcal{A}}$, $n^{\mathcal{B}}$, and λ_0 held fixed, the various SPP-wave modes can be tracked in relation to $\Omega^{\mathcal{A}}$ and organized into branches [63,159]. Figure 6.5 shows the real and imaginary parts of several but not all SPP wavenumbers q as functions of $\Omega^{\mathcal{A}}/\lambda_0$ organized into 19 branches, for the planar interface of bulk aluminum ($n^{\mathcal{B}} = 1.38 + 7.61i$) and a rugate filter described by Eq. (6.10) with $n_{\text{avg}}^{\mathcal{A}} = 1.885$ and $\Delta n^{\mathcal{A}} = 0.87$,

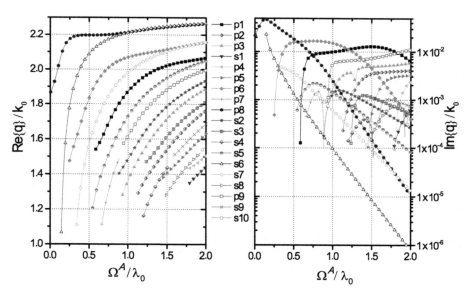

Figure 6.5 Several solutions q of the canonical boundary-value problem as functions of $\Omega^{\mathcal{A}}/\lambda_0$ for propagation guided by the planar interface of bulk aluminum and a rugate filter described by Eq. (6.10); $n_{\text{avg}}^{\mathcal{A}} = 1.885$, $\Delta n^{\mathcal{A}} = 0.87$, $n^{\mathcal{B}} = 1.38 + 7.61i$, and $\lambda_0 = 633$ nm [63,159]. These solutions are organized into 19 branches. Some solutions are not shown for clarity [159]. If q represents an SPP wave propagating in the $\underline{u}_{\text{prop}}$ direction, $-q$ represents an SPP wave propagating in the $-\underline{u}_{\text{prop}}$ direction.

when $\lambda_0 = 633$ nm. The branches labeled s1–s10 represent s-polarized SPP waves and the branches labeled p1–p9 represent p-polarized SPP waves. A few of these branches extend into the high-phase-speed regime $\mathrm{Re}\{q\}/k_0 < 1.0$ (i.e. $v_p > c_0$) [159], but the extensions are not shown in the figure for clarity.

None of the s branches intersects with a p branch, which indicates that only linearly polarized SPP waves can be guided by metal/rugate-filter interfaces. As $\Omega^{\mathcal{A}}/\lambda_0$ increases, the branches come closer to each other. At some very high value of $\Omega^{\mathcal{A}}/\lambda_0$, only one solution should exist, since the rugate filter would then be essentially homogeneous over the region in which the fields of the SPP wave have significant magnitudes [63]. Extending the calculations to very large values of $\Omega^{\mathcal{A}}$, however, leads to computational difficulties—which are absent when $\sin(\pi z/\Omega^{\mathcal{A}})$ is replaced by zero in Eq. (6.10).

Although the degree of localization varies for different SPP-wave modes, calculations show that the wave on the rugate-filter side of the interface is confined to a few periods of the rugate filter. This is good news for practical implementations because the thickness of the rugate filter is necessarily limited in practice. Figure 6.6 shows the field profiles for two different solutions on the branch labeled p8 in Figure 6.5, one at $\Omega/\lambda_0 = 0.1$ and the other at $\Omega/\lambda_0 = 1$. Likewise, Figure 6.7 shows the field profiles for two different solutions—one at $\Omega/\lambda_0 = 1$ and the other at $\Omega/\lambda_0 = 1.5$—on the branch labeled s2. In both figures, localization to the interface on the rugate-filter side is weaker at the lower value than at the higher value of Ω/λ_0.

Two types of field profiles appear to exist, as exemplified in Figures 6.6 and 6.7. For both types, the fields decrease monotonically with distance from the interface inside the metal, just as for the simple SPP wave guided by the interface of a metal and a homogeneous isotropic dielectric material. One type also has a monotonic decrease of the field magnitudes on the rugate-filter side with distance from the interface, exactly like the simple SPP wave. However, the other type has an oscillatory decrease on the rugate-filter side. Both types of field profiles are consistent with Floquet theory. Profiles of the time-average Poynting vector also available in Figures 6.6 and 6.7 provide corroboration. The different spatial profiles of different SPP-wave modes offer promise for multi-analyte SPR sensing without spatial multiplexing.

The lowest refractive index in the rugate filter equals $n_{\mathrm{avg}}^{\mathcal{A}} - \Delta n^{\mathcal{A}}/2$. Therefore, three different regimes of the phase speed can be delineated:

i. low-phase-speed regime $v_p < (n_{\mathrm{avg}}^{\mathcal{A}} - \Delta n^{\mathcal{A}}/2)^{-1}c_0$,

ii. intermediate-phase-speed regime $(n_{\mathrm{avg}}^{\mathcal{A}} - \Delta n^{\mathcal{A}}/2)^{-1}c_0 < v_p < c_0$, and

iii. high-phase-speed regime $v_p > c_0$.

Most SPP wavenumbers in Figure 6.5 lie in the low-phase-speed regime, but others lie in the intermediate-phase-speed regime. Still others lie in the high-phase-speed regime [159]—just like SPP waves guided by the interface of a metal and a Bragg mirror whose unit cell comprises two homogeneous dielectric layers of quarter-wave thickness [155, 157, 160].

The high-phase-speed SPP waves can be either p or s polarized, and can be excited by direct illumination of a metal-backed rugate filter of finite thickness. SPP waves with extremely high phase speeds have been found [159], but these do not *propagate*

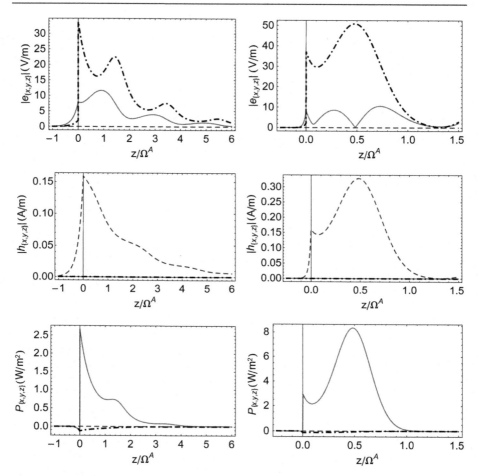

Figure 6.6 Variations with z of the magnitudes of the Cartesian components of $\underline{e}(z)$ and $\underline{h}(z)$ of two p-polarized SPP waves lying on branch p8 in Figure 6.5, when $\underline{u}_{prop} = \underline{u}_x$. Also shown are the variations of Cartesian components of the time-averaged Poynting vector $\underline{P}(x, y, z)$ with z on the line $\{x = 0, y = 0\}$. The components parallel to $\underline{u}_x, \underline{u}_y,$ and $\underline{u}_z$ are represented by red solid, blue dashed, and black chain-dashed lines, respectively. The data were computed by setting $n_{avg}^A = 1.885, \Delta n^A = 0.87, n^B = 1.38 + 7.61i$ (bulk aluminum), $\lambda_0 = 633$ nm, and $b_p/n^B = 1$ V m^{-1}. (Left) $\Omega/\lambda_0 = 0.1$ and $q/k_0 = 2.00943 + 0.04468i$, and (right) $\Omega/\lambda_0 = 1.0$ and $q/k_0 = 2.21456 + 0.00246i$ [63]. (For interpretation of the references to color in this figure legend, the reader is referred to the web version of this book.)

in the interface plane because $\text{Re}\{q\} \approx 0$. Hence, these SPP waves are useless for communications.

Just as the excitation of multiple low-phase-speed SPP waves at a fixed frequency is promising for optical sensing applications [338], the excitation of multiple SPP waves in the intermediate- and the high-phase-speed regimes is likely to be similarly useful. This

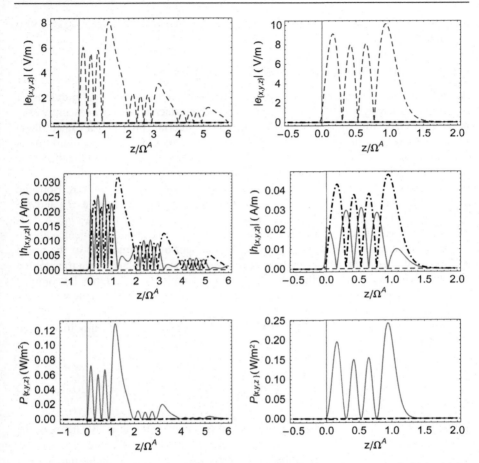

Figure 6.7 Variations with z of the magnitudes of the Cartesian components of $\underline{e}(z)$ and $\underline{h}(z)$ of two s-polarized SPP waves lying on branch s2 in Figure 6.5, when $\underline{u}_{\text{prop}} = \underline{u}_x$. Also shown are the variations of Cartesian components of the time-averaged Poynting vector $\underline{P}(x, y, z)$ with z on the line $\{x = 0, y = 0\}$. The components parallel to $\underline{u}_x, \underline{u}_y$, and $\underline{u}_z$ are represented by red solid, blue dashed, and black chain-dashed lines, respectively. The data were computed by setting $n_{\text{avg}}^{\mathcal{A}} = 1.885$, $\Delta n^{\mathcal{A}} = 0.87$, $n^{\mathcal{B}} = 1.38 + 7.61i$ (bulk aluminum), $\lambda_0 = 633$ nm, and $b_s = 1$ V m^{-1}. (left) $\Omega/\lambda_0 = 1.0$ and $q/k_0 = 1.4864 + 0.0013203i$, and (right) $\Omega/\lambda_0 = 1.5$ and $q/k_0 = 1.7873 + 0.0007801i$ [63]. (For interpretation of the references to color in this figure legend, the reader is referred to the web version of this book.)

would require the rugate filter to have sufficient porosity to be infiltrated by the analytes to be sensed. Neither the prism-coupled configuration nor the periodic corrugation of the metal/rugate-filter interface would be necessary, leading to greater integration of chemical sensors based on SPP waves in optical circuits. For harvesting solar energy, if the rugate filter is made of an appropriate semiconductor, the excitation of high-phase-speed SPP waves is expected to enhance absorption of the solar flux and therefore the generation of electron-hole pairs, as has already been noted for low-phase-speed SPP waves [84,207,208].

6.4.1.2   Turbadar-Kretschmann-Raether Configuration

An SPP wave is excited in the Turbadar-Kretschmann-Raether configuration of Figure 2.4, when light incident on the prism/metal interface excites evanescent waves in the thin metal film and its partnering dielectric material (presently a rugate filter) of finite thickness. The prism in the Turbadar-Kretschmann-Raether configuration is extremely thick compared to both the free-space wavelength of light and the thickness of the metal film, and can therefore be modeled as a semi-infinite medium.

The essence of this prism-coupled configuration is shown in Figure 6.8. Both the rugate filter (material $\mathcal{A}$) and the metal (material $\mathcal{B}$) are of finite thicknesses $d_{\mathcal{A}}$ and $d_{\mathcal{B}}$, respectively. The metal film is also in contact with a prism of refractive index $n_{\text{prism}} > 1$, whereas the rugate filter $\mathcal{A}$ is also in contact with air ($n_{\text{air}} = 1$). For theoretical analysis, a linearly polarized plane wave is taken to be incident on the interface of the prism and material $\mathcal{B}$ at an angle θ_{inc} with respect to the z axis. The plane wave transmitted into air has a wave vector inclined to the z axis at an angle θ_{tr}, which may be complex valued. As all materials are isotropic, the wave vector of the incident plane wave may be taken to lie wholly in the xz plane so that $\underline{u}_{\text{prop}} = \underline{u}_x$.

The expression for the permittivity of the rugate filter is modified from Eq. (6.10) to

$$\epsilon^{\mathcal{A}}(z) = \epsilon_0 \left\{ n_{\text{avg}}^{\mathcal{A}} + \frac{\Delta n^{\mathcal{A}}}{2} \sin\left[\frac{\pi(z - d_{\mathcal{B}})}{\Omega^{\mathcal{A}}} \right] \right\}^2, \quad z \in (d_{\mathcal{B}}, d_{\mathcal{A}} + d_{\mathcal{B}}), \quad (6.11)$$

whereas $n^{\mathcal{B}}$ still denotes the refractive index of the metal. The refractive index of the prism should be larger than the maximum refractive index of the rugate filter (i.e. $n_{\text{prism}} > n_{\text{avg}}^{\mathcal{A}} + \Delta n^{\mathcal{A}}/2$), both materials being assumed to be nondissipative. For

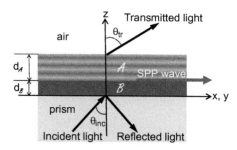

Figure 6.8 Schematic of the Turbadar-Kretschmann-Raether configuration to excite multiple SPP-wave modes guided by the planar interface of a periodically nonhomogeneous dielectric material $\mathcal{A}$ and a metal $\mathcal{B}$. Material $\mathcal{A}$ can be isotropic or anisotropic, and is periodically non-homogeneous along the z axis. For analytical simplicity, air in the half-space $z > d_{\mathcal{A}} + d_{\mathcal{B}}$ can be replaced by the prism material, provided that the layer of material $\mathcal{A}$ has a sufficiently large number of periods (i.e. it is sufficiently thick).

theoretical studies, it is convenient to always choose d_A to be an integral multiple of $2\Omega^A$; that choice is implicit throughout this chapter.

The mathematical technique described in Section 3.7 is straightforward to apply for analysis and prediction. As the integer $d_A/2\Omega^A$ increases, the boundary-value problem may yield solutions that violate the principle of conservation of energy due to the limited precision available on digital computers. The emergence of these erroneous solutions can often be circumvented by adapting the RCWA algorithm of Sections 2.3.4.5 and 3.8.6 for the rugate filter [333].

In order to investigate the excitation of SPP waves, the angle of incidence θ_{inc} has to be varied from $0°$ toward $90°$, and the linear absorptances A_p and A_s of Eqs. (3.128) are computed as functions of that angle. When a linear absorptance is plotted against θ_{inc}, several peaks may be evident. An absorptance peak whose location on the θ_{inc} axis does not vary significantly with respect to $d_A/2\Omega^A$ and d_B beyond some thresholds for both thicknesses very likely indicates the excitation of an SPP wave. Confirmation is provided by the solution of the underlying canonical boundary-value problem.

Although in an actual experiment the material that the light encounters on exiting the rugate filter would be air, this material is often replaced, for simplicity of calculation, with the same material that the prism is composed of. Then, $\theta_{tr} = \theta_{inc}$ is definitely a real angle. In spite of this replacement, the essential characteristics of SPP-wave propagation are captured. For direct comparison with experimental results, however, the exit region should be taken to be occupied by air.

Figure 6.9 shows typical results for absorptance A_p as a function of θ_{inc} for the interface of an aluminum thin film ($n^B = 0.75 + 3.9i$) and a rugate filter described by Eq. (6.11) with $n_{avg}^A = 1.885$, $\Delta n^A = 0.87$, and $\Omega^A = 1.5\lambda_0$, when $\lambda_0 = 633$ nm. The prism material is zinc selenide ($n_{prism} = 2.58$), which is taken to occupy both half-spaces $z < 0$ and $z > d_A + d_B$. The thickness $d_B = 30$ nm of the metal film is typical. Data are presented for two different thicknesses of the rugate filter. The angular

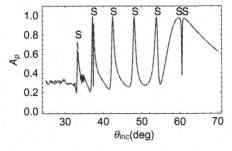

Figure 6.9 Calculated linear absorptance A_p at $\lambda_0 = 633$ nm as a function of the angle of incidence θ_{inc} for the Turbadar-Kretschmann-Raether configuration incorporating (i) a rugate filter described by Eq. (6.11) with $n_{avg}^A = 1.885$, $\Delta n^A = 0.87$, $\Omega^A = 1.5\lambda_0$, and $d_A = 2N_{per}\Omega^A$ and (ii) an aluminum thin film with $n^B = 0.75 + 3.9i$ and $d_B = 30$ nm, encased by the half-spaces $z < 0$ and $z > d_A + d_B$ occupied by zinc selenide ($n_{prism} = 2.58$). Red solid line is for $N_{per} = 3$ and blue dashed line is for $N_{per} = 4$ [333]. The letter S identifies the absorptance peaks that represent the excitation of SPP waves. (For interpretation of the references to color in this figure legend, the reader is referred to the web version of this book.)

Table 6.2 Values of θ_{inc} and $\kappa/k_0 = n_{prism} \sin \theta_{inc}$ for those absorptance peaks in Figure 6.9 that do not shift when the thickness of the rugate filter is changed [333].

θ_{inc} (deg)	κ/k_0
33.23	1.4138
37.20	1.5599
42.41	1.7400
48.01	1.9176
53.86	2.0836
59.66	2.2266
61.01	2.2567

Table 6.3 Normalized SPP wavenumbers q/k_0 of p-polarized SPP-wave modes obtained by solving the canonical boundary-value problem for the metal/rugate-filter interface with same partnering materials as in Figure 6.9 [333]. If q represents an SPP wave propagating in the $\underline{u}_{prop}$ direction, $-q$ represents an SPP wave propagating in the $-\underline{u}_{prop}$ direction.

Polarization state	q/k_0	q/k_0	q/k_0
p-pol	1.4125 + 0.0004i	1.5600 + 0.0008i	1.7401 + 0.0013i
	1.9175 + 0.0015i	2.0836 + 0.0018i	2.2302 + 0.0173i
	2.2498 + 0.0079i		

locations of seven peaks in this figure are not affected by the rugate filter having either three or four periods, suggesting that only these peaks represent excitation of SPP-wave modes rather than waveguide modes [299] which must depend on d_A.

Table 6.2 lists the angle of incidence θ_{inc} for each of these seven absorptance peaks and the corresponding wavenumber $\kappa = k_0 n_{prism} \sin \theta_{inc}$. Table 6.3 lists the SPP wavenumbers q obtained from the solution of the underlying canonical boundary-value problem for p-polarized SPP-wave modes guided by the interface of the chosen rugate filter and the aluminum thin film. The close agreement of κ and Re$\{q\}$ in these two tables confirms that the seven absorptance peaks identified in Figure 6.9 do indeed indicate the excitation of seven distinct SPP-wave modes.

The counterpart of Figure 6.9 for the excitation of s-polarized SPP-wave modes is Figure 6.10, wherein the absorptance A_s is plotted against θ_{inc}. The angular locations of five peaks in this figure are not affected by the number of periods of the rugate filter, which indicates that these peaks represent excitation of SPP-wave modes. The values of θ_{inc} and the corresponding wavenumber $\kappa = k_0 n_{prism} \sin \theta_{inc}$ are listed in Table 6.4. Table 6.5 lists the SPP wavenumbers q obtained from the solution of the underlying canonical boundary-value problem for s-polarized SPP-wave modes guided by the interface of the chosen rugate filter and the aluminum thin film. Again, the close agreement of κ and Re$\{q\}$ in these two tables confirms that the five absorptance peaks identified in Figure 6.10 do indeed indicate the excitation of five distinct s-polarized SPP-wave modes.

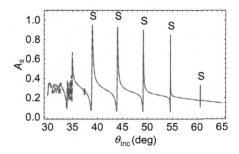

Figure 6.10 Same as Figure 6.9, except A_s is plotted instead of A_p. Although the absorptance peak at $\theta_{inc} \simeq 35°$ appears to be independent of the number of periods N_{per} of the rugate filter, this peak shifts significantly when N_{per} increases beyond 4; therefore, it does not represent the excitation of an SPP-wave mode [333].

Table 6.4 Values of θ_{inc} and $\kappa/k_0 = n_{prism} \sin \theta_{inc}$ for those absorptance peaks in Figure 6.10 that do not shift when the thickness of the rugate filter is changed [333].

θ_{inc} (deg)	κ/k_0
38.97	1.6226
44.01	1.7925
49.22	1.9536
54.63	2.1038
60.66	2.2491

Table 6.5 Normalized SPP wavenumbers q/k_0 of s-polarized SPP-wave modes obtained by solving the canonical boundary-value problem for the metal/rugate-filter interface with the same partnering materials as in Figure 6.10 [333]. If q represents an SPP-wave mode propagating in the $\underline{u}_{prop}$ direction, $-q$ represents the same SPP-wave mode propagating in the $-\underline{u}_{prop}$ direction.

Polarization state	q/k_0	q/k_0	q/k_0
s-pol	$1.6226 + 0.0010i$	$1.7924 + 0.0008i$	$1.9534 + 0.0004i$
	$2.1038 + 0.0001i$	$2.2490 + 1.0136 \times 10^{-5}i$	

Thus, all 12 SPP-wave modes found from examining the absorptance peaks in Figures 6.9 and 6.10 for the Turbadar-Kretschmann-Raether configuration could have been predicted after solving the canonical boundary-value problem. Nevertheless, some SPP-wave modes are so weakly bound to the metal/rugate-filter interface that it is not possible to observe them as absorptance peaks when the number of periods of the rugate filter is not large [159]. Thick rugate filters are, however, impractical. Furthermore, almost all SPP-wave modes are excited with maximal energy when the metal film is approximately equal to the skin depth $[k_0 \text{Im}\{n^B\}]^{-1}$ of the metal [333]. This thickness of the metal film would also be sufficient to minimize the coupling of any

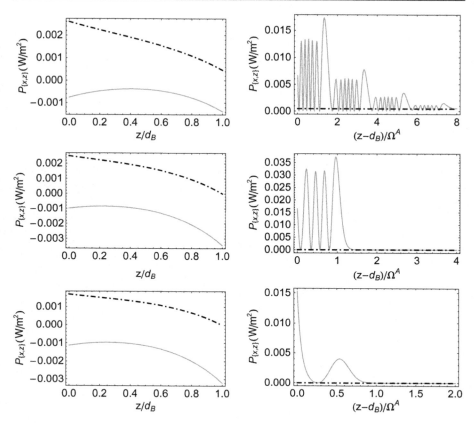

Figure 6.11 Variations with z of the non-zero Cartesian components of the time-averaged Poynting vector in (left) the metal film and (right) the rugate filter for three absorptance peaks in Figure 6.9 identified with the excitation of p-polarized SPP-wave modes at (top) $\theta_{inc} = 32.23°$, (middle) $\theta_{inc} = 42.42°$, and (bottom) $\theta_{inc} = 59.66°$, when $\psi = 0°$ and $\lambda_0 = 633$ nm. The components parallel to $\underline{u}_x$ and $\underline{u}_z$ are represented by red solid and black chain-dashed lines, respectively. The amplitude of the incident electric field is 1 V m^{-1} [333]. (For interpretation of the references to color in this figure legend, the reader is referred to the web version of this book.)

SPP-wave modes guided solely by the metal/prism interface with the ones guided solely by the metal/rugate-filter interface [297,333].

Different SPP-wave modes have different field profiles in the canonical problem, as evidenced by Figures 6.6 and 6.7. The same is true in any prism-coupled configuration. As an example, Figures 6.11 and 6.12 show the variations of the the non-zero Cartesian components of the time-averaged Poynting vector as functions of distance from the metal/rugate-filter interface on either side of the interface for three p-polarized and two s-polarized SPP-wave modes, respectively. In the bottom half of Figure 6.12, the rise in magnitude of both P_x and P_z as z approaches the metal/prism interface $z = 0$

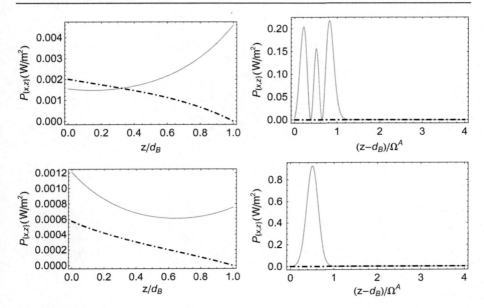

Figure 6.12 Variations with z of the non-zero Cartesian components of the time-averaged Poynting vector in (left) the metal film and (right) the rugate filter for two absorptance peaks in Figure 6.10 identified with the excitation of s-polarized SPP-wave modes at (top) $\theta_{\mathrm{inc}} = 49.22°$, and (bottom) $\theta_{\mathrm{inc}} = 60.66°$, when $\psi = 0°$ and $\lambda_0 = 633$ nm. The components parallel to $\underline{u}_x$ and $\underline{u}_z$ are represented by red solid and black chain-dashed lines, respectively. The magnitude of the incident electric field is 1 V m^{-1} [333]. (For interpretation of the references to color in this figure legend, the reader is referred to the web version of this book.)

indicates coupling between SPP-wave modes guided independently by the metal/prism and the metal/rugate-filter interfaces. But the coupling is weak because the metal film is 30-nm thick whereas the skin depth in the metal is 24.91 nm [333].

6.4.1.3   Grating-Coupled Configuration

As discussed in Section 1.3.2.2, surface-relief gratings are sometimes used for exciting SPP waves [3,4], thereby avoiding the use of a prism. The partnering dielectric material has been reported to be homogeneous in experimental situations so far. But very recently, this configuration has been successfully employed with a periodic multilayer in that role, for the excitation of multiple p- and s-polarized SPP-wave modes over the entire visible regime [339].

As shown in Figure 6.13, the half-spaces $z < 0$ and $z > d_3$ are occupied by air ($n_{\mathrm{air}} = 1$), the region $0 < z \leqslant d_1$ is occupied by a dielectric material $\mathcal{A}$ of permittivity $\underline{\epsilon}^{\mathcal{A}}(z)$, and the region $d_2 \leqslant z < d_3$ by a metal $\mathcal{B}$ of permittivity $\epsilon^{\mathcal{B}}$. The region $d_1 < z < d_2$ contains a periodically undulating interface with period L_x along the x axis. The permittivity $\underline{\underline{\epsilon}}_g(x, z) = \underline{\underline{\epsilon}}_g(x \pm L_x, z)$ in the reference unit cell $0 < x < L_x$

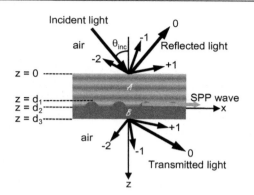

Figure 6.13 Schematic of the grating-coupled configuration to excite SPP-wave propagation guided by the periodically corrugated interface of a periodically nonhomogeneous dielectric material $\mathcal{A}$ and a metal $\mathcal{B}$. Material $\mathcal{A}$ can be isotropic or anisotropic, and is periodically nonhomogeneous along the z axis. The incident plane wave propagates in the xz plane, so that the method of Section 3.8 can be applied with $\psi = 0°$ and $n \in \{0\}$.

of this region is expressed as

$$\underline{\underline{\epsilon}}_g(x, z) = \begin{cases} \epsilon^{\mathcal{B}}\underline{\underline{I}} - \left[\epsilon^{\mathcal{B}}\underline{\underline{I}} - \underline{\underline{\epsilon}}^{\mathcal{A}}(z)\right] \mathcal{U}[d_2 - z - w(x)], & x \in (0, L_1), \\ \underline{\underline{\epsilon}}^{\mathcal{A}}(z), & x \in (L_1, L_x), \end{cases} \tag{6.12}$$

with $\underline{\underline{I}}$ as the identity dyadic, $\mathcal{U}(\sigma)$ as the unit step function, and

$$w(x) = (d_2 - d_1) \sin\left(\frac{\pi x}{L_1}\right), \quad L_1 \in (0, L_x). \tag{6.13}$$

Functions other than the sinusoid can be used for $w(x)$, subject to the restrictions mentioned in Section 3.8. Thus the grating depicted in Figure 6.13 is singly periodic and the grating plane is the xz plane. The ratio $d_1/2\Omega^{\mathcal{A}}$ is taken to be an integer throughout this chapter, unless otherwise specified.

A linearly polarized plane wave, propagating in the half-space $z < 0$ at an angle $\theta_{\text{inc}} \in [0°, 90°)$ to the z axis and at an angle $\psi = 0°$ to the x axis in the xy plane, is taken to be incident on the plane $z = 0$. The field phasors of the incident plane wave are described by Eqs. (3.141) with $n \in \{0\}$ and $n_1 = n_{\text{air}}$. Linear absorptances A_p and A_s defined in Eqs. (3.151) can be calculated after using the RCWA described in Section 3.8.5.

Illustrative numerical results are presented in Figure 6.14 at $\lambda_0 = 633$ nm for the periodically corrugated interface of bulk aluminum ($n^{\mathcal{B}} = 1.38 + 7.61i$) and a rugate filter described by

$$\underline{\underline{\epsilon}}^{\mathcal{A}}(z) = \epsilon_0 \underline{\underline{I}} \left\{ n_{\text{avg}}^{\mathcal{A}} + \frac{\Delta n^{\mathcal{A}}}{2} \sin\left[\frac{\pi(d_2 - z)}{\Omega^{\mathcal{A}}}\right] \right\}^2, \tag{6.14}$$

with $n_{\text{avg}}^{\mathcal{A}} = 1.885$, $\Delta n^{\mathcal{A}} = 0.87$, and $\Omega^{\mathcal{A}} = \lambda_0$ [153]. Calculated linear absorptances A_p and A_s are plotted as functions of θ_{inc}, for three values of the thickness of the rugate

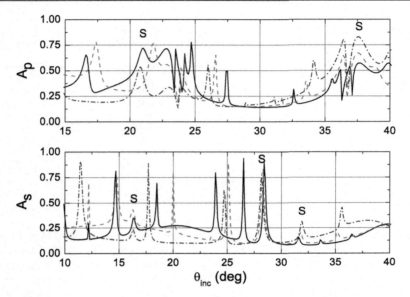

Figure 6.14 Calculated linear absorptances A_p and A_s as functions of the angle of incidence θ_{inc}, when $\psi = 0°$ and $\lambda_0 = 633$ nm, for the grating-coupled configuration of Figure 6.13. Material $\mathcal{A}$ is the rugate filter described by Eq. (6.14) with $n_{\text{avg}}^{\mathcal{A}} = 1.885$, $\Delta n^{\mathcal{A}} = 0.87$, and $\Omega^{\mathcal{A}} = \lambda_0$. Material $\mathcal{B}$ is bulk aluminum with refractive index $n^{\mathcal{B}} = 1.38 + 7.61i$. The grating is described via Eqs. (6.12) and (6.13), with $d_3 - d_2 = 30$ nm, $d_2 - d_1 = 50$ nm, $L_1 = 0.5L_x$, and $L_x = \Omega^{\mathcal{A}}$. Three values of d_1 are used: $4\Omega^{\mathcal{A}}$ (blue chain-dashed lines), $5\Omega^{\mathcal{A}}$ (red dashed lines), and $6\Omega^{\mathcal{A}}$ (black solid lines). The letter S identifies the absorptance peaks that represent the excitation of SPP waves. (For interpretation of the references to color in this figure legend, the reader is referred to the web version of this book.)

filter above the grating: $d_1 \in \{4\Omega^{\mathcal{A}}, 5\Omega^{\mathcal{A}}, 6\Omega^{\mathcal{A}}\}$. Although numerous absorptance peaks are evident, most do not represent the excitation of an SPP wave, since they shift to different values of θ_{inc} as the integer $d_1/2\Omega^{\mathcal{A}}$ varies. Any peak which represents the excitation of an SPP wave does not shift significantly on the θ_{inc} axis as the integer $d_1/2\Omega^{\mathcal{A}}$ increases beyond a threshold, and is identified by the letter S in the figure.

The wavenumbers $k_{xy}^{(m,0)}/k_0$ of linear Floquet harmonics of order $(m, 0)$, at the absorptance peaks marked by S in Figure 6.14, are tabulated in Table 6.6 for $m \in [-2, 2]$. Table 6.7 lists the SPP wavenumbers q obtained from the solution of the underlying canonical boundary-value problem for SPP-wave modes guided by the interface of the chosen rugate filter and bulk aluminum.

The A_p-peak at $\theta_{\text{inc}} = 37.7°$ in Figure 6.14 has $k_x^{(1,0)}/k_0 = 1.6115$, according to Table 6.6. This number is quite close to $\text{Re}\{q\}/k_0 = 1.61782$ in Table 6.7. Hence, this A_p-peak represents the excitation of a p-polarized SPP wave as a Floquet harmonic of order $n = 1$. Every peak identified by S in Figure 6.14 is able to satisfy a similar test, thereby indicating that every SPP-wave mode is excited as a linear Floquet harmonic in the grating-coupled configuration.

Table 6.6 Relative wavenumbers $k_{xy}^{(m,n)}/k_0$ of linear Floquet harmonics of order (m, n), $m \in [-2, 2]$ and $n = 0$, at the absorptance peaks marked by S in Figure 6.14. Boldface entries signify SPP waves [153].

θ_{inc} (deg)	$k_{xy}^{(m,0)}/k_0$				
	$m = -2$	$m = -1$	$m = 0$	$m = 1$	$m = 2$
16.3	**−1.7210**	−0.7210	0.2790	1.2790	2.2790
21.0	−1.6416	−0.6416	0.3584	**1.3584**	2.3584
28.4	−1.5244	−0.5244	0.4756	**1.4756**	2.4756
31.6	**−1.4760**	−0.4760	0.5240	1.5240	2.5240
37.7	−1.3885	−0.3885	0.6115	**1.6115**	2.6115

Table 6.7 Normalized SPP wavenumbers q/k_0 of SPP-wave modes obtained by solving the underlying canonical boundary-value problem for the metal/rugate-filter interface withe same partnering materials as in Figure 6.14 [153]. If q represents an SPP-wave mode propagating in the $\underline{u}_{prop}$ direction, $-q$ represents the same SPP-wave mode propagating in the $-\underline{u}_{prop}$ direction.

Polarization state	q/k_0	q/k_0	q/k_0
p-pol	$1.36479 + 0.00169i$	$1.61782 + 0.00548i$	$1.87437 + 0.00998i$
	$2.06995 + 0.01526i$	$2.21456 + 0.00246i$	
s-pol	$1.48639 + 0.00132i$	$1.7324 + 0.0014i$	

Indeed, it is the panoply of linear Floquet harmonics that facilitates the excitation of multiple SPP-wave modes in the grating-coupled configuration. This abundance is manifested, in some cases, by the excitation of a single SPP-wave mode by light incident at two different angles θ_{inc}, but the angle of incidence for which the linear Floquet harmonic is of a lower order leads to the more efficient excitation. An example is furnished by Figure 6.14. The A_s-peak at $\theta_{inc} = 28.4°$ represents the excitation of an SPP-wave mode as the linear Floquet harmonic of order $(1, 0)$ because $k_{xy}^{(1,0)} = 1.4756k_0$ (Table 6.6) is then close to $\text{Re}\{q\} = 1.48639$ (Table 6.7), while the A_s-peak at $\theta_{inc} = 31.6°$ represents the excitation of the same SPP-wave mode as the linear Floquet harmonic of order $(-2, 0)$ because $k_{xy}^{(-2,0)} = -1.4760k_0$ (Table 6.6) is then close to $\text{Re}\{q\} = -1.48639$ (Table 6.7). Although the field profiles for the two absorptance peaks are scaled mirror images of each other, the excitation of the s-polarized SPP-wave mode at $\theta_{inc} = 31.6°$ is less efficient—i.e. it has a lower A_s-peak—than the excitation of the same mode at $\theta_{inc} = 28.4°$ [153].

The number of available SPP-wave modes actually excitable in the grating-coupled configuration depends crucially on the grating period L_x. For that reason, a compound surface-relief grating can help excite many more SPP-wave modes than a single simple surface-relief grating [86]. Each period of a compound grating comprises two or more simple gratings, each several periods long.

Finally, a comparison of absorptance peaks leads to two useful conclusions: (i) A_p-peaks are wider than A_s-peaks on the θ_{inc} axis, and (ii) SPP-wave modes with higher phase speeds have narrower absorptance peaks.

6.4.2 Metal/SNTF Interface

The morphology of an SNTF [42] is engineered during PVD simply by rocking the substrate in order to change χ_v dynamically. By varying the way in which the substrate is rocked, a seemingly endless variety of SNTFs, some unidirectionally periodic but not others, could be fabricated. An example of simple SNTFs with periodic nonhomogeneity along the thickness direction (i.e. the z axis) is the one with chevronic morphology shown in Figure 1.9. Maintenance of a fixed orientation of the substrate for half of the temporal period and the opposite orientation for the remainder of the temporal period during fabrication results in a periodic variation of the tilt angle χ between two values, one of which is the negative of the other. Continuous rocking of the substrate would make χ vary smoothly.

Thus, χ becomes $\chi(z)$ for SNTFs, and it is also appropriate to consider the vapor flux angle as a function of z. Since the permittivity of a CTF depends on the vapor flux angle, as discussed in Section 6.3.2, the permittivity of an SNTF is a straightforward modification of Eq. (6.5); thus,

$$\underline{\underline{\epsilon}}_{SNTF}(z) = \underline{\underline{S}}_y[\chi(z)] \cdot [\epsilon_a(z)\underline{u}_z\underline{u}_z + \epsilon_b(z)\underline{u}_x\underline{u}_x + \epsilon_c(z)\underline{u}_y\underline{u}_y] \cdot \underline{\underline{S}}_y^{-1}[\chi(z)],$$

(6.15)

where $\chi(z)$ and $\epsilon_{a,b,c}(z)$ must depend on $\chi_v(z)$.

Experimental data on $\chi(z)$, $\epsilon_a(z)$, $\epsilon_b(z)$, and $\epsilon_c(z)$ of SNTFs do not exist. Therefore, although expressions such as Eqs. (6.6) and (6.7) characterize the permittivity of CTFs, it may be assumed for initial studies that such relationships are reasonable estimates of the local permittivity of an SNTF. Recent evidence with optical filters now indicates that a refined optical characterization of SNTFs should be a new research avenue [340].

Calculations have been performed to determine the characteristics of SPP-wave propagation guided by the interface of a metal and a periodically nonhomogeneous SNTF for: the canonical boundary-value problem [334], the Turbadar-Kretschmann-Raether configuration [54,55], and the grating-coupled configuration [259]. All of these theoretical reports used

$$\left. \begin{array}{l} \epsilon_a(z) = \epsilon_0 \left\{ n_{a0} + n_{a1}\chi_v^{norm}(z) + n_{a2}\left[\chi_v^{norm}(z)\right]^2 \right\}^2 \\[2mm] \epsilon_b(z) = \epsilon_0 \left\{ n_{b0} + n_{b1}\chi_v^{norm}(z) + n_{b2}\left[\chi_v^{norm}(z)\right]^2 \right\}^2 \\[2mm] \epsilon_c(z) = \epsilon_0 \left\{ n_{c0} + n_{c1}\chi_v^{norm}(z) + n_{c2}\left[\chi_v^{norm}(z)\right]^2 \right\}^2 \\[2mm] \chi(z) = \tan^{-1}[a\tan\chi_v(z)] \end{array} \right\},$$

(6.16)

with $\chi_v^{norm}(z) = 2\chi_v(z)/\pi$ and Table 6.1 providing the parameters $n_{j\ell}$ and a for SNTFs made from three different oxides. Moreover, only SNTFs resulting from a sinusoidal

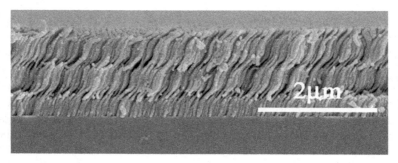

Figure 6.15 Cross-sectional image of an SNTF on a scanning electron microscope. This SNTF was fabricated by evaporating silicon oxide with a sinusoidal variation of the vapor flux angle. Courtesy: Yi-Jun Jen, National Taipei University of Technology.

variation of $\chi_v(z)$ were considered. In particular,

$$\chi_v(z) = \tilde{\chi}_v + \delta_v \sin\left(\frac{\pi z}{\Omega}\right), \tag{6.17}$$

where $\tilde{\chi}_v$ is the average value, δ_v is the modulation amplitude, and 2Ω is the period. Without loss of generality, the SNTF was oriented so that its morphologically significant plane is the xz plane.

The sole experimental investigation reported was conducted on the interface of an aluminum thin film and a silicon-oxide SNTF in the Turbadar-Kretschmann-Raether configuration [57]. The dynamically varying vapor flux angle followed Eq. (6.17). Figure 6.15 shows an SEM image of an SNTF of the type used.

6.4.2.1 Canonical Boundary-Value Problem

The formulation and solution of the canonical boundary-value problem [334] for the metal/SNTF interface follows the 4×4-matrix methodology of Section 3.5, with $\underline{u}_{\text{prop}} = \underline{u}_x \cos \psi + \underline{u}_y \sin \psi$. However, when the SPP wave propagates along the

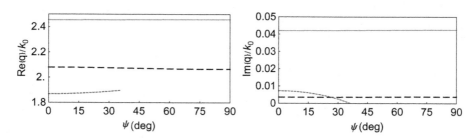

Figure 6.16 Solutions q of the canonical boundary-value problem as functions of ψ for SPP-wave propagation at $\lambda_0 = 633$ nm guided by the interface of bulk aluminum ($n^{\mathcal{B}} = 1.38 + 7.61i$) and a periodically nonhomogeneous titanium-oxide SNTF. The SNTF is described by Eqs. (6.16) and (6.18), along with data from Table 6.1, $\Omega^{\mathcal{A}} = 200$ nm, $\tilde{\chi}_v^{\mathcal{A}} = 45°$, and $\delta_v^{\mathcal{A}} = 30°$ [334].

x axis, two autonomous 2×2-matrix methodologies for p- and s-polarized SPP waves can be devised from the 4×4-matrix methodology.

In the configuration shown in Figure 6.4, let material $\mathcal{A}$ be an SNTF and material $\mathcal{B}$ a metal. Calculations for the interface of bulk aluminum ($n^{\mathcal{B}} = 1.38 + 7.61i$) and a

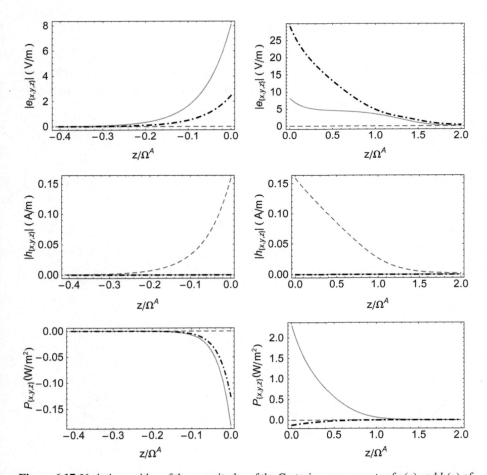

Figure 6.17 Variations with z of the magnitudes of the Cartesian components of $\underline{e}(z)$ and $\underline{h}(z)$ of a p-polarized SPP-wave mode guided by the planar interface $z = 0$ of a metal and a periodically nonhomogeneous titanium-oxide SNTF, when $\psi = 0°$ so that $\underline{u}_{\text{prop}} = \underline{u}_x$. Also shown are the variations of Cartesian components of the time-averaged Poynting vector $\underline{P}(x, y, z)$ with z on the line $\{x = 0, y = 0\}$. The components parallel to $\underline{u}_x, \underline{u}_y$, and $\underline{u}_z$ are represented by red solid, blue dashed, and black chain-dashed lines, respectively. Plots in the left column are for variations on the metal side of the interface, whereas those on the right are for variations on the SNTF side of the interface. The metal is bulk aluminum ($n^{\mathcal{B}} = 1.38 + 7.61i$). The SNTF is described by Eqs. (6.16) and (6.18), along with data from Table 6.1, $\Omega^{\mathcal{A}} = 200$ nm, $\tilde{\chi}_v^{\mathcal{A}} = 45°$, and $\delta_v^{\mathcal{A}} = 30°$. The data were computed by setting $b_p/n^{\mathcal{B}} = 1$ V m^{-1} and $q/k_0 = 2.455 + 0.04208i$ [334]. (For interpretation of the references to color in this figure legend, the reader is referred to the web version of this book.)

titanium-oxide SNTF revealed three different SPP-wave modes [334] at $\lambda_0 = 633$ nm. The SNTF was described by

$$\left.\begin{aligned}
\underline{\underline{\epsilon}}^{\mathcal{A}}(z) &= \underline{\underline{S}}_y[\chi^{\mathcal{A}}(z)] \cdot \left[\epsilon_a^{\mathcal{A}}(z)\underline{u}_z\underline{u}_z + \epsilon_b^{\mathcal{A}}(z)\underline{u}_x\underline{u}_x \right.\\
&\left. + \epsilon_c^{\mathcal{A}}(z)\underline{u}_y\underline{u}_y\right] \cdot \underline{\underline{S}}_y^{-1}[\chi^{\mathcal{A}}(z)] \\
\chi_v^{\mathcal{A}}(z) &= \tilde{\chi}_v^{\mathcal{A}} + \delta_v^{\mathcal{A}} \sin\left(\frac{\pi z}{\Omega^{\mathcal{A}}}\right)
\end{aligned}\right\}, \quad z > 0, \tag{6.18}$$

and Eqs. (6.16) were used to express $\epsilon_{a,b,c}^{\mathcal{A}}(z)$ and $\chi^{\mathcal{A}}(z)$ in terms of $\chi_v^{\mathcal{A}}(z)$, along with $\Omega^{\mathcal{A}} = 200$ nm, $\tilde{\chi}_v^{\mathcal{A}} = 45°$, and $\delta_v^{\mathcal{A}} = 30°$. Of course, the SNTF is periodically nonhomogeneous along the z axis.

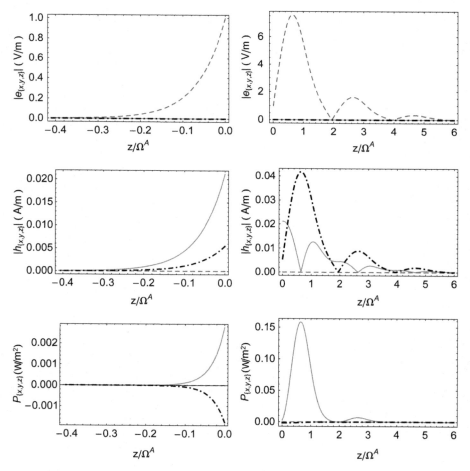

Figure 6.18 Same as Figure 6.17, except that an s-polarized SPP-wave mode is represented with $b_s = 1$ V m^{-1} and $q/k_0 = 2.080 + 0.003538i$ [334].

Figure 6.16 shows the real and imaginary parts of the SPP wavenumber q as a function of $\psi \in [0°, 90°]$. There is twofold symmetry, so that the same SPP wavenumbers exist for $\pm\psi$. For $0° \leqslant \psi \lesssim 36°$ there are three SPP-wave modes, while for $36° \lesssim \psi \leqslant 90°$ only two SPP-wave modes exist. Thus, SPP-wave propagation occurs in all possible directions parallel to the interface plane and the AED is the maximum possible. The imaginary part of the wavenumber determines the propagation length Δ_{prop} according to Eq. (2.38). Information on Δ_{prop} cannot be obtained from calculations for the prism-coupled and grating-coupled configurations.

Spatial profiles of the fields and the time-averaged Poynting vector for the three modes when $\psi = 0°$ are available in Figures 6.17–6.19. Clearly, the SPP-wave modes in Figures 6.17 and 6.19 are p polarized, whereas the mode in Figure 6.18 is s polarized. While the fields of one SPP-wave mode have maximum magnitudes at the plane $z = 0+$ on the SNTF side of the interface, the fields of the two other SPP-wave modes have

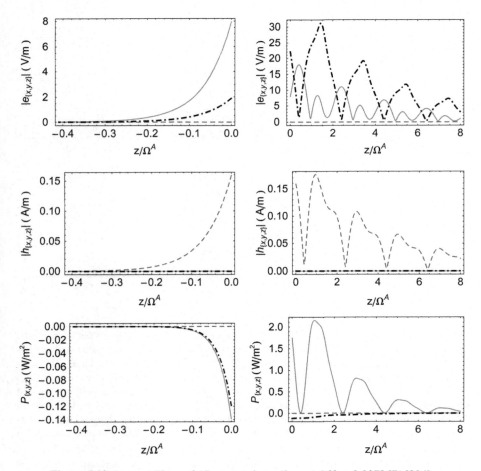

Figure 6.19 Same as Figure 6.17, except that $q/k_0 = 1.868 + 0.007267i$ [334].

maximum magnitudes at different distances into the SNTF. These characteristics are
the same as in Section 6.4.1 for the interface of a metal and a rugate filter. The different
spatial profiles of different SPP-wave modes are very promising for multi-analyte SPR
sensing without spatial multiplexing.

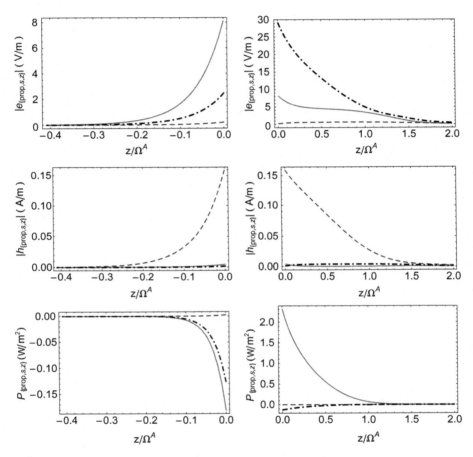

Figure 6.20 Variations with z of the magnitudes of the Cartesian components of $\underline{e}(z)$ and
$\underline{h}(z)$ of an SPP-wave mode guided by the planar interface $z = 0$ of a metal and a periodically nonhomogeneous titanium-oxide SNTF, when $\psi = 75°$. Also shown are the variations of Cartesian components of the time-averaged Poynting vector $\underline{P}(x, y, z)$ with z on
the line $\{x = 0, y = 0\}$. The components parallel to $\underline{u}_{prop}$, $\underline{u}_s$, and $\underline{u}_z$ are represented by red
solid, blue dashed, and black chain-dashed lines, respectively. The metal is bulk aluminum
($n^{\mathcal{B}} = 1.38 + 7.61i$). The SNTF is described by Eqs. (6.16) and (6.18), along with data from
Table 6.1, $\Omega^{\mathcal{A}} = 200$ nm, $\tilde{\chi}_v^{\mathcal{A}} = 45°$, and $\delta_v^{\mathcal{A}} = 30°$. The data were computed by setting
$b_p/n^{\mathcal{B}} = 1$ V m^{-1}, $b_s = 0.1919 - 0.0429i$ V m^{-1}, and $q/k_0 = 2.459 + 0.04247i$ [334]. (For
interpretation of the references to color in this figure legend, the reader is referred to the web
version of this book.)

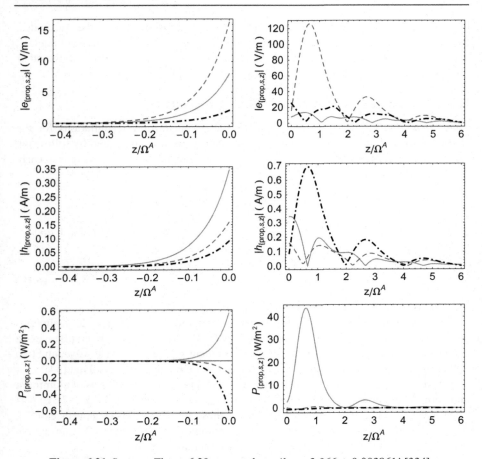

Figure 6.21 Same as Figure 6.20, except that $q/k_0 = 2.066 + 0.003861i$ [334].

Distinct polarization states cannot be defined for SPP-wave modes when $\psi \in (0°, 90°]$. This can be deduced from the spatial profiles shown in Figures 6.20 and 6.21 for the two possible modes when $\psi = 75°$. However, while the fields of one SPP-wave mode have maximum magnitudes on the SNTF side of the interface right at the interface itself, the fields of the other SPP-wave mode have maximum magnitudes quite deep into the SNTF. Again, this difference is important for multi-analyte SPR sensing.

Even more SPP-wave modes are possible [297] when a thin metal film is inserted between two identical SNTFs which are periodically nonhomogeneous and semi-infinitely thick. Such problems can be theoretically treated using the methodology of Section 3.6. If the metal film is much thicker than the skin depth in the metal, SPP-wave modes exist on each metal/SNTF interface independently and their number is the same as for a single metal/SNTF interface. When the metal film is made thinner, however, the number of SPP-wave modes increases as the two metal/SNTF interfaces begin to interact with each other. In addition to an increase in the number of modes, the phase speeds of the SPP-wave modes decrease as the metal film is made thinner. If, additionally, one of the two SNTFs is rotated about the z axis with respect to the other SNTF, the number of SPP-wave modes can be affected further [298].

6.4.2.2   Turbadar-Kretschmann-Raether Configuration

Theoretical analysis and results. As the SNTF is anisotropic, calculations for the Turbadar-Kretschmann-Raether configuration of Figure 6.8 must generally follow the full 4×4-matrix methodology of Section 3.7, even if the incident plane wave is linearly polarized [55]. Instead of Eqs. (6.18),

$$
\left.
\begin{aligned}
\underline{\underline{\epsilon}}^{\mathcal{A}}(z) &= \underline{\underline{S}}_y \left[\chi^{\mathcal{A}}(z - d_{\mathcal{B}}) \right] \cdot \left[\epsilon_a^{\mathcal{A}}(z - d_{\mathcal{B}}) \underline{u}_z \underline{u}_z \right. \\
&\quad + \epsilon_b^{\mathcal{A}}(z - d_{\mathcal{B}}) \underline{u}_x \underline{u}_x + \epsilon_c^{\mathcal{A}}(z - d_{\mathcal{B}}) \underline{u}_y \underline{u}_y \right] \\
&\quad \cdot \underline{\underline{S}}_y^{-1} \left[\chi^{\mathcal{A}}(z - d_{\mathcal{B}}) \right] \\
\chi_v^{\mathcal{A}}(z) &= \tilde{\chi}_v^{\mathcal{A}} + \delta_v^{\mathcal{A}} \sin\left(\frac{\pi z}{\Omega^{\mathcal{A}}} \right)
\end{aligned}
\right\}, \quad z \in (d_{\mathcal{B}}, d_{\mathcal{A}} + d_{\mathcal{B}}),
$$

(6.19)

must be used for the periodically nonhomogeneous SNTF.

An exception is possible only for propagation parallel to the morphologically significant plane of the SNTF, when the treatment for an incident p-polarized plane wave can be completely separated from that for an incident s-polarized plane wave, and two autonomous systems involving 2×2 matrixes can be independently solved [54].

Linear absorptances A_p and A_s have been calculated for the interfaces of aluminum and periodically nonhomogeneous SNTFs for four different SNTFs [54,55,57]. Experimental results have been reported for one of these four metal/SNTF interfaces [57]. All investigations are in accord with the main finding from Section 6.4.2.1: more than one SPP-wave modes can be guided by a single metal/SNTF interface.

Figures 6.22 and 6.23 show plots of the linear absorptances A_p and A_s as functions of the angle of incidence θ_{inc}, when $\psi = 0°$ and $\lambda_0 = 633$ nm. Material $\mathcal{A}$ is a titanium-oxide SNTF described by Eqs. (6.16) and (6.19), along with data from Table 6.1, $\Omega^{\mathcal{A}} = 200$ nm, $\tilde{\chi}_v^{\mathcal{A}} = \delta_v^{\mathcal{A}} = 30°$, and $d_{\mathcal{A}} = 4\Omega^{\mathcal{A}}$. Material $\mathcal{B}$ is bulk aluminum of thickness $d_{\mathcal{B}} \in \{1, 5, 10, 20, 30, 40\}$ nm. The prism material is zinc selenide ($n_{\text{prism}} = 2.58$), which is taken to occupy both half-spaces $z < 0$ and $z > d_{\mathcal{A}} + d_{\mathcal{B}}$.

Three SPP-wave modes are evident in Figures 6.22 and 6.23 as two A_p-peaks at $\theta_{\text{inc}} \simeq 37°$ and $\theta_{\text{inc}} \simeq 66°$, and one A_s-peak at $\theta_{\text{inc}} \simeq 50°$. Calculations with higher values of the integer $d_{\mathcal{A}}/2\Omega^{\mathcal{A}}$ did not appreciably change the locations of these peaks on the θ_{inc} axis [55], thereby discounting the possibilities of the excitation of waveguide modes [299]. Furthermore, for each peak the wavenumber $k_0 n_{\text{prism}} \sin \theta_{\text{inc}}$ turned out to be very close to the real part of the SPP wavenumber q emerging from the solution of the underlying canonical boundary-value problem.

The two SPP-wave modes in Figure 6.22 are excitable with incident p-polarized light, while the sole SPP-wave mode in Figure 6.23 is excitable with incident s-polarized light. The efficiency of excitation, as gauged by the maximum absorptance, varies with the thickness $d_{\mathcal{B}}$ of the metal film. The optimum value of $d_{\mathcal{B}}$ depends on the choice

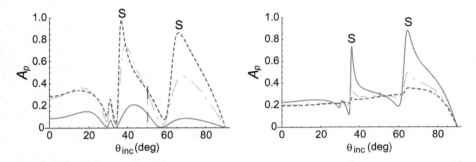

Figure 6.22 Calculated linear absorptance A_p as a function of the angle of incidence θ_{inc} and $\psi = 0°$, at $\lambda_0 = 633$ nm, when bulk aluminum ($n^B = 1.38 + 7.61i$) and a periodically nonhomogeneous titanium-oxide SNTF are incorporated in the Turbadar-Kretschmann-Raether configuration of Figure 6.8. Materials $\mathcal{A}$ and $\mathcal{B}$ are encased by the half-spaces $z < 0$ and $z > d_\mathcal{A} + d_\mathcal{B}$ occupied by zinc selenide ($n_{prism} = 2.58$). The SNTF is described by Eqs. (6.16) and (6.19), along with data from Table 6.1, $\Omega^\mathcal{A} = 200$ nm, and $\tilde{\chi}_v^\mathcal{A} = \delta_v^\mathcal{A} = 30°$. Whereas $d_\mathcal{A} = 4\Omega^\mathcal{A}$, $d_\mathcal{B}$ is variable. (left) $d_\mathcal{B} = 1$ nm (red solid line), $d_\mathcal{B} = 5$ nm (green irregularly dashed line), and $d_\mathcal{B} = 10$ nm (blue dashed line); (right) $d_\mathcal{B} = 20$ nm (red solid line), $d_\mathcal{B} = 30$ nm (green irregularly dashed line line), and $d_\mathcal{B} = 40$ nm (blue dashed line) [54]. The letter S identifies the absorptance peaks that represent the excitation of SPP waves. (For interpretation of the references to color in this figure legend, the reader is referred to the web version of this book.)

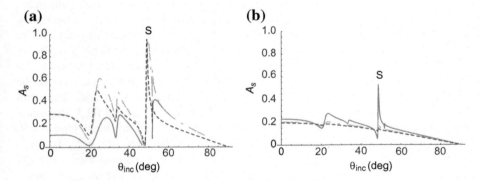

Figure 6.23 Same as Figure 6.22, except that the linear absorptance A_s is plotted [54].

of the material evaporated to fabricate the SNTF as well as on the substrate rocking dynamics. In the two figures, the optimum thickness is in the vicinity of 10 nm.

Just as in Section 6.4.1 for the interface of a metal and a rugate filter, two types of field profiles exist: those with a monotonic decrease in the fields with distance from the interface on the SNTF side of the interface, and ones with an oscillatory decrease with distance. Figure 6.24 shows an example of an oscillatory profile for an SNTF incorporated in the Turbadar-Kretschmann-Raether configuration. Additionally, the

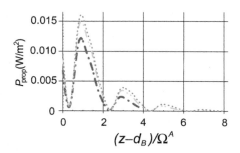

Figure 6.24 Variations with z of $\underline{u}_{prop} \cdot \underline{P}(x, y, z)$ on the line $\{x = 0, y = 0\}$ in a periodically nonhomogeneous titanium-oxide SNTF, which partners a layer of bulk aluminum in the Turbadar-Kretschmann-Raether configuration of Figure 6.8. Both materials are encased by the half-spaces $z < 0$ and $z > d_A + d_B$ occupied by zinc selenide ($n_{prism} = 2.58$). All parameters are the same as in Figure 6.22, except that $\psi = 30°$, $\theta_{inc} = 47.78°$, $d_B = 10$ nm, $\tilde{\chi}_v^A = 45°$, and $d_A \in \{4\Omega^A, 6\Omega^A, 8\Omega^A\}$. The incident plane wave is p polarized and its electric field has an amplitude of 1 V m^{-1}. The blue dash-dotted line is for $d_A = 4\Omega^A$, the green dash-dot-dotted line for $d_A = 6\Omega^A$, and the red dotted line for $d_A = 8\Omega^A$ [55]. (For interpretation of the references to color in this figure legend, the reader is referred to the web version of this book.)

identical nature of the curves for the three different thicknesses of the SNTF in this figure precludes an attribution to a waveguide mode.

Experimental observations. The existence of multiple SPP-wave modes guided by the interface of a metal and a periodically nonhomogeneous SNTF was demonstrated experimentally in 2009 [57], using the Turbadar-Kretschmann-Raether configuration. Two SNTFs were fabricated of silicon oxide and both were described using Eqs. (6.19). Sixth-order polynomial expressions were devised to connect the principal permittivity scalars of the SNTFs to the vapor flux angle. The substrates were rocked so that $\tilde{\chi}_v^A = 25°$, $\delta_v^A = 20°$, and $\Omega^A = 310$ nm. One SNTF was three periods in thickness, the other four. The three-period SNTF was deposited on top of a 30-nm thick aluminum film, while the four-period SNTF was deposited on top of a 34-nm thick aluminum film. The aluminum thin films had been deposited on slides of borosilicate glass BK-7. During these depositions in a low-pressure chamber, the base pressure was 3 μTorr.

A 45°-90°-45° prism made of borosilicate glass BK-7 glass was used when the Turbadar-Kretschmann-Raether configuration was implemented. The slide was affixed to the hypotenuse of the prism with an index-matching liquid. Figure 6.25a shows the measured linear absorptances A_p and A_s at $\lambda_0 = 632.8$ nm as functions of θ_{inc}, for the sample with the three-period SNTF. These measurements were made with the wave vector of the incident light oriented wholly in the morphologically significant plane of the SNTF so that $\psi = 0°$. Three peaks in the absorptance plots were identified as indicating the excitation of SPP-wave modes, because their locations on the θ_{inc} axis were almost the same as that of the peaks for the sample with the four-period SNTF: two p-polarized modes, one at $\theta_{inc} = 50.1°$ and another at $\theta_{inc} = 72.7°$, and one s-polarized mode at $\theta_{inc} = 57.0°$. Other absorptance peaks most likely represent waveguide modes.

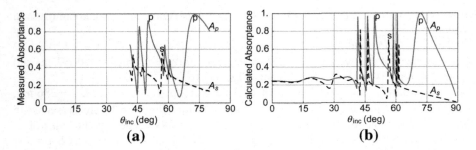

Figure 6.25 Experimental and simulated linear absorptances A_p and A_s for the interface of an aluminum thin film and a three-period SNTF of silicon oxide in the Turbadar-Kretschmann-Raether configuration. The letter p identifies a p-polarized SPP-wave mode, the letter s an s-polarized SPP-wave mode [57]. (a) Experimental and (b) Simulated.

Simulations were performed using the methodology of Section 3.7, but there was a discrepancy between theory and experiment for the A_p-peak observed experimentally at $\theta_{\text{inc}} = 72.7°$. The replacement of the first 30 nm of the SNTF by an isotropic homogeneous dielectric layer with a refractive index of $\sqrt{1.4}$—in order to simulate the early-stage growth of an STF—eliminated the discrepancy. The simulated linear absorptances, shown in Figure 6.25b as functions of θ_{inc}, agree reasonably well with their experimental counterparts.

6.4.2.3   Grating-Coupled Configuration

The grating-coupled configuration is described in Section 6.4.1.3 and illustrated in Figure 6.13. It was adapted for material $\mathcal{A}$ being a periodically nonhomogeneous SNTF by setting

$$
\begin{rcases}
\underline{\underline{\epsilon}}^{\mathcal{A}}(z) = \underline{\underline{A}}_z(\gamma^{\mathcal{A}}) \cdot \underline{\underline{S}}_y\left[\chi^{\mathcal{A}}(d_2 - z)\right] \cdot \left[\epsilon_a^{\mathcal{A}}(d_2 - z)\underline{u}_z\underline{u}_z\right. \\
\qquad \left. + \epsilon_b^{\mathcal{A}}(d_2 - z)\underline{u}_x\underline{u}_x + \epsilon_c^{\mathcal{A}}(d_2 - z)\underline{u}_y\underline{u}_y\right] \\
\qquad \cdot \underline{\underline{S}}_y^{-1}\left[\chi^{\mathcal{A}}(d_2 - z)\right] \cdot \underline{\underline{A}}_z^{-1}(\gamma^{\mathcal{A}}) \\
\chi_v^{\mathcal{A}}(z) = \tilde{\chi}_v^{\mathcal{A}} + \delta_v^{\mathcal{A}}\sin\left(\frac{\pi z}{\Omega^{\mathcal{A}}}\right) \\
\underline{\underline{A}}_z(\gamma) = \underline{u}_z\underline{u}_z + (\underline{u}_x\underline{u}_x + \underline{u}_y\underline{u}_y)\cos\gamma + (\underline{u}_y\underline{u}_x - \underline{u}_x\underline{u}_y)\sin\gamma
\end{rcases},
\tag{6.20}
$$

and using Eqs. (6.16) along with data from Table 6.1 [259]. The angle $\gamma^{\mathcal{A}}$ is the angle between the morphologically significant plane of the SNTF and the grating plane (i.e. the xz plane).

Calculations have been reported for $\psi = 0°$ and various values of $\gamma^{\mathcal{A}}$ [259]. A twofold symmetry exists with respect to this angle: results for $\gamma^{\mathcal{A}}$ and $-\gamma^{\mathcal{A}}$ are identical. Both p- and s-polarized SPP-wave modes are excited by incident p- and s-polarized plane waves, respectively, when the morphologically significant plane, the grating plane, and the plane of incidence are the same, i.e. $\gamma^{\mathcal{A}} = 0°$. For $\gamma^{\mathcal{A}} \in (0°, 90°]$, a distinct polarization state cannot be assigned to an SPP-wave mode, because the relative magnitudes and phases of the Cartesian components of $\underline{e}(z)$ are not uniform in material $\mathcal{A}$.

For $\gamma^{\mathcal{A}} = 75°$ and the same titanium-oxide SNTF as in Section 6.4.2.1, the SPP wavenumbers of the two allowed SPP-wave modes are $q = (2.0664 + 0.0039i)k_0$ and $q = (2.4588 + 0.0425i)k_0$, according to Figure 6.16.

As an example, the linear absorptances A_p and A_s for $L_1 = 0.5L_x$, $L_x = 286$ nm, and $\gamma^{\mathcal{A}} = 75°$ in the grating-coupled configuration of Figure 6.13 are presented in Figure 6.26 as functions of the angle of incidence θ_{inc}. Independent of the value of d_1 and the polarization state of the incident plane wave, an absorptance peak is present at $\theta_{\text{inc}} \simeq 9.2°$. This peak represents the excitation of an SPP-wave mode as a linear Floquet harmonic of order $(-1, 0)$ because $k_{xy}^{(-1,0)}/k_0 = -2.0534$ is very close to $-\text{Re}(2.0664 + 0.0039i)$. Confirmation is provided by the field profiles. This SPP-wave mode transports energy mainly along the $-\underline{u}_x$ direction, which is reasonable because this mode is excited in the grating-coupled configuration as a linear Floquet harmonic of negative order. A notable characteristic of this SPP-wave mode is that it can be excited by a plane wave of either polarization state; however, the excitation is more efficient if the incident plane wave is s polarized [259].

At $\theta \simeq 15.5°$ in Figure 6.26, a peak is present regardless of the value of d_1 (above a threshold) in the plots of A_p, but not of A_s. This A_p-peak represents the excitation of an

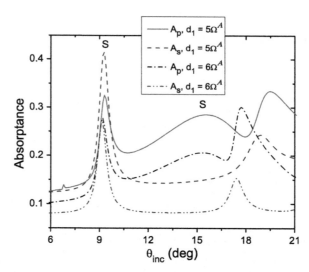

Figure 6.26 Calculated linear absorptances A_p and A_s as functions of the angle of incidence θ_{inc}, when $\psi = 0°$ and $\lambda_0 = 633$ nm, for the grating-coupled configuration of Figure 6.13. Material $\mathcal{A}$ is a titanium-oxide SNTF described by Eqs. (6.16) and (6.20), along with data from Table 6.1, $\Omega^{\mathcal{A}} = 200$ nm, $\tilde{\chi}_v^{\mathcal{A}} = 45°$, $\delta_v^{\mathcal{A}} = 30°$, and $\gamma^{\mathcal{A}} = 75°$. Material $\mathcal{B}$ is bulk aluminum with refractive index $n^{\mathcal{B}} = 1.38 + 7.61i$. The grating is described via Eqs. (6.12) and (6.13), with $d_3 - d_2 = 30$ nm, $d_2 - d_1 = 20$ nm, $L_1 = 0.5L_x$, and $L_x = 286$ nm [259]. The letter S identifies the absorptance peaks that represent the excitation of SPP waves.

SPP-wave mode as a linear Floquet harmonic of order $(1, 0)$ because $k_{xy}^{(1,0)}/k_0 = 2.4805$ is very close to $\mathrm{Re}(2.4588+0.0425i)$. Contrary to the SPP wave excited at $\theta_{\mathrm{inc}} = 9.2°$, the absence of the peak in the curves of A_s shows that this SPP-wave mode is excited only by a p-polarized incident plane wave.

The A_p- and A_s-peaks at $\theta_{\mathrm{inc}} = 9.2°$ are narrower than the A_p-peak at $\theta_{\mathrm{inc}} = 15.5°$, thereby supporting the conclusion that the absorptance peak representing the excitation of an SPP wave with smaller phase speed is broader.

6.4.3   Metal/Chiral-STF Interface

Chiral STFs are fabricated by PVD. Vapor is directed in a low-pressure chamber at a fixed angle relative to the surface of a substrate while the substrate is rotated steadily about an axis perpendicular to its surface [42]. Unlike SNTFs but like CTFs, the vapor flux angle χ_v is constant and χ is not a function of z for chiral STFs. The periodic nonhomogeneity of a chiral STF comes from its helical morphology, as depicted in Figure 6.27. Furthermore, a chiral STF possesses structural handedness, as it is either left handed or right handed.

Therefore, the permittivity of a chiral STF is written as

$$\underline{\underline{\epsilon}}_{\text{chiral STF}}(z) = \underline{\underline{S}}_z(z) \cdot \underline{\underline{S}}_y(\chi) \cdot (\epsilon_a \underline{u}_z \underline{u}_z + \epsilon_b \underline{u}_x \underline{u}_x + \epsilon_c \underline{u}_y \underline{u}_y) \cdot \underline{\underline{S}}_y^{-1}(\chi) \cdot \underline{\underline{S}}_z^{-1}(z), \tag{6.21}$$

where

$$\underline{\underline{S}}_z(z) = \underline{u}_z \underline{u}_z + (\underline{u}_x \underline{u}_x + \underline{u}_y \underline{u}_y) \cos\left(h\frac{\pi z}{\Omega}\right) + (\underline{u}_y \underline{u}_x - \underline{u}_x \underline{u}_y) \sin\left(h\frac{\pi z}{\Omega}\right) \tag{6.22}$$

indicates rotation about the z axis with period 2Ω. The structural handedness parameter h is equal to either $+1$ for right handedness or -1 for left handedness.

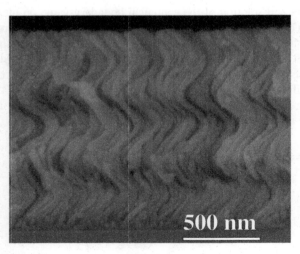

500 nm

Figure 6.27 Cross-sectional image of a chiral STF on a scanning electron microscope. This chiral STF was fabricated by evaporating lanthanum fluoride.

Experimental data on χ, ϵ_a, ϵ_b, and ϵ_c of chiral STFs in relation to χ_v do not exist for any material. Empirical relations that characterize the permittivity of a CTF relative to the vapor flux angle are expected to be more suitable for chiral STFs than for SNTFs because the substrate is not rocked during the fabrication of chiral STFs [99]. Nevertheless, a refined optical characterization of chiral STFs ought to be undertaken by specialist researchers.

Theoretical research at the present time is limited to chiral STFs made of the same small set of materials described in Section 6.3.2, for which CTFs have been empirically characterized. Equations (6.6) and (6.7) along with the Table 6.1 provided the principal permittivity scalars $\epsilon_{a,b,c}$ and the tilt angle χ for calculations in order to determine the characteristics of SPP-wave propagation guided by a metal/chiral-STF interface [56,161,336]. These calculations have revealed the existence of multiple SPP-wave modes with widely differing characteristics. The canonical boundary-value problem was solved and the Turbadar-Kretschmann-Raether configuration was also investigated.

The first experimental confirmation of the theoretically predicted multiple SPP-wave modes [56] came in 2009 [58] from measurement of the reflectance of a magnesium-fluoride chiral STF backed by an aluminum thin film in the Turbadar-Kretschmann-Raether configuration with the prism illuminated by p-polarized light. Further experimental evidence was provided a couple of years later [341] by a similar experiment involving a silicon-oxide chiral STF and a silver thin film. In the meantime, illumination by both p- and s-polarized light was experimentally demonstrated to be effective in exciting multiple SPP-wave modes guided by the interface of a magnesium-fluoride chiral STF and a gold thin film in the Turbadar-Kretschmann-Raether configuration [59].

6.4.3.1 Canonical Configuration

The formulation and solution of the canonical boundary-value problem [56] for the metal/chiral-STF interface follows the 4×4-matrix methodology of Section 3.5. Regardless of the angle ψ in $\underline{u}_{\mathrm{prop}} = \underline{u}_x \cos \psi + \underline{u}_y \sin \psi$, further simplification is not possible and a definite polarization state cannot be assigned to any SPP-wave mode found—because the relative magnitudes and phases of the Cartesian components of $\underline{e}(z)$ are not uniform in the chiral STF.

As shown in Figure 6.4, material $\mathcal{A}$ is the chiral STF and material $\mathcal{B}$ is the metal. The chiral STF is described by

$$
\left.
\begin{aligned}
\underline{\underline{\epsilon}}^{\mathcal{A}}(z) &= \underline{\underline{S}}_{z}^{\mathcal{A}}(z) \cdot \underline{\underline{S}}_{y}(\chi^{\mathcal{A}}) \cdot \left(\epsilon_a^{\mathcal{A}} \underline{u}_z \underline{u}_z + \epsilon_b^{\mathcal{A}} \underline{u}_x \underline{u}_x \right. \\
&\quad \left. + \epsilon_c^{\mathcal{A}} \underline{u}_y \underline{u}_y \right) \cdot \left[\underline{\underline{S}}_{z}^{\mathcal{A}}(z) \cdot \underline{\underline{S}}_{y}(\chi^{\mathcal{A}}) \right]^{-1} \\
\underline{\underline{S}}_{z}^{\mathcal{A}}(z) &= \underline{u}_z \underline{u}_z + (\underline{u}_x \underline{u}_x + \underline{u}_y \underline{u}_y) \cos \left(h^{\mathcal{A}} \frac{\pi z}{\Omega^{\mathcal{A}}} \right) \\
&\quad + (\underline{u}_y \underline{u}_x - \underline{u}_x \underline{u}_y) \sin \left(h^{\mathcal{A}} \frac{\pi z}{\Omega^{\mathcal{A}}} \right)
\end{aligned}
\right\}, \quad z > 0.
\tag{6.23}
$$

The refractive index of the metal in the half-space $z < 0$ is denoted by $n^{\mathcal{B}}$.

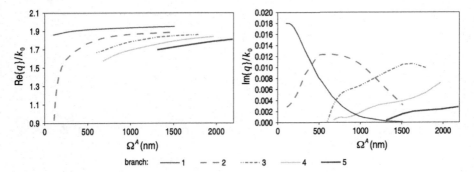

Figure 6.28 Solutions q of the canonical boundary-value problem as functions of Ω^A for SPP-wave propagation guided by the planar interface of bulk aluminum ($n^B = 1.38 + 7.61i$) and a titanium-oxide chiral STF, when $\psi = 0°$ and $\lambda_0 = 633$ nm. The chiral STF is described by Eqs. (6.6), (6.7), and (6.23), data from Table 6.1, and $\chi_v^A = 20°$ [56]. The solutions are organized in five branches labeled 1–5, and do not depend on the structural handedness parameter $h^A \in \{-1, 1\}$.

Equations (6.6) and (6.7) were used along with the Table 6.1 to theoretically show that the interface of bulk aluminum ($n^B = 1.38 + 7.61i$) and a titanium-oxide chiral STF can guide as many as five SPP-wave modes at $\lambda_0 = 633$ nm, depending on the ratio Ω^A/λ_0 [56, 161, 336]. The solutions of the canonical boundary-value problem are organized in five branches in Figure 6.28. As the ratio Ω^A/λ_0 increases beyond a certain value, the number of SPP-wave modes decreases—just as for the metal/rugate-filter interface in Section 6.4.1.1—and will eventually reduce to just unity. Only the branch labeled 5 should survive as $\Omega^A/\lambda_0 \to \infty$.

The data in Figure 6.28 are reinterpreted in terms of the phase speed v_p and the propagation length Δ_{prop} in Figure 6.29. The phase speed is clearly a decreasing function of Ω^A for all SPP-wave modes. Computational difficulties prevented exploration of the intermediate-phase-speed regime—wherein c_0 exceeds v_p by a small margin—and the high-phase-speed regime ($v_p > c_0$). Not only does the range of propagation length Δ_{prop} exceed two orders of magnitude, but the values of Δ_{prop} for some SPP-wave modes are quite large on an absolute scale. The largest value of of Δ_{prop} on branch 1 is about 3 mm.

The field phasors of any SPP-wave mode decay in the metal with distance from the interface, because the condition (2.13)$_1$ in $\exp(-i\alpha_{met}z)$ ensures exponential decay as $z \to -\infty$. The typical penetration depth Δ_{met} into the metal is a few tens of nanometers, and it depends very weakly through q on the specific SPP-wave mode as well as the constitution of the chiral STF. The field phasors have both s- and p-polarized components, in accordance with Eqs. (6.4), because the partnering dielectric material is anisotropic. The relative magnitudes of the Cartesian components of the field phasors do depend on the constitutive properties of the chiral STF and vary from mode to mode.

In contrast, the manner of the decay with distance from the interface on the chiral-STF side of the interface of an SPP-wave mode depends not only on the chiral STF

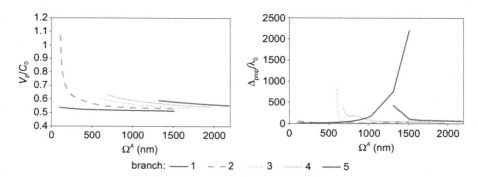

Figure 6.29 Same as Figure 6.28, except that the phase speed v_p and propagation length Δ_{prop} are plotted as functions of $\Omega^{\mathcal{A}}$.

but also strongly on the specific mode. The variations with z of $|\underline{e}(z)|$ and $\underline{P}(0, 0, z)$ depicted in Figure 6.30 for modes on branches labeled 1, 2, and 3 in Figure 6.28 for $\Omega^{\mathcal{A}} = 657$ nm and $\chi_v^{\mathcal{A}} = 20°$ clearly show differences in field profiles. Some SPP-wave modes have the highest magnitude of the electric field close to the interface, while other modes—such as the one on branch 3 in Figure 6.28—display peak magnitudes at some distance from the interface into the chiral STF. Furthermore, if decay constants

$$\left.\begin{array}{l}\beta_1^{\mathcal{A}} = \exp\left(-2\Omega^{\mathcal{A}}\mathrm{Im}\left\{\alpha_1^{\mathcal{A}}\right\}\right) \in (0, 1)\\\beta_2^{\mathcal{A}} = \exp\left(-2\Omega^{\mathcal{A}}\mathrm{Im}\left\{\alpha_2^{\mathcal{A}}\right\}\right) \in (0, 1)\end{array}\right\} \tag{6.24}$$

are defined for the half-space $z > 0$ in line with Eqs. (5.33), they will vary from mode to mode. The smaller the value of a decay constant, the more tightly is the associated electromagnetic field bound to the interface.

The field profile for a specific branch can also vary considerably from one value of $\Omega^{\mathcal{A}}$ to another. Figure 6.31 shows the profiles for the SPP-wave mode for $\Omega^{\mathcal{A}} = 131$ nm on branch 2 in Figure 6.28. In comparison to Figure 6.30b for $\Omega^{\mathcal{A}} = 657$ nm, the decay constants are larger and the fields decay much more slowly as $z/\Omega^{\mathcal{A}} \to \infty$. Even when the fields were examined with respect to z and not $z/\Omega^{\mathcal{A}}$, the SPP-wave mode in Figure 6.31 would be far more tightly bound to the metal/chiral-STF interface than the one in Figure 6.30b.

The direction of propagation in the interface plane can also have a dramatic influence on SPP-wave propagation. This influence can be gauged by choosing different values of ψ in $\underline{u}_{prop} = \underline{u}_x \cos\psi + \underline{u}_y \sin\psi$ [336]. Alternatively, one can keep $\psi = 0°$ fixed, but replace $\underline{\underline{\epsilon}}^{\mathcal{A}}(z)$ by $\underline{\underline{A}}_z^{-1}(\gamma^{\mathcal{A}}) \cdot \underline{\underline{\epsilon}}^{\mathcal{A}}(z) \cdot \underline{\underline{A}}_z(\gamma^{\mathcal{A}})$ with $\underline{\underline{A}}_z(\cdot)$ defined in Eq. (6.20)$_3$.

Thus, the metal/chiral-STF interface offers several modalities for control of SPP-wave propagation, a capability likely to attract the attention of those designing practical SPP-wave devices.

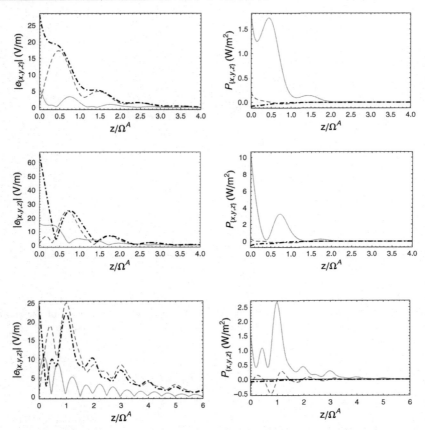

Figure 6.30 Variations with z of the magnitudes of the Cartesian components of $\underline{e}(z)$ of SPP-wave modes guided by the planar interface of bulk aluminum ($n^{\mathcal{B}} = 1.38 + 7.61i$) and a titanium-oxide chiral STF, when $\psi = 0°$, $\lambda_0 = 633$ nm, and $b_s = 1$ V m^{-1}. The chiral STF is described by Eqs. (6.6), (6.7), and (6.23), data from Table 6.1, $\Omega^{\mathcal{A}} = 657$ nm, $\chi_v^{\mathcal{A}} = 20°$, and $h^{\mathcal{A}} = 1$ [56]. Also shown are the variations of Cartesian components of the time-averaged Poynting vector $\underline{P}(x, y, z)$ with z on the line $\{x = 0, y = 0\}$. The components parallel to $\underline{u}_x$, $\underline{u}_y$, and $\underline{u}_z$ are represented by red solid, blue dashed, and black chain-dashed lines, respectively. The three modes lie on branches labeled 1, 2, and 3 in Figure 6.28, and the respective SPP wavenumbers are $q = (1.9246 + 0.0047753i)k_0$, $q = (1.8199 + 0.01238i)k_0$, and $q = (1.6824 + 0.002560i)k_0$. (For interpretation of the references to color in this figure legend, the reader is referred to the web version of this book.)

6.4.3.2   Turbadar-Kretschmann-Raether Configuration

Theoretical analysis and results. In order to incorporate a chiral STF as material $\mathcal{A}$ in the Turbadar-Kretschmann-Raether configuration of Figure 6.8,

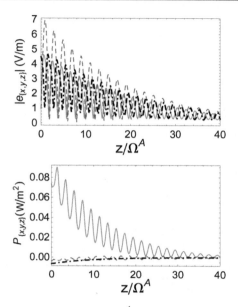

Figure 6.31 Same as Figure 6.30, except that $\Omega^A = 131$ nm and only the mode on the branch labeled 2 is considered. The SPP wavenumber is $q = (1.2154 + 0.003075i)k_0$.

$$\left.\begin{aligned}
\underline{\underline{\epsilon}}^A(z) &= \underline{\underline{S}}_z^A(z - d_B) \bullet \underline{\underline{S}}_y(\chi^A) \bullet \left(\epsilon_a^A \underline{u}_z \underline{u}_z + \epsilon_b^A \underline{u}_x \underline{u}_x\right. \\
&\quad + \epsilon_c^A \underline{u}_y \underline{u}_y\Big) \bullet \left[\underline{\underline{S}}_z^A(z - d_B) \bullet \underline{\underline{S}}_y(\chi^A)\right]^{-1} \\
\underline{\underline{S}}_z^A(z) &= \underline{u}_z \underline{u}_z + (\underline{u}_x \underline{u}_x + \underline{u}_y \underline{u}_y) \cos\left(h^A \frac{\pi z}{\Omega^A}\right) \\
&\quad + (\underline{u}_y \underline{u}_x - \underline{u}_x \underline{u}_y) \sin\left(h^A \frac{\pi z}{\Omega^A}\right)
\end{aligned}\right\}, \quad z \in (d_B, d_A + d_B),$$

$$(6.25)$$

must be used. The refractive index of the metal in the region $0 < z < d_B$ is denoted by n^B.

The excitation of an SPP-wave mode in the Turbadar-Kretschmann-Raether configuration is manifested in terms of a well-defined peak in the plots of linear absorptances A_p and A_s as functions of the angle of incidence θ_{inc}, provided that the thicknesses of both partnering materials exceed threshold values. This has been theoretically demonstrated for multiple SPP-wave modes guided by the planar interface of an aluminum thin film and a titanium-oxide chiral STF [336]. Some SPP-wave modes obtained by solving the underlying canonical boundary-value problem of Section 6.4.3.1 are evident with chiral STFs only two periods thick, while others require the chiral STF to be thicker. Still others do not show up in the Turbadar-Kretschmann-Raether configuration at all, because they are so loosely bound on the chiral-STF side to the interface as to require very thick chiral STFs.

Excitation of SPP-wave modes is evident with both s- and p-polarized incident light [56,336]. However, excitation with incident p-polarized light seems to be more efficient. Furthermore, the efficiency of excitation via incident s-polarized light in comparison to incident p-polarized light can be strongly influenced by the number of periods in the chiral STF. All three factors χ_v^A, Ω^A, and ψ affect the shape and location of the absorptance peaks on the θ_{inc} axis, though χ_v^A has the greatest influence by far. The largest number of SPP-wave modes occur at intermediate values of χ_v^A. This is reasonable, because a chiral STF shall be (i) mostly air if deposited at a very low value of χ_v and (ii) dense and almost homogeneous like a CTF if deposited at a very high value of χ_v.

Illustrative plots of A_p and A_s against θ_{inc} are shown in Figures 6.32 and 6.33 for a range of values of χ_v^A, when the number of periods of the chiral STF is either two or three. Thereby, the problem of distinguishing between SPP-wave modes and waveguide modes is clearly highlighted. Comparison of an absorptance plot with a two-period chiral STF against the one for a three-period chiral STF shows that the majority of absorptance peaks shift when the number of periods changes. Only those peaks which do not shift significantly represent SPP-wave modes; the other peaks represent waveguide modes.

In Figures 6.32 and 6.33, the absorptance peaks indicating the excitation of SPP-wave modes on branches 1, 2, and 3 are indicated with blue S, red S, and black S, respectively. The mode on branch 3 appears only when $\chi_v^A = 60°$. Close examination of the plots for $\chi_v^A = 5°$ reveals that peaks in the curves of both A_p and A_s at $\theta = 22.6°$ remain stationary as the number of periods of the chiral STF is changed from two to three. This SPP-wave mode is excitable with both s- and p-polarized incident light.

Experimental observations. The existence of multiple SPP-wave modes guided by the interface of a metal and a chiral STF was demonstrated experimentally in 2009 [58], using the Turbadar-Kretschmann-Raether configuration. Two chiral STFs were fabricated of magnesium fluoride, both with period $2\Omega^A \sim 425$ nm. During deposition in a low-pressure chamber with 2.1 μTorr base pressure, the vapor flux angle $\chi_v^A = 15°$ was kept fixed. One chiral STF was three periods in thickness, the other four. Each was deposited on top of a 25-nm thick aluminum film, which itself had been deposited on a slide of borosilicate glass BK-7.

When the Turbadar-Kretschmann-Raether configuration was implemented, a 45°-90°-45° prism made of borosilicate glass BK-7 was used. The slide was affixed to the hypotenuse of the prism with an index-matching liquid. Light from a 633-nm He-Ne laser was first filtered to eliminate the s-polarized component and then directed toward one slanted face of the prism. Although the incident light was only p polarized, the reflectance was measured without regard to the polarization state of the light reflected through the other slanted face of the prism.

Figure 6.34 shows the experimental curves of reflectance in relation to θ_{inc} [58] when this interface is illuminated by p-polarized light. As the transmittance across the metal-coated chiral STF was definitely absent for $\theta_{inc} \gtrsim 45°$, absorptance peaks would be manifested as reflectance dips. The reflectance dips in Figure 6.34 at $\theta_{inc} \sim 54°$ and $\sim 64°$, marked by S, indicate the excitation of SPP-wave modes, because their locations

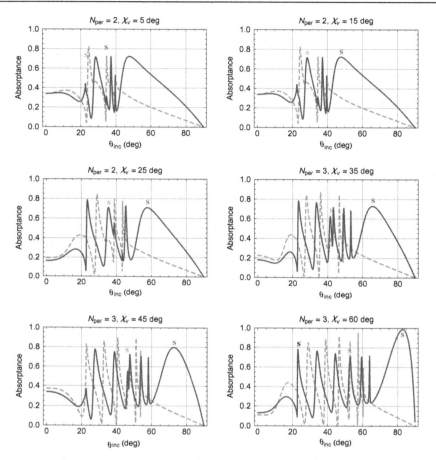

Figure 6.32 Calculated linear absorptances A_p (blue solid lines) and A_s (red dashed lines), when $\psi = 0°$ and $\lambda_0 = 633$ nm, as functions of the angle of incidence θ_{inc} for the Turbadar-Kretschmann-Raether configuration incorporating (i) a titanium-oxide chiral STF described by Eqs. (6.6), (6.7), and (6.25), data from Table 6.1, $\Omega^{\mathcal{A}} = 150$ nm, $h^{\mathcal{A}} = 1, d_{\mathcal{A}} = 2N_{\text{per}}\Omega^{\mathcal{A}}, \chi_v^{\mathcal{A}} \in \{5°, 15°, 25°\}$, and $N_{\text{per}} \in 2, 3$, and (ii) an aluminum thin film with $n^{\mathcal{B}} = 0.75 + 3.9i$ and $d_{\mathcal{B}} = 15$ nm, encased by the half-spaces $z < 0$ and $z > d_{\mathcal{A}} + d_{\mathcal{B}}$ occupied by rutile ($n_{\text{prism}} = 2.6$). The blue letter S identifies the absorptance peaks that represent the excitation of SPP-wave modes on branch 1, the red letter S is for SPP-wave modes on branch 2, and the black letter S is the SPP-wave modes on branch 3 [336]. (For interpretation of the references to color in this figure legend, the reader is referred to the web version of this book.)

are the same whether the chiral STF has three periods or four. The other reflectance dips shift with the number of periods and, therefore, represent waveguide modes.

Later experimental measurements for the interface of a gold thin film and a magnesium-fluoride chiral STF provided evidence for four distinct SPP-wave modes, of which two were excited by incident s-polarized light and two by p-polarized light [59].

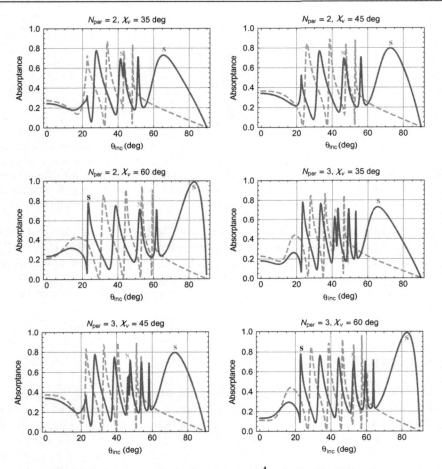

Figure 6.33 Same as Figure 6.32, except $\chi_v^{\mathcal{A}} \in \{35°, 45°, 60°\}$ [336].

6.4.3.3 Grating-Coupled Configuration

The grating-coupled configuration described in Section 6.4.1.3 and illustrated in Figure 6.13 can be adapted for material $\mathcal{A}$ being a chiral STF by setting

$$
\left.
\begin{aligned}
\underline{\underline{\epsilon}}^{\mathcal{A}}(z) &= \underline{\underline{S}}_z^{\mathcal{A}}\left(d_2 - z - \Omega^{\mathcal{A}}\frac{\gamma^{\mathcal{A}}}{\pi}\right) \bullet \underline{\underline{S}}_y(\chi^{\mathcal{A}}) \bullet \left(\epsilon_a^{\mathcal{A}}\underline{u}_z\underline{u}_z + \epsilon_b^{\mathcal{A}}\underline{u}_x\underline{u}_x\right. \\
&\quad \left. + \epsilon_c^{\mathcal{A}}\underline{u}_y\underline{u}_y\right) \bullet \left[\underline{\underline{S}}_z^{\mathcal{A}}\left(d_2 - z - \Omega^{\mathcal{A}}\frac{\gamma^{\mathcal{A}}}{\pi}\right) \bullet \underline{\underline{S}}_y(\chi^{\mathcal{A}})\right]^{-1} \\
\underline{\underline{S}}_z^{\mathcal{A}}(z) &= \underline{u}_z\underline{u}_z + (\underline{u}_x\underline{u}_x + \underline{u}_y\underline{u}_y)\cos\left(h^{\mathcal{A}}\frac{\pi z}{\Omega^{\mathcal{A}}}\right) \\
&\quad + (\underline{u}_y\underline{u}_x - \underline{u}_x\underline{u}_y)\sin\left(h^{\mathcal{A}}\frac{\pi z}{\Omega^{\mathcal{A}}}\right)
\end{aligned}
\right\},
$$

$$
(6.26)
$$

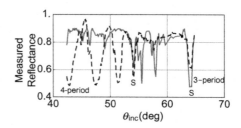

Figure 6.34 Measured linear reflectance as a function of θ_{inc} for the interface of an aluminum thin film and a chiral STF in the Turbadar-Kretschmann-Raether configuration. The chiral STF has either three or four periods. The incident light was only p polarized, but the reflected light was not broken up into its linear-polarization components before measuring the reflectance. Measurements were made at $\lambda_0 = 633$ nm. The letter S identifies an SPP-wave mode [58].

and using Eqs. (6.6) and (6.7) along with data from Table 6.1. The angle γ^A denotes a twist about the z axis with respect to the grating plane (i.e. the xz plane).

Theoretical results obtainable for a periodically corrugated metal/chiral-STF in the grating-coupled configuration are expected to be consistent with those for the canonical boundary-value problem of Section 6.4.3.1 and the prism-coupled configuration of Section 6.4.3.2.

6.5 Optical Sensing

Exploiting only interfaces between homogeneous isotropic dielectric materials and metals, the practical application of SPP-wave-based optical sensors has blossomed in the past two decades [3,4]. The previous sections in this chapter have demonstrated the degree of control over SPP-wave propagation that CTFs and STFs allow—not to mention the multiplicity of SPP-wave modes guided by interfaces formed with periodically nonhomogeneous STFs. At least three attributes make CTFs and STFs attractive for SPP-wave-based optical sensing:

   i. control of permittivity,
  ii. control of porosity, and
 iii. multiplicity of SPP-wave modes in the case of periodic STFs.

The flexibility offered by CTFs and STFs and the possibility of multiple SPP-wave modes will spur new research for optical sensing.

The ability to change the permittivity dyadic of a CTF simply by reorienting the vapor in relation to the substrate in any PVD technique is evident from Section 6.3.2. However, the permittivity dyadics of CTFs of only a handful of materials have been related to the vapor flux angle—that too, at just one or two values of the free-space wavelength. CTFs produced by evaporating other materials await optical characterization. The permittivity dyadics of SNTFs, chiral STFs, and other STFs have not been measured yet.

Were it not for another property of CTFs and STFs—namely, porosity—the ability to design and fabricate these films with a particular permittivity dyadic would not be of much use for optical sensing [342,343]. In order for an analyte to affect SPP-wave propagation, and thus be sensed, it must be able to migrate close to the metal/dielectric interface. Unlike many other solid materials, thin films have the porosity which will allow infiltration by analytes. Commercial SPR detectors use a very thin layer of recognition molecules attached to the metal surface in order to bind specific analytes for sensing. It might be possible to embed the recognition molecules into the protective environment of the pores of CTFs and STFs, while simultaneously protecting the metal film from environmental degradation. The porosity, like the permittivity dyadic, can be set simply by choosing the vapor flux angle appropriately.

The porosity of a CTF or an STF is significant in two ways. First, it determines the amount of analyte that might be infused into the thin film. This determines the degree to which the characteristics of SPP-wave propagation are changed and thus the sensitivity of the sensor. Second, the diffusion of analytes would also be affected by porosity. This might allow the differentiation of various chemical species because of different diffusion rates. Additionally, it may allow a peek into the kinetics of a chemical reaction.

Periodically nonhomogeneous STFs, enabling the propagation of multiple SPP-wave modes, may provide a way to detect multiple analytes simultaneously. With each SPP-wave mode having a distinct SPP wavenumber and field profile, the effect of each analyte on each SPP-wave mode would be different. For the sensing of a single analyte, multiple SPP-wave modes would provide multiple channels for more reliable sensing.

The path toward the exploitation of the multiple SPP-wave modes for optical sensing has been opened by the development of a theoretical model for CTFs and chiral STFs [98,338]. Very recently, a proof-of-concept experiment has also been performed [344]. The sensing configuration chosen in these studies is the Turbadar-Kretschmann-Raether configuration, but the Turbadar-Otto configuration may be useful in some situations. Both of these configurations employ a prism. Grating-coupled sensors exploiting multiple SPP-wave modes remain to be investigated.

6.5.1   Metal/CTF Interface

6.5.1.1   Nominal Model of a CTF

Theoretical analysis requires a model to connect the nanostructure to the continuum. In a nominal model of CTFs [42, Section 6.4] described in Appendix E, the columns of a CTF are modeled as strings of identical, highly elongated, and electrically small ellipsoids, as shown in Figure 6.35. Each inclusion is described by the shape dyadic $\underline{\underline{U}} = \underline{u}_n \underline{u}_n + \gamma_\tau \underline{u}_\tau \underline{u}_\tau + \gamma_b \underline{u}_b \underline{u}_b$, where the unit vectors $\underline{u}_n = -\underline{u}_x \sin \chi + \underline{u}_z \cos \chi$, $\underline{u}_\tau = \underline{u}_x \cos \chi + \underline{u}_z \sin \chi$, and $\underline{u}_b = -\underline{u}_y$ are defined in terms of the tilt angle $\chi \in (0°, 90°]$. Since the morphology is highly aciculate, the shape parameters $\gamma_b \gtrsim 1$ and $\gamma_\tau \gg 1$.

Let the ellipsoidal particles be made of an isotropic dielectric material of refractive index n_s. Let $v_s \in (0, 1)$ be the volume fraction occupied by the ellipsoidal particles, while the void region is filled with air (or vacuum). Thus, the porosity of the CTF equals $1 - v_s$.

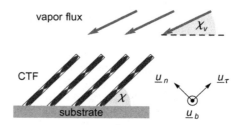

Figure 6.35 The columns of a CTF are represented as a set of elongated ellipsoidal particles, strung together end-to-end. The columns grow tilted at angle χ, along the direction of the unit vector $\underline{u}_\tau = \underline{u}_x \cos\chi + \underline{u}_z \sin\chi$, due to a vapor flux incident at an angle $\chi_v \leqslant \chi$.

Knowledge of the nanoscale parameters n_s, v_s, γ_b, and γ_τ is not easy to get for a CTF for at least three reasons, as discussed in Appendix E.2. One reason is the variability that exists due to differences in deposition conditions, so that the bulk material that is evaporated may be quite different from the material that is actually deposited as a thin film. In other words, the columns of a CTF comprise a material that can be different from the material that was evaporated to fabricate the CTF. Therefore, n_s cannot be presumed known. Second, the porosity is often inaccurately determined. Third, even when $\gamma_\tau > 10$ is fixed, γ_b cannot be measured as it is a notional quantity in a nominal model rather than a physical quantity.

Provided that the three principal permittivity scalars ϵ_a, ϵ_b, and ϵ_c of an uninfiltrated CTF have been measured by suitable optical experiments [94,345,346] and $\gamma_\tau > 10$ has been fixed [347], an inverse Bruggeman formalism can be used to determine n_s, v_s, and γ_b [308], as discussed in Appendix E.2. The angle χ can be measured from SEM images of the uninfiltrated CTF.

Finally, when the CTF is uniformly infiltrated by a fluid of refractive index $n_{f\ell}$, the value of ϵ_a, ϵ_b, and ϵ_c of the infiltrated CTF can be predicted using n_s, v_s, and γ_b in a forward Bruggeman formalism [98]. The details are provided in Appendix E.3.

The calculated data in Table 6.8 for a titanium-oxide CTF shows that the porosity $1 - v_s$ decreases and the cross section of the columns tends towards the circular shape, as the vapor flux angle increases. Both of these trends are in accord with numerous observations [42,327]. The refractive index n_s of the material also decreases toward

Table 6.8 The dimensionless quantities n_s, v_s, and γ_b computed using the inverse Bruggeman formalism for a titanium-oxide CTF described by Eqs. (6.5)–(6.7), along with data from Table 6.1 and $\chi_v \in \{15°, 30°, 60°, 90°\}$. The values of n_s should hold at $\lambda_0 = 633$ nm [98].

χ_v (deg)	n_s	v_s	γ_b
15	3.2510	0.3614	2.2793
30	3.0517	0.5039	1.8381
60	2.9105	0.6956	1.4054
90	2.8828	0.7859	1.0020

the refractive index (i.e. 2.58) of rutile. That n_s exceeds the bulk refractive index of the material evaporated to fabricate the CTF must be carefully noted.

All three principal permittivity scalars of a CTF increase roughly linearly with the refractive index $n_{f\ell}$ of the infiltrating fluid [98]. A more rapid change of values with $n_{f\ell}$ is observed for lower values of χ_v, which is consistent with the greater porosity of CTFs fabricated at lower values of χ_v.

6.5.1.2   Canonical Boundary-Value Problem

SPP-wave propagation guided by the interface of bulk aluminum and a titanium-oxide CTF has been studied in detail, when the CTF is infiltrated by a liquid of refractive index $n_{f\ell} \in [1, 1.5]$ [98]. The real part of the SPP wavenumber q increases approximately linearly as $n_{f\ell}$ increases, and it also increases as χ_v increases. Similarly, $\mathrm{Im}\{q\}$ also increases as both $n_{f\ell}$ and χ_v increase. Clearly then, both v_p and Δ_{prop} are decreasing functions of both $n_{f\ell}$ and χ_v. Not only that but the depth of penetration into the CTF is also a decreasing function of both $n_{f\ell}$ and χ_v.

6.5.1.3   Turbadar-Kretschmann-Raether Configuration

Although the results of the canonical boundary-value problem give an indication of how SPP-wave propagation is affected by infiltrating a CTF with a fluid, a real appreciation of the performance of a metal/CTF interface as an SPR sensor comes from an examination of absorptance as a function of the angle of incidence θ_{inc} in the Turbadar-Kretschmann-Raether configuration illustrated in Figure 6.8.

This prism-coupled configuration is very popular for optical sensing [3,4]. The mathematical technique described in Section 3.7 is straightforward to apply for analysis and prediction. Care must be taken to ensure during modeling and analysis that not only are the pores of the CTF in the region $d_B < z < d_A + d_B$ occupied by the infiltrant fluid, but the half-space $z > d_A + d_B$ is occupied by the same fluid as well.

When $\psi = 0°$, only incident p-polarized light can launch an SPP wave guided by the metal/CTF interface. Calculations for the interface of bulk aluminum and a titanium-oxide CTF in the Turbadar-Kretschmann-Raether configuration yield the value $\theta_{\mathrm{inc}}^{\mathrm{SPR}}$ of the angle of incidence of the A_p-peak which is insensitive to d_A beyond a certain threshold. Optical sensing is based on the detection of the shift of the resonance angle $\theta_{\mathrm{inc}}^{\mathrm{SPR}}$ with $n_{f\ell}$. Calculations have shown that the SPR sensing is very sensitive in a practical situation when both $n_{f\ell}$ and χ_v^A are small [98].

A sensitivity measure can be devised as

$$\rho = \frac{\theta_{\mathrm{inc}}^{\mathrm{SPR}}(n_{f\ell}) - \theta_{\mathrm{inc}}^{\mathrm{SPR}}\left(n_{f\ell}^{\mathrm{std}}\right)}{n_{f\ell} - n_{f\ell}^{\mathrm{std}}}, \tag{6.27}$$

where the notation $\theta_{\mathrm{inc}}^{\mathrm{SPR}}(n_{f\ell})$ indicates the dependence of $\theta_{\mathrm{inc}}^{\mathrm{SPR}}$ on $n_{f\ell}$, and $n_{f\ell}^{\mathrm{std}}$ is the refractive index of a standard fluid. Figure 6.36 shows plots of ρ as a function of $n_{f\ell}$ at $\lambda_0 = 633$ nm, when the prism is made of rutile, the 10-nm thick metal film (material B) is assumed to have the refractive index of bulk aluminum, the 1000-nm thick CTF (material A) is made of titanium oxide with $\chi_v^A \in \{15°, 30°, 60°, 90°\}$. Except for a

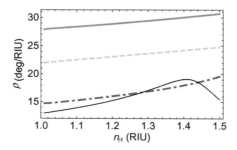

Figure 6.36 Calculated sensitivity ρ (in deg/RIU) as a function of $n_{f\ell}$ at $\lambda_0 = 633$ nm, for an aluminum film and a titanium-oxide CTF in the Turbadar-Kretschmann-Raether configuration. The prism is made of rutile ($n_{\text{prism}} = 2.6$); the aluminum film has a thickness $d_B = 10$ nm and refractive index $n^B = 1.38 + 7.61i$; and the CTF described by Eqs. (6.5), (6.6), and (6.7), with data from Table 6.1, has a thickness $d_A = 1000$ nm. The half-space $z > d_A + d_B$ is occupied by the infiltrant fluid. Air is used as the standard: $n_{f\ell}^{\text{std}} = 1$. Data are presented for $\chi_v^A = 15°$ (red thick solid line), $\chi_v^A = 30°$ (green dashed line), $\chi_v^A = 60°$ (blue chain-dashed line), and $\chi_v^A = 90°$ (black thin solid line) [98]. (For interpretation of the references to color in this figure legend, the reader is referred to the web version of this book.)

bump in the plot for $\chi_v^A = 90°$ at large values of $n_{f\ell}$, the sensitivity decreases with increasing χ_v^A, and increases with increasing $n_{f\ell}$.

The values of ρ in Figure 6.36 are less than what may be obtained using a conventional SPR sensor with the thin metal film in direct contact with a solution of the analyte. As an example, $\rho = 79$ deg/RIU has been obtained with an aluminum film at $\lambda_0 = 653$ nm [318]. This is to be expected since $v_s = 0$ in effect for a conventional SPR sensor, whereas $v_s > 0$ when a CTF is used.

The refractive-index sensitivity defined as

$$\rho_{\text{RI}} = \frac{d\theta_{\text{inc}}^{\text{SPR}}(n_{f\ell})}{dn_{f\ell}} \tag{6.28}$$

is useful when small fluctuations in the concentration of an analyte have to be sensed. Figure 6.37 shows the refractive-index sensitivity for the same aluminum/CTF interface as in Figure 6.36. For $n_{f\ell} \in (1, 1.15)$, this figure indicates that an SPR sensor employing a metal/CTF interface shall be most sensitive to changes in $n_{f\ell}$ when χ_v^A is small. For larger values of $n_{f\ell}$, plots of the refractive-index sensitivities appear to slowly converge as $n_{f\ell}$ increases for $\chi_v^A \in \{15°, 30°, 60°\}$. In contrast, the refractive-index sensitivity for $\chi_v^A = 90°$ decreases abruptly as $n_{f\ell}$ increases beyond 1.4.

The sensitivities should be expected to decrease as χ_v^A increases because, as has been quantified in Table 6.8, the porosity $1 - v_s$ decreases. Furthermore, the sensitivities should be expected to decrease as $n_{f\ell}$ comes closer to n_s, because the spatial variations of the dielectric properties inside an infiltrated thin film would be weaker than when it was not infiltrated. Although Figures 6.36 and 6.37 indicate that the interaction of the CTF and the infiltrant is complicated, these simple expectations are borne out in part by data in both figures.

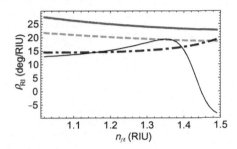

Figure 6.37 Same as Figure 6.36, except the calculated refractive-index sensitivity ρ_{RI} is plotted [98].

6.5.2   Metal/SNTF Interface

As discussed in Section 6.4.2, experimental data on $\chi(z)$, $\epsilon_a(z)$, $\epsilon_b(z)$, and $\epsilon_c(z)$ in relation to $\chi_v(z)$ do not exist for any SNTF, and the appropriation of data from Table 6.1 for use in Eqs. (6.16) is, at best, an initial approximation for uninfiltrated SNTFs.

Furthermore, $\epsilon_a(z)$, $\epsilon_b(z)$, and $\epsilon_c(z)$ have not been measured for any particular SNTF. Although the procedure involving the inverse and the forward Bruggeman formalisms has been successfully used for infiltrated CTFs, conceptual difficulties are insurmountable for the same procedure when applied to infiltrated SNTFs.

6.5.3   Metal/Chiral-STF Interface

6.5.3.1   Theoretical Analysis and Results

The use of a metal/chiral-STF bilayer in the Turbadar-Kretschmann-Raether configuration can be theoretically investigated following the procedure outlined in Section 6.4.3.2 with the modifications described in Section 6.5.1 to incorporate infiltration by a fluid. Now, multiple SPP-wave modes can be exploited for optical sensing—which is, of course, very attractive.

Calculations for the combination of the metal and the chiral STF similar to the one chosen for illustrative results in Section 6.4.3.1 show that, although details vary from one SPP-wave mode to the next, all available SPP-wave modes could be used in SPR sensors [338]. The sensitivity and the refractive-index sensitivity calculated as functions of $n_{f\ell}$ for two SPP-wave modes are shown in Figures 6.38 and 6.39, respectively. Both sensitivities are similar in magnitude and trend to those in Figures 6.36 and 6.37 when a CTF is used as the partnering dielectric material.

Of course, the use of a chiral STF offers multiple SPR shifts in contrast to a single SPR shift afforded by the use of a CTF. Moreover, with judicious choices of geometric and constitutive parameters, the circular Bragg phenomenon exhibited by chiral STFs offers yet another sensing mode which could be employed in parallel with SPR sensing [343].

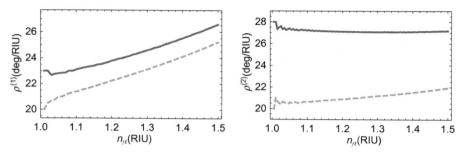

Figure 6.38 Calculated sensitivity ρ (in deg/RIU) as a function of $n_{f\ell}$ at $\lambda_0 = 633$ nm, for an aluminum film and a titanium-oxide chiral STF in the Turbadar-Kretschmann-Raether configuration. The prism is made of zinc selenide ($n_{\text{prism}} = 2.58$); the aluminum film has a thickness $d_B = 15$ nm and refractive index $n^B = 1.38 + 7.61i$; and the chiral STF described by Eqs. (6.6), (6.7), and (6.23), and data from Table 6.1, has a thickness $d_A = 4\Omega^A$ nm, and half-period $\Omega^A = 328.33$ nm. The half-space $z > d_A + d_B$ is occupied by the infiltrant fluid. Air is used as the standard: $n_{f\ell}^{\text{std}} = 1$. The chosen SPP-wave mode lies on either (left) branch 1 or (right) branch 2, when $n_{f\ell} = 1$. Data are presented for $\chi_v^A = 15°$ (red solid line) and $\chi_v^A = 30°$ (green dashed line), and are the same whether the chiral STF is structurally right handed ($h^A = 1$) or left handed ($h^A = -1$) [338]. (For interpretation of the references to color in this figure legend, the reader is referred to the web version of this book.)

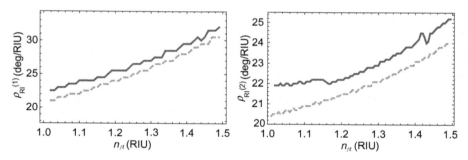

Figure 6.39 Same as Figure 6.38, except the calculated refractive-index sensitivity ρ_{RI} is plotted [338].

6.5.3.2 Experimental Observations

Experimental verification that the multiple SPP-wave modes guided by the interface of a metal and a periodically nonhomogeneous dielectric material offer multiple channels for optical sensing became available recently. A 30-nm thick aluminum film was partnered with a three-period-thick chiral STF of lanthanum fluoride in the Turbadar-Kretschmann-Raether configuration. A prism made of dense flint glass was used in the apparatus depicted in Figure 2.5. A laser diode ($\lambda_0 = 635$ nm) was used as the light source. Light incident on a slanted face of the prism was p polarized. The intensity

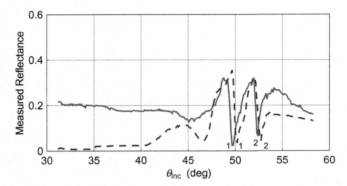

Figure 6.40 Linear reflectance measured as a function of the angle of incidence θ_{inc} at $\lambda_0 = 635$ nm in the Turbadar-Kretschmann-Raether configuration of Figure 2.5. The incident light was p polarized. The prism was made of dense flint glass ($n_{prism} = 1.78471$), the $\sim$30-nm thick metal film of aluminum was deposited on a substrate made of dense flint glass, and the dielectric partnering material was a lanthanum-fluoride chiral STF infiltrated either by water (red solid line) or a sucrose solution of refractive index 1.34 (blue dashed line). The two SPP-wave modes are identified by 1 and 2. (For interpretation of the references to color in this figure legend, the reader is referred to the web version of this book.)

of light leaving the second slanted face of the prism was measured and divided by the intensity of light incident on the first slanted face of the prism to obtain the reflectance. Transmittance being measured to be negligibly small over the θ_{inc}-range of interest, a sharp dip in reflectance plotted as a function of θ_{inc} implied an A_p-peak.

Figure 6.40 shows the measured reflectance as a function of θ_{inc}. The infiltrant fluid is either water or a sucrose solution. Two reflectance dips, both identified as indicative of the excitation of two different SPP-wave modes, shift on the θ_{inc} axis when water of refractive index 1.33 is replaced by sucrose solution of refractive index 1.34 [344]. Thus, a multiple SPP-wave-based optical sensor with a single metal/dielectric interface has been demonstrated and confirms theoretical predictions [338,343] of the viability of such sensors.

6.6 Harvesting of Solar Energy

Whereas prism-coupled configurations are widely used for optical sensing, the grating-coupled configuration has been known for at least three decades to be useful in enhancing the absorptance of light in thin-film solar cells [207,208].

The electric current produced is directly proportional to the energy absorbed from the light incident on the solar cell. Therefore, to maximize the electrical output of a solar cell, the absorptance of light over the spectral regime of the solar flux (roughly, $\lambda_0 \in$ [400, 1100] nm) needs to be maximized. A semiconductor such as silicon functions as the partnering dielectric material. This material is in intimate contact with a metallic back-reflector. One way to increase the absorptance is to periodically corrugate the

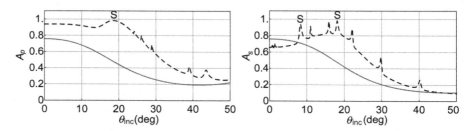

Figure 6.41 Calculated linear absorptances as functions of the angle of incidence θ_{inc}, when $\psi = 0°$ and $\lambda_0 = 827$ nm, for the grating-coupled configuration of Figure 6.13. Material $\mathcal{A}$ is a rugate filter made of a-Si$_{1-b}$C$_b$:H described by Eq. (6.14) with $n_{\text{avg}}^{\mathcal{A}} = 3.16+0.0008i$, $\Delta n^{\mathcal{A}} = 0.2n_{\text{avg}}^{\mathcal{A}}$, and $\Omega^{\mathcal{A}} = 300$ nm. Material $\mathcal{B}$ is bulk aluminum with refractive index $n^{\mathcal{B}} = 2.74 + 8.31i$. The grating is described via Eqs. (6.12) and (6.13), with $L_1 = 0.5L_x$, $d_3 - d_2 = 30$ nm, and $d_1 = 8\Omega^{\mathcal{A}}$. (left) A_p when $L_x = 244.5$ nm; (right) A_s when $L_x = 282$ nm. The red solid line is for metal/rugate-filter interface being planar ($d_2 = d_1$), and the black dashed line is for that interface being periodically corrugated with $d_2 - d_1 = 20$ nm. The letter S identifies the absorptance peaks that represent the excitation of SPP waves. (For interpretation of the references to color in this figure legend, the reader is referred to the web version of this book.)

metallic back-reflector of the solar cell—just as in the grating-coupled configuration. This periodic corrugation may lead to the excitation of an SPP wave and thereby increase the absorptance of light [206].

However, direct sunlight is almost completely unpolarized [348, p. 377], i.e. at any given instant of time, it has components of both linear polarization states in almost equal proportions. If the semiconductor is homogeneous, only one SPP wave at a given wavelength can be excited. As only p-polarized light would be effective in exciting SPP waves, this strategy cannot enhance the absorptance of one half of the incident sunlight.

If the semiconductor were to be periodically nonhomogeneous normal to the mean plane of the periodic corrugation of the metal/semiconductor interface, multiple SPP-wave modes of both p- and s-polarization states could be excited at every wavelength in the solar spectrum. Figure 6.41 shows the absorptances A_p and A_s as functions of θ_{inc}, incident on a finitely thick isotropic semiconductor backed by a finitely thick metallic film, the metal/semiconductor interface being either planar or periodically corrugated.

When the metal/semiconductor interface is planar, no SPP wave can be excited. This is because $\kappa_{xy}^{(m,n)} \equiv \kappa_{xy}^{(0,0)}$ for all $m \in (-\infty, \infty)$ and $n \in (-\infty, \infty)$, and no angle of incidence $\theta_{\text{inc}} \in [0°, 90°)$ can then satisfy Eq. (3.216).

Periodic corrugation of the metal/semiconductor interface leads to enhancement of absorptance of light of either polarization state, and the absorptance increases significantly if multiple SPP-wave modes are excited as compared to the situation when no SPP wave-modes are excited [84]. The enhancement in absorptance for both the p- and s-polarized components of the incident sunlight is an important advantage over conventional plasmonic solar cells in which no s-polarized SPP-wave modes are excited

[206–208]. However, careful optimization of the grating is necessary to obtain high absorptance of unpolarized light over a broad range of free-space wavelengths [349].

6.7 Outlook

Multiple SPP-wave modes—all at the same frequency and propagating in the same direction in the interface plane, but with distinct phase speeds, propagation lengths, and field profiles—can be guided by the interface of a metal and a dielectric material that is periodically nonhomogeneous in the direction normal to the interface. This phenomenon—now dubbed *surface multiplasmonics* [350]—was theoretically predicted and subsequently confirmed by experimentation. A proof-of-concept experiment has demonstrated the availability of multiple channels for more reliable sensing of a single analyte, and for simultaneous sensing of multiple analytes. Surface multiplasmonics is also expected to be highly useful for boosting the quantum efficiencies of tandem solar cells. A tandem solar cell is a cascade of multiple *p-i-n* solar cells, each comprising a layer of a doped semiconductor of the *p* type, a layer of an intrinsic semiconductor, and a layer of a doped semiconductor of the *n* type. Absorption of light for conversion to electricity occurs only in the intrinsic semiconductor layers when the solar cell is made of amorphous silicon. Thus, broad avenues have opened up for both theoretical and experimental research along with significant applications.

7 Dyakonov-Tamm Waves

7.1 Introduction

Having considered in detail Dyakonov waves in Chapter 4 and Tamm waves in Chapter 5, let us now turn to what is essentially an amalgamation of these two types of surface waves: Dyakonov-Tamm waves. By definition, at least one of the two partnering materials is required to be both anisotropic and periodically nonhomogeneous. Periodic nonhomogeneity of an anisotropic partnering material distinguishes Dyakonov–Tamm waves from Dyakonov waves.

Dyakonov-Tamm waves offer two distinct advantages over Dyakonov waves for optical applications: First, Dyakonov-Tamm-wave propagation has a much larger AED than Dyakonov waves. Second, more than one Dyakonov-Tamm-wave mode may be guided by a particular interface in a specific direction in the interface plane, whereas only one Dyakonov-wave mode is supported at a given interface as shown in Chapter 4. However, more than one Dyakonov-wave mode may be possible if one of the partnering materials supports NPV [69,170].

Development of the theory of Dyakonov-Tamm waves (like that of Dyakonov waves) has preceded experimental research. Indeed, the term *Dyakonov-Tamm wave* was coined in 2007 when the existence of such a surface wave guided by the interface of a chiral sculptured thin film and a homogeneous isotropic dielectric material was predicted [46]. While this work provided a rigorous basis for Dyakonov-Tamm waves, other approaches to surface waves in the context of anisotropic periodically nonhomogeneous materials had also been reported [351–353]. Experimental studies of Dyakonov-Tamm waves are only now being undertaken, but no definitive results have emerged yet.

In this chapter we focus on the excitation and propagation of Dyakonov-Tamm waves for certain relatively simple cases that have been reported in the recent literature. A selection of illustrative numerical results is provided. The theoretical treatment of the general case for Dyakonov-Tamm waves, wherein the two partnering materials are both bianisotropic and periodically nonhomogeneous normal to their interface, is comprehensively covered in Chapter 3.

7.2 Canonical Boundary-Value Problem

As in the preceding three chapters, the canonical boundary-value problem introduced in Section 3.5 underlies the description of Dyakonov-Tamm waves. Given the complexity

Electromagnetic Surface Waves. http://dx.doi.org/10.1016/B978-0-12-397024-4.00007-4

of the partnering materials involved, it is reasonable to distinguish the case when one partnering material is homogeneous from the other when neither of the two is.

7.2.1 Interface of a Homogeneous Isotropic Material and a Periodically Nonhomogeneous Anisotropic Material

Suppose that material $\mathcal{A}$, occupying the half-space $z > 0$, is a periodically nonhomogeneous, anisotropic dielectric material while material $\mathcal{B}$, occupying the half-space $z < 0$, is a homogeneous, isotropic dielectric material, as illustrated schematically in Figure 7.1. Various forms of material $\mathcal{A}$ may be considered. It may be continuously nonhomogeneous like a chiral STF or it may be piecewise homogeneous such as a Reusch pile. Even with a given form of material $\mathcal{A}$, different implementations may be considered. As examples, SNTFs and chiral STFs have permittivity dyadics identical in form to some nematic liquid crystals and cholesteric liquid crystals, respectively. In the following sections, we consider three particular types of periodically nonhomogeneous dielectric materials which have been studied in some detail: chiral STFs, SNTFs, and Reusch piles.

7.2.1.1 Interface of an Isotropic Dielectric Material and a Chiral Sculptured Thin Film

Let material $\mathcal{A}$ be a chiral sculptured thin film specified by the permittivity dyadic

$$
\left.
\begin{aligned}
\underline{\underline{\epsilon}}^{\mathcal{A}}(z) &= \underline{\underline{S}}_z^{\mathcal{A}}\left(z + \Omega^{\mathcal{A}}\frac{\gamma^{\mathcal{A}}}{\pi}\right) \cdot \underline{\underline{S}}_y(\chi^{\mathcal{A}}) \cdot \left(\epsilon_a^{\mathcal{A}}\underline{u}_z\underline{u}_z + \epsilon_b^{\mathcal{A}}\underline{u}_x\underline{u}_x \right. \\
&\quad \left. + \epsilon_c^{\mathcal{A}}\underline{u}_y\underline{u}_y\right) \cdot \left[\underline{\underline{S}}_z^{\mathcal{A}}\left(z + \Omega^{\mathcal{A}}\frac{\gamma^{\mathcal{A}}}{\pi}\right) \cdot \underline{\underline{S}}_y(\chi^{\mathcal{A}})\right]^{-1} \\
\underline{\underline{S}}_z^{\mathcal{A}}(z) &= \underline{u}_z\underline{u}_z + (\underline{u}_x\underline{u}_x + \underline{u}_y\underline{u}_y)\cos\left(h^{\mathcal{A}}\frac{\pi z}{\Omega^{\mathcal{A}}}\right) \\
&\quad + (\underline{u}_y\underline{u}_x - \underline{u}_x\underline{u}_y)\sin\left(h^{\mathcal{A}}\frac{\pi z}{\Omega^{\mathcal{A}}}\right)
\end{aligned}
\right\}, \quad z > 0,
$$

$$(7.1)$$

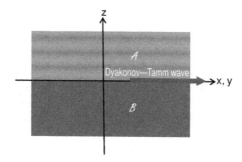

Figure 7.1 Schematic of the canonical boundary-value problem for Dyakonov-Tamm waves guided by the planar interface of a periodically nonhomogeneous anisotropic material $\mathcal{A}$ and a homogeneous isotropic material $\mathcal{B}$.

wherein the dyadic $\underline{\underline{S}}_y$ is defined by Eq. (1.30), $h^{\mathcal{A}} \in \{-1, 1\}$ is the structural handedness parameter, $2\Omega^{\mathcal{A}}$ is the period, and the angle $h^{\mathcal{A}}\gamma^{\mathcal{A}} > 0°$ denotes the orientation of the chiral STF about the z axis. The permittivity of the homogeneous partnering dielectric material $\mathcal{B}$ is denoted by $\epsilon^{\mathcal{B}}$ and its refractive index by $n^{\mathcal{B}} = \sqrt{\epsilon^{\mathcal{B}}/\epsilon_0}$.

As material $\mathcal{B}$ is isotropic, either one can keep $\gamma^{\mathcal{A}} = 0°$ fixed and vary $\psi \in [0°, 360°)$, or one can keep $\psi = 0°$ fixed and vary $\gamma^{\mathcal{A}} \in [0°, 360°)$. Either scheme is fine to determine the AED of Dyakonov-Tamm propagation for the chosen interface, the AED being the range of the variable angle over which solutions of the dispersion equation (3.87) can be found.

Suppose that the Dyakonov-Tamm wave is taken to propagate parallel to the x axis; i.e. $\psi = 0°$ and $\underline{u}_{prop} = \underline{u}_x$. As in Section 4.2.1.1, we can then write[1]

$$\left. \begin{array}{l} \underline{e}(z) = C_s^{\mathcal{B}}\underline{u}_y + C_p^{\mathcal{B}}\left(\frac{-\alpha^{\mathcal{B}}\underline{u}_x + q\underline{u}_z}{k_0 n^{\mathcal{B}}}\right)\exp\left(i\alpha^{\mathcal{B}}z\right) \\[3mm] \underline{h}(z) = \frac{n^{\mathcal{B}}}{\eta_0}\left[C_s^{\mathcal{B}}\left(\frac{-\alpha^{\mathcal{B}}\underline{u}_x + q\underline{u}_z}{k_0 n^{\mathcal{B}}}\right) - C_p^{\mathcal{B}}\underline{u}_y\right]\exp\left(i\alpha^{\mathcal{B}}z\right) \end{array} \right\}, \quad z < 0, \quad (7.2)$$

where $C_s^{\mathcal{B}}$ and $C_p^{\mathcal{B}}$ are amplitudes of the s- and p-polarized field components, respectively; the wavenumber $q > 0$ is purely real for unattenuated propagation in the interface plane; and the wavenumber

$$\alpha^{\mathcal{B}} = \sqrt{\omega^2 \mu_0 \epsilon^{\mathcal{B}} - q^2} \qquad (7.3)$$

has to obey the inequality $\text{Im}\{\alpha^{\mathcal{B}}\} < 0$ for attenuation as $z \to -\infty$. Accordingly,

$$[\underline{f}(0-)] = \begin{bmatrix} 0 & -\dfrac{\alpha^{\mathcal{B}}}{k_0 n^{\mathcal{B}}} \\[3mm] 1 & 0 \\[3mm] -\dfrac{\alpha^{\mathcal{B}}}{\omega\mu_0} & 0 \\[3mm] 0 & -\dfrac{n^{\mathcal{B}}}{\eta_0} \end{bmatrix} \cdot \begin{bmatrix} C_s^{\mathcal{B}} \\[3mm] C_p^{\mathcal{B}} \end{bmatrix}. \qquad (7.4)$$

The procedure to express the field phasors in the region $z > 0$ provided in Section 3.5.1 yields Eq. (3.80) with $\psi = 0°$. Therefrom the boundary value

---

[1]In Chapters 1, 2, and 6, the half-space $z < 0$ in the canonical boundary-value problem for SPP-wave propagation is occupied by a metal, and the spatial dependence of the field phasors is taken to be $\exp[i(q\underline{u}_{prop} - \alpha_{met}\underline{u}_z) \cdot \underline{r}]$ with $\text{Im}\{\alpha_{met}\} > 0$. In Chapters 4, 5, and 7, sometimes the half-space $z < 0$ in the canonical boundary-value problem for surface-wave propagation is occupied by an isotropic homogeneous dielectric material. In consonance with Chapter 3, we then adopt the spatial dependence $\exp[i(q\underline{u}_{prop} + \alpha^{\mathcal{B}}\underline{u}_z) \cdot \underline{r}]$ with $\text{Im}\{\alpha^{\mathcal{B}}\} < 0$. Both representations are identical, as can be ascertained by setting $\alpha_{met} = -\alpha^{\mathcal{B}}$.

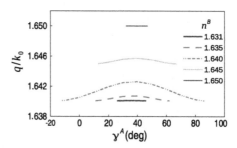

Figure 7.2 Solutions q of the canonical boundary-value problem illustrated in Figure 7.1 as functions of $\gamma^{\mathcal{A}}$, when $\psi = 0°$ and $\lambda_0 = 633$ nm. The Dyakonov-Tamm waves are guided by the planar interface of a homogeneous isotropic dielectric material of refractive index $n^{\mathcal{B}} \in \{1.631, 1.635, 1.64, 1.645, 1.65\}$, and a titanium-oxide chiral STF described by Eqs. (6.6), (6.7), and (7.1), data from Table 6.1, $h^{\mathcal{A}} = 1$, $\Omega^{\mathcal{A}} = 197$ nm, and $\chi_v^{\mathcal{A}} = 7.2°$. If q is a solution for $\gamma^{\mathcal{A}}$, then it is also a solution for $\gamma^{\mathcal{A}} + 180°$ [46].

$$
\left[\underline{f}(0+) \right] = \begin{bmatrix} v_{11}^{\mathcal{A}} & v_{12}^{\mathcal{A}} \\ v_{21}^{\mathcal{A}} & v_{22}^{\mathcal{A}} \\ v_{31}^{\mathcal{A}} & v_{32}^{\mathcal{A}} \\ v_{41}^{\mathcal{A}} & v_{42}^{\mathcal{A}} \end{bmatrix} \cdot \begin{bmatrix} C_1^{\mathcal{A}} \\ C_2^{\mathcal{A}} \end{bmatrix} \tag{7.5}
$$

emerges with unknown coefficients $C_1^{\mathcal{A}}$ and $C_2^{\mathcal{A}}$. Finally, imposition of the boundary condition (3.84) leads to the dispersion equation (3.87) from which the wavenumber q for Dyakonov-Tamm-wave propagation may be numerically determined.

Figure 7.2 presents the solutions q of the dispersion equation for $\gamma^{\mathcal{A}} \in [-20°, 100°]$ and $n^{\mathcal{B}} \in \{1.631, 1.635, 1.64, 1.645, 1.65\}$ at $\lambda_0 = 633$ nm, when material $\mathcal{A}$ is a titanium-oxide chiral STF described by Eqs. (6.6), (6.7), and (7.1), data from Table 6.1, $h^{\mathcal{A}} = 1$, $\Omega^{\mathcal{A}} = 197$ nm, and $\chi_v^{\mathcal{A}} = 7.2°$. The chosen values for $n^{\mathcal{B}}$ roughly span the range for which the dispersion equation (3.87) has at least one solution for this particular example [46]. Twofold symmetry exists with respect to $\gamma^{\mathcal{A}}$: for any $\gamma^{\mathcal{A}}$ for which a solution q can be found, the same solution holds at $\gamma^{\mathcal{A}} + 180°$.

As the dispersion equation for Dyakonov-Tamm-wave propagation does not admit a solution for every $\gamma^{\mathcal{A}} \in [-90°, 90°]$, each curve in Figure 7.2 is drawn over the continuous $\gamma^{\mathcal{A}}$-range over which a solution exists. Every curve is smooth with a maximum at some $\gamma^{\mathcal{A}} \in (35°, 40°)$, and the widest ranges of q arise for mid-range values of $n^{\mathcal{B}}$. The widest $\gamma^{\mathcal{A}}$-range for Dyakonov-Tamm-wave propagation for the example considered in Figure 7.2 is approximately 98°, which arises for $n^{\mathcal{B}} = 1.64$. Clearly, the AEDs for Dyakonov-Tamm waves are much larger in comparison to those in Figure 4.2 for Dyakonov waves, when the anisotropic partnering materials in two cases have principal permittivity scalars, and birefringences of comparable magnitudes.

Close examination of Figure 7.2 shows that the phase speed $v_p = \omega/q$ of a Dyakonov-Tamm wave is lower than the phase speed $\omega/k_0 n^{\mathcal{B}}$ in the bulk isotropic partnering material. As $n^{\mathcal{B}} \geqslant 1$ is commonplace for optical materials, high-phase-speed Dyakonov-Tamm waves ($v_p \geqslant c_0$) shall be difficult to observe experimentally.

For a specific choice of the chiral STF $\mathcal{A}$ and the homogeneous dielectric material $\mathcal{B}$, Figure 7.2 indicates that at most one Dyakonov-Tamm-wave mode can propagate along a given direction in the interface plane at a specific frequency. This attribute is similar to that for Dyakonov-wave propagation as exemplified by Figure 4.4, but is in contrast to multiple Tamm-wave modes indicated in Figures 5.2 and 5.3. However, multiple Dyakonov-Tamm-wave modes are possible, as may be gathered from Figure 7.6 in Section 7.2.1.2.

The decay constants

$$
\left.\begin{aligned}
\beta_1^{\mathcal{A}} &= \exp\left(-2\Omega^{\mathcal{A}}\mathrm{Im}\left\{\alpha_1^{\mathcal{A}}\right\}\right) \in (0, 1) \\
\beta_2^{\mathcal{A}} &= \exp\left(-2\Omega^{\mathcal{A}}\mathrm{Im}\left\{\alpha_2^{\mathcal{A}}\right\}\right) \in (0, 1)
\end{aligned}\right\} \tag{7.6}
$$

indicate the decay of the two components of the Dyakonov-Tamm wave over one period of the chiral STF, and thus provide a measure of how tightly the Dyakonov-Tamm wave is bound to the plane $z = 0+$ for $z > 0$. The smaller a decay constant is, the stronger is the field associated with it bound to the interface. Likewise, the decay rate $-\mathrm{Im}\left\{\alpha^{\mathcal{B}}\right\}$ indicates the degree of localization to the plane $z = 0-$ for $z < 0$. The larger that a decay rate is, the more strongly is its associated field bound to the interface.

All three quantities are plotted in Figure 7.3 against $\gamma^{\mathcal{A}}$ for the same parameters as in Figure 7.2. Note that $\alpha_1^{\mathcal{A}}$, $\alpha_2^{\mathcal{A}}$, and $\alpha^{\mathcal{B}}$ turned out to be purely imaginary for this particular example. The Dyakonov-Tamm waves are bound most strongly to both sides of the interface for $\gamma^{\mathcal{A}} \in [35°, 40°]$, i.e. in the central part of the $\gamma^{\mathcal{A}}$-range in which propagation is allowed. Regardless of the specific value of $\gamma^{\mathcal{A}}$, in the chiral STF the

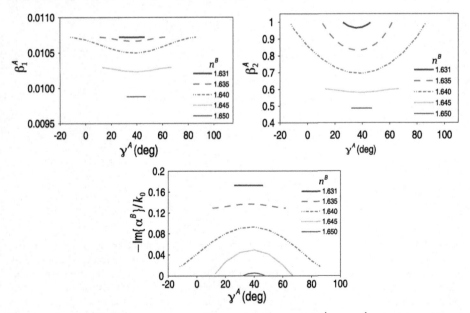

Figure 7.3 Same as Figure 7.2, except that the decay constants $\beta_1^{\mathcal{A}}$ and $\beta_2^{\mathcal{A}}$ and the normalized decay rate $-\mathrm{Im}\{\alpha^{\mathcal{B}}\}/k_0$ are plotted [46].

components of field phasors associated with α_1^A are much more strongly bound to the plane $z = 0+$ than the components associated with α_2^A.

The surface waves are bound very strongly to the interface for mid-range values of n^B, according to Figure 7.3. Delocalization occurs on the chiral-STF side ($z > 0$) of the interface at low values of n^B, but on the isotropic-material side ($z < 0$) of the interface for high values of n^B.

The spatial profiles of the field phasors $\underline{e}(z)$ and $\underline{h}(z)$, as well as the time-averaged Poynting vector $\underline{P}(x, y, z)$ in the vicinity of the interface $z = 0$, are provided in Figure 7.4 as functions of z for $n^B = 1.64$ and $\gamma^A = 40°$. The spatial profiles have a decaying periodic characteristic in material $\mathcal{A}$ in consonance with Floquet theory, but exhibit an exponential decay in material $\mathcal{B}$ because of Eq. (7.2). Also, the surface wave is considerably more tightly bound to the interface in the region occupied by the periodically nonhomogeneous partnering material than it is in the region occupied by the homogeneous partnering material. Also, the spatial profiles show that no polarization state can be assigned to a Dyakonov-Tamm wave, because the relative magnitudes and phases of the Cartesian components of $\underline{e}(z)$ are not uniform in the anisotropic partnering material.

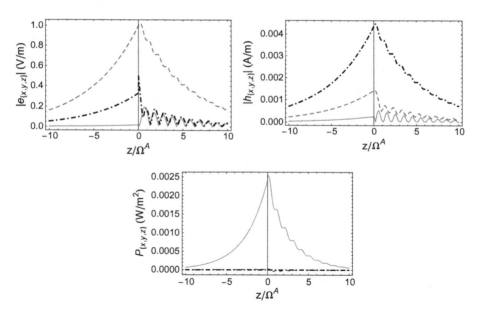

Figure 7.4 Magnitudes of the Cartesian components of $\underline{e}(z)$, $\underline{h}(z)$, and $\underline{P}(x, y, z)$ plotted against z/Ω^A for a Dyakonov-Tamm wave guided by the interface $z = 0$ of a chiral STF labeled $\mathcal{A}$ and a homogeneous dielectric material ($n^B = 1.64$), when $\lambda_0 = 633$ nm. All parameters of the chiral STF are specified in Figure 7.2, except that $\gamma^A = 40°$. The components parallel to $\underline{u}_x$, $\underline{u}_y$, and $\underline{u}_z$ are represented by red solid, blue dashed, and black chain-dashed lines, respectively. The data were computed by setting $C_s^B = 1$ V m^{-1}. (For interpretation of the references to color in this figure legend, the reader is referred to the web version of this book.)

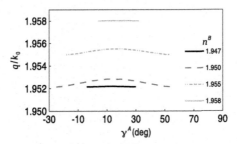

Figure 7.5 Same as Figure 7.2 but for $\chi_v^{\mathcal{A}} = 25°$ and $n^{\mathcal{B}} \in \{1.947, 1.95, 1.955, 1.958\}$ [46].

Qualitatively similar phase speeds, decay rates, and decay constants are obtained for other values of vapor flux angle $\chi_v^{\mathcal{A}}$. For example, in Figure 7.5 the solutions q of the dispersion equation are presented for the same set of partnering materials as for Figure 7.2 except that $\chi_v^{\mathcal{A}} = 25°$. Although the dependencies on $\gamma^{\mathcal{A}}$ and $n^{\mathcal{B}}$ in both figures are qualitatively similar, the phase speeds are lower for the higher value of $\chi_v^{\mathcal{A}}$. Furthermore, the $n^{\mathcal{B}}$-range for $\chi_v^{\mathcal{A}} = 25°$ is roughly half the width as that for $\chi_v^{\mathcal{A}} = 7.2°$; the AED for $\chi_v^{\mathcal{A}} = 25°$ is slightly less than that for $\chi_v^{\mathcal{A}} = 7.2°$; and the maximums of q with respect to $\gamma^{\mathcal{A}}$ occur at smaller values of $\gamma^{\mathcal{A}}$ for $\chi_v^{\mathcal{A}} = 25°$ than for $\chi_v^{\mathcal{A}} = 7.2°$. Additionally, Im $\{\alpha_1^{\mathcal{A}}\}$, Im $\{\alpha_2^{\mathcal{A}}\}$, and $-$Im $\{\alpha^{\mathcal{B}}\}$ are typically smaller for $\chi_v^{\mathcal{A}} = 25°$ than for $\chi_v^{\mathcal{A}} = 7.2°$, indicating the fields of a Dyakonov-Tamm wave are more weakly bound for the larger value of vapor flux angle used to fabricate the chiral STF.

Calculations at $\lambda_0 = 533$ and 733 nm, based on the same constitutive and geometric parameters as used for Figures 7.2–7.4, yield qualitatively similar results to the results for $\lambda_0 = 633$ nm [46]. Two general conclusions have been drawn: as the ratio $\lambda_0/\Omega^{\mathcal{A}}$ increases, the phase speed of the Dyakonov-Tamm wave tends to increase and it becomes less strongly bound to the interface.

Finally, a reversal in the structural handedness parameter $h^{\mathcal{A}}$ of the chiral STF, along with a concomitant reversal of the direction of the angle $\gamma^{\mathcal{A}}$, results in no change to the numerical results presented in this section.

7.2.1.2 Interface of an Isotropic Dielectric Material and a Sculptured Nematic Thin Film

Suppose next that the chiral STF in Section 7.2.1.1 is replaced by a periodically non-homogeneous SNTF described by

$$
\left.
\begin{aligned}
\underline{\underline{\epsilon}}^{\mathcal{A}}(z) &= \underline{\underline{A}}_z(\gamma^{\mathcal{A}}) \cdot \underline{\underline{S}}_y \left[\chi^{\mathcal{A}}(z)\right] \cdot \left[\epsilon_a^{\mathcal{A}}(z)\underline{u}_z\underline{u}_z \right. \\
&\quad + \epsilon_b^{\mathcal{A}}(z)\underline{u}_x\underline{u}_x + \epsilon_c^{\mathcal{A}}(z)\underline{u}_y\underline{u}_y \Big] \cdot \underline{\underline{S}}_y^{-1} \left[\chi^{\mathcal{A}}(z)\right] \cdot \underline{\underline{A}}_z^{-1}(\gamma^{\mathcal{A}}) \\
\chi_v^{\mathcal{A}}(z) &= \tilde{\chi}_v^{\mathcal{A}} + \delta_v^{\mathcal{A}} \sin\left(\frac{\pi z}{\Omega^{\mathcal{A}}}\right) \\
\underline{\underline{A}}_z(\gamma) &= \underline{u}_z\underline{u}_z + (\underline{u}_x\underline{u}_x + \underline{u}_y\underline{u}_y)\cos\gamma + (\underline{u}_y\underline{u}_x - \underline{u}_x\underline{u}_y)\sin\gamma
\end{aligned}
\right\}, \quad (7.7)
$$

whose provenance is discussed at some length in Section 6.4.2.

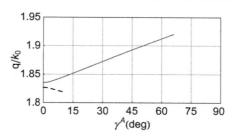

Figure 7.6 Solutions q of the canonical boundary-value problem as functions of $\gamma^{\mathcal{A}}$ for Dyakonov-Tamm-wave propagation at $\lambda_0 = 633$ nm guided by the interface of an isotropic dielectric material ($n^{\mathcal{B}} = 1.8$) and a periodically nonhomogeneous titanium-oxide SNTF. The SNTF is described by Eqs. (6.16) and (7.7), along with data from Table 6.1, $\Omega^{\mathcal{A}} = 197$ nm, $\tilde{\chi}_v^{\mathcal{A}} = 19.1°$, and $\delta_v^{\mathcal{A}} = 16.2°$. The solutions are organized in two branches [182].

Because the angle $\gamma^{\mathcal{A}}$ plays the same role as in Eq. (7.1), one can set $\psi = 0°$ so that $\underline{u}_{prop} = \underline{u}_x$. The formulation and solution of the canonical boundary-value problem [48, 182] for the isotropic-dielectric/SNTF interface follows the 4×4-matrix methodology of Section 3.5. Just as in Section 7.2.1.1 for the isotropic-dielectric/chiral-STF interface, no polarization state can be assigned to a Dyakonov-Tamm wave guided by the isotropic-dielectric/SNTF interface, except when $\sin \gamma^{\mathcal{A}} = 0$; then, two autonomous 2×2-matrix methodologies for p- and s-polarized Dyakonov-Tamm waves can be devised from the 4×4-matrix methodology.

A twofold symmetry exists for Dyakonov-Tamm waves guided by the isotropic-dielectric/SNTF interface. All results obtained for $\gamma^{\mathcal{A}} \in [0°, 90°]$ also hold for $-\gamma^{\mathcal{A}}$ [182].

Figure 7.6 provides solutions q of the dispersion equation (3.87) in relation to $\gamma^{\mathcal{A}}$ for Dyakonov-Tamm-wave propagation guided by the interface of an isotropic dielectric material ($n^{\mathcal{B}} = 1.8$) and a periodically nonhomogeneous SNTF made of titanium oxide and described by Eqs. (6.16) and (7.7), along with data from Table 6.1, $\Omega^{\mathcal{A}} = 197$ nm, $\tilde{\chi}_v^{\mathcal{A}} = 19.1°$, and $\delta_v^{\mathcal{A}} = 16.2°$. The solutions are organized into two branches. Two Dyakonov-Tamm waves are supported for $\gamma^{\mathcal{A}} \in [0°, 11°]$, one for $\gamma^{\mathcal{A}} \in [11°, 67°]$, but none for $\gamma^{\mathcal{A}} \in [67°, 90°]$. This figure demonstrates that multiple Dyakonov-Tamm-wave modes can exist.

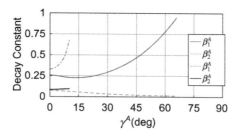

Figure 7.7 Same as Figure 7.6, except that decay constants $\beta_1^{\mathcal{A}}$ and $\beta_2^{\mathcal{A}}$ are plotted [182].

The decay constants β_1^A and β_2^A are plotted in Figure 7.7 as functions of γ^A for both solution branches in Figure 7.6. On both branches, the decay constants become quite small when γ^A approaches $0°$. Hence, both Dyakonov-Tamm waves are quite strongly bound to the interface on the SNTF side when γ^A is small. However, the decay constant β_1^A increases markedly as γ^A increases, which indicates that the Dyakonov-Tamm wave is more weakly bound to the interface on the SNTF side as γ^A increases toward its maximum value on the specific branch.

7.2.1.3 Interface of an Isotropic Dielectric Material and an Electro-Optic Reusch Pile

Due to the incorporation of periodic nonhomogeneity into material $\mathcal{A}$, Dyakonov-Tamm-wave propagation exhibits a much wider AED than Dyakonov-wave propagation. However, the ranges of values of n^B necessary for the existence of Dyakonov-Tamm waves in the examples presented in Sections 7.2.1.1 and 7.2.1.2 are small.

One way to widen the n^B-range (as well as the AED) in a controllable manner is to use an electro-optic material as material $\mathcal{A}$. Instead of a periodically nonhomogeneous STF, suppose that material $\mathcal{A}$ is the electro-optic version of a Reusch pile introduced in Section 1.4.5. This is an attractive proposition because a Reusch pile has many parameters—including the tilt angle with respect to the xy plane, number of homogeneous layers per period, and the thickness of each layer—which may be chosen to tailor the characteristics of Dyakonov-Tamm waves to specific applications. Furthermore, a Reusch pile made from an electro-optic material may be readily envisioned to allow the characteristics to be fine-tuned dynamically through an applied DC (or quasistatic) electric field [271]. The fabrication of an electro-optic Reusch pile seems a more practicable prospect than the fabrication of an electro-optic chiral STF or SNTF.

Instead of Eq. (1.39), the permittivity dyadic of the ℓth layer occupying the region $(\ell - 1)d_{\ell yr} < z < \ell d_{\ell yr}$ in the electro-optic Reusch pile is given by

$$\underline{\underline{\epsilon}}^A(z) = \underline{\underline{A}}_\ell \cdot \underline{\underline{S}}_y(\chi^A) \cdot \underline{\underline{\epsilon}}_{EO}^A \cdot \underline{\underline{S}}_y^{-1}(\chi^A) \cdot \underline{\underline{A}}_\ell^{-1},$$
$$(\ell - 1)d_{\ell yr} < z < \ell d_{\ell yr}, \quad \ell \in [1, \infty), \tag{7.8}$$

wherein

$$\underline{\underline{A}}_\ell = (\underline{u}_x\underline{u}_x + \underline{u}_y\underline{u}_y) \cos\left[\frac{h^A(\ell - 1)\pi}{Q} + \gamma^A\right]$$

$$+ (\underline{u}_y\underline{u}_x - \underline{u}_x\underline{u}_y) \sin\left[\frac{h^A(\ell - 1)\pi}{Q} + \gamma^A\right] + \underline{u}_z\underline{u}_z \tag{7.9}$$

allows for an offset angle γ^A in the same way as in Sections 7.2.1.1 and 7.2.1.2 and the structural handedness parameter $h^A \in \{-1, 1\}$. The parameter $Q > 1$ is an integer, and the permittivity dyadic of a Reusch pile approaches that of a chiral STF as $Q \to \infty$ while $Qd_{\ell yr}$ remains fixed. The permittivity dyadic $\underline{\underline{\epsilon}}_{EO}^A$ in Eq. (7.8) has the electro-optic form presented in Eq. (4.54).

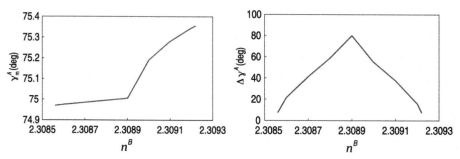

Figure 7.8 Midpoint $\gamma_m^{\mathcal{A}}$ and the width $\Delta\gamma^{\mathcal{A}}$ of the AED for Dyakonov-Tamm waves as functions of the refractive index $n^{\mathcal{B}}$ of material $\mathcal{B}$, when material $\mathcal{A}$ is an ambichiral and electro-optic Reusch pile consisting of layers of orthorhombic potassium niobate. Whereas $\chi^{\mathcal{A}} = 20°$, $h^{\mathcal{A}} = 1$, $Q = 3$, $d_{\ell yr} = 50$ nm, and $\underline{E}^{\mathrm{DC}} = 10^8 \underline{u}_z$ V m^{-1}, the constitutive parameters of potassium niobate entering Eqs. (4.54) and (7.8) are [354]: $\epsilon_1^{A0} = 4.70326\epsilon_0$, $\epsilon_2^{A0} = 5.19886\epsilon_0$, $\epsilon_3^{A0} = 5.42604\epsilon_0$, $r_{13} = 34 \times 10^{-12}$ V m^{-1}, $r_{23} = 6 \times 10^{-12}$ V m^{-1}, $r_{33} = 63.4 \times 10^{-12}$ V m^{-1}, $r_{42} = 450 \times 10^{-12}$ V m^{-1}, and $r_{51} = 120 \times 10^{-12}$ V m^{-1}, with all other electro-optic coefficients being null-valued [49].

Care must be exercised in choosing the thickness of each layer of the Reusch pile. If $d_{\ell yr}$ is too large in comparison to λ_0 then any surface waves which are excited will effectively be of the Dyakonov type, because the fields will decay to insignificant values within the first layer of the Reusch pile. If $d_{\ell yr}$ is too small in comparison to λ_0 then again any surface waves which are excited will be of the Dyakonov type as the Reusch pile will then be effectively homogeneous.

The angle $\gamma^{\mathcal{A}}$ plays the same role as in Eq. (7.1). Hence, one can set $\psi = 0°$ so that $\underline{u}_{\mathrm{prop}} = \underline{u}_x$. The formulation and solution of the canonical boundary-value problem [49] for the interface of isotropic dielectric material and the electro-optic Reusch pile follows the 4×4-matrix methodology of Section 3.5.

Let $\gamma_m^{\mathcal{A}}$ denote the midpoint of the $\gamma^{\mathcal{A}}$-range which supports Dyakonov-Tamm-wave propagation while $\Delta\gamma^{\mathcal{A}}$ denotes the width of this range. In Figure 7.8, $\gamma_m^{\mathcal{A}}$ and $\Delta\gamma^{\mathcal{A}}$ are plotted against $n^{\mathcal{B}}$, when $\lambda_0 = 633$ nm. The electro-optic Reusch pile is supposed to be made of potassium niobate, $\chi^{\mathcal{A}} = 20°$, $h^{\mathcal{A}} = 1$, $Q = 3$, $d_{\ell yr} = 50$ nm, and a DC field $\underline{E}^{\mathrm{DC}} = 10^8 \underline{u}_z$ V m^{-1} is supposed to be applied. The width of the $n^{\mathcal{B}}$-range that permits the propagation of Dyakonov-Tamm waves is just 0.00066, but the width of the AED is definitely much larger than for Dyakonov waves. Indeed, the maximum of $\Delta\gamma^{\mathcal{A}}$ in Figure 7.8 is approximately 80° which occurs at $n^{\mathcal{B}} = 2.3089$.

The range of values of $n^{\mathcal{B}}$ that supports the propagation of Dyakonov-Tamm waves and the maximum value of $\Delta\gamma^{\mathcal{A}}$ change only by small amounts if the calculations of Figure 7.8 are repeated for $\underline{E}^{\mathrm{DC}} = \underline{0}$. If, however, these calculations are repeated for larger magnitudes of $\underline{E}^{\mathrm{DC}}$ than 10^8 V m^{-1} then the range of values of $n^{\mathcal{B}}$ that supports Dyakonov-Tamm waves increases dramatically. This may be observed in Figure 7.9 wherein $\gamma_m^{\mathcal{A}}$ and $\Delta\gamma^{\mathcal{A}}$ are plotted against $n^{\mathcal{B}}$ for $\underline{E}^{\mathrm{DC}} = 10^9 \underline{u}_z$ V m^{-1}, with all other parameters being the same as for Figure 7.8. The $n^{\mathcal{B}}$-range for Dyakonov-Tamm-wave

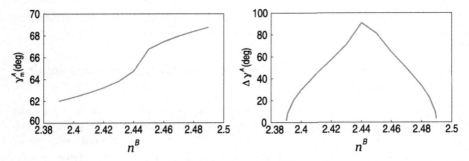

Figure 7.9 Same as Figure 7.8 except that $\underline{E}^{DC} = 10^9 \underline{u}_z$ V m^{-1} [49].

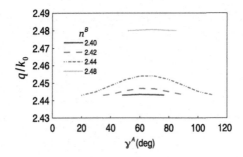

Figure 7.10 Solutions q of the canonical boundary-value problem illustrated in Figure 7.1 as functions of $\gamma^{\mathcal{A}}$, when $\psi = 0°$ and $\lambda_0 = 633$ nm. The Dyakonov-Tamm waves are guided by the planar interface of a homogeneous isotropic dielectric material of refractive index $n^{\mathcal{B}} \in \{2.4, 2.42, 2.44, 2.48\}$ and an electro-optic Reusch pile which is the same as for Figure 7.8 except that $\underline{E}^{DC} = 10^9 \underline{u}_z$ V m^{-1} [49].

propagation is approximately two orders larger in magnitude than in Figure 7.8. In addition, $\gamma_m^{\mathcal{A}}$ varies over ~7° in Figure 7.9 but over only ~0.4° in Figure 7.8. There is also a modest increase in the width of the AED.

The wavenumbers q of the Dyakonov-Tamm waves represented in Figure 7.9 are plotted against $\gamma^{\mathcal{A}}$ in Figure 7.10 for $n^{\mathcal{B}} \in \{2.4, 2.42, 2.44, 2.48\}$ and $\underline{E}^{DC} = 10^9 \underline{u}_z$ V m^{-1}. Clearly, the phase speed of the Dyakonov-Tamm wave decreases as the refractive index $n^{\mathcal{B}}$ increases. For each value of $n^{\mathcal{B}}$, q has a maximum value when $\gamma^{\mathcal{B}} \simeq 65°$ and has local minimum values at the two ends of the AED. For the $n^{\mathcal{B}}$-range depicted, the range of q is widest for $n^{\mathcal{B}} = 2.44$. This value of $n^{\mathcal{B}}$ also delivers the largest value of $\Delta\gamma^{\mathcal{A}}$. Qualitatively similar results are obtained when the DC electric field is of lower magnitude.

The magnitude of $\underline{E}^{DC}$ in Figures 7.9 and 7.10 corresponds to voltages comparable to the half-wave voltages of electro-optic materials. The half-wave voltage is the voltage required to bring about a 180° phase change in transmitted light [309]. Also, characteristic atomic electric field strengths are much larger than 10^9 V m^{-1}

[271]. Although electric breakdown is a possibility, its likelihood may be reduced by reducing the duration that the DC electric field is switched on.

7.2.2   Interface of Two Periodically Nonhomogeneous and Anisotropic Materials

More generally for Dyakonov-Tamm-wave propagation than in Section 7.2.1, both materials $\mathcal{A}$ and $\mathcal{B}$ are periodically nonhomogeneous along the z axis, and one of them is anisotropic. In the most general case, both materials are periodically nonhomogeneous and ansiotropic, as illustrated schematically in Figure 7.11. In this section, two examples are considered.

7.2.2.1   Interface of Two Sculptured Nematic Thin Films

Suppose that two SNTFs occupy the half-spaces $z > 0$ and $z < 0$. The applicable permittivity dyadics then are

$$
\left.
\begin{aligned}
\underline{\underline{\epsilon}}^{\mathcal{A}}(z) &= \underline{\underline{A}}_z(\gamma^{\mathcal{A}}) \cdot \underline{\underline{S}}_y\left[\chi^{\mathcal{A}}(z)\right] \cdot \left[\epsilon_a^{\mathcal{A}}(z)\underline{u}_z\underline{u}_z\right. \\
&\quad + \left. \epsilon_b^{\mathcal{A}}(z)\underline{u}_x\underline{u}_x + \epsilon_c^{\mathcal{A}}(z)\underline{u}_y\underline{u}_y\right] \cdot \underline{\underline{S}}_y^{-1}\left[\chi^{\mathcal{A}}(z)\right] \cdot \underline{\underline{A}}_z^{-1}(\gamma^{\mathcal{A}}) \\
\chi_v^{\mathcal{A}}(z) &= \tilde{\chi}_v^{\mathcal{A}} + \delta_v^{\mathcal{A}} \sin\left(\frac{\pi z}{\Omega^{\mathcal{A}}} + \tau^{\mathcal{A}}\right)
\end{aligned}
\right\}, \quad z > 0,
$$

$$(7.10)$$

and

$$
\left.
\begin{aligned}
\underline{\underline{\epsilon}}^{\mathcal{B}}(z) &= \underline{\underline{A}}_z(\gamma^{\mathcal{B}}) \cdot \underline{\underline{S}}_y\left[\chi^{\mathcal{B}}(z)\right] \cdot \left[\epsilon_a^{\mathcal{B}}(z)\underline{u}_z\underline{u}_z\right. \\
&\quad + \left. \epsilon_b^{\mathcal{B}}(z)\underline{u}_x\underline{u}_x + \epsilon_c^{\mathcal{B}}(z)\underline{u}_y\underline{u}_y\right] \cdot \underline{\underline{S}}_y^{-1}\left[\chi^{\mathcal{B}}(z)\right] \cdot \underline{\underline{A}}_z^{-1}(\gamma^{\mathcal{B}}) \\
\chi_v^{\mathcal{B}}(z) &= \tilde{\chi}_v^{\mathcal{B}} + \delta_v^{\mathcal{B}} \sin\left(\frac{\pi z}{\Omega^{\mathcal{B}}} - \tau^{\mathcal{B}}\right)
\end{aligned}
\right\}, \quad z < 0,
$$

$$(7.11)$$

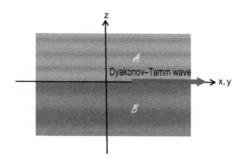

Figure 7.11 Schematic of the canonical boundary-value problem for Dyakonov-Tamm waves guided by the planar interface of two periodically nonhomogeneous and anisotropic materials $\mathcal{A}$ and $\mathcal{B}$.

where $\underline{\underline{A}}_z(\gamma)$ is defined in Eq. (7.7)$_3$, γ^A and γ^B are two angular offsets from the xz plane for the morphologically significant planes, and τ^A and τ^B are two phase angles for the rocking needed to fabricate SNTFs.

Since the angles γ^A and γ^B are independent of each other, there is no loss in generality taking the direction of Dyakonov-Tamm-wave propagation to be along the x axis, in keeping with Section 7.2.1. Then, $\psi = 0°$ and $\underline{u}_{\text{prop}} = \underline{u}_x$. The procedure to express the field phasors in the region $z \geqslant 0$ provided in Section 3.5.1 yields Eq. (3.80). Therefrom the boundary value

$$[\underline{f}(0+)] = \begin{bmatrix} v_{11}^A & v_{12}^A \\ v_{21}^A & v_{22}^A \\ v_{31}^A & v_{32}^A \\ v_{41}^A & v_{42}^A \end{bmatrix} \cdot \begin{bmatrix} C_1^A \\ C_2^A \end{bmatrix} \tag{7.12}$$

emerges with unknown coefficients C_1^A and C_2^A. The same procedure for the half-space $z \leqslant 0$ gives Eq. (3.82), which provides the boundary value

$$[\underline{f}(0-)] = \begin{bmatrix} v_{13}^B & v_{14}^B \\ v_{23}^B & v_{24}^B \\ v_{33}^B & v_{34}^B \\ v_{43}^B & v_{44}^B \end{bmatrix} \cdot \begin{bmatrix} C_3^B \\ C_4^B \end{bmatrix} \tag{7.13}$$

with unknown coefficients C_3^B and C_4^B. Finally, imposition of the boundary condition (3.84) leads to the dispersion equation (3.87) from which the wavenumber q for Dyakonov-Tamm-wave propagation may be numerically determined.

Equations (7.10) and (7.11) contain a large number of parameters to describe the two partnering SNTFs. Although a comprehensive program to investigate surface waves guided by the interface of two distinct SNTFs has yet to be undertaken, some progress can be made by examining the interface of two identical SNTFs that differ only in the orientations of their morphologically significant planes and the phase angles for substrate rocking. This pair of SNTFs can be fabricated as follows: During the PVD process, the substrate is rocked back and forth sinusoidally about an axis parallel to the plane $z = 0$, in consonance with Eq. (7.11)$_2$. When the switch to Eq. (7.10)$_2$ has to be implemented, the sinusoidal rocking is halted, the vapor flux is switched off, and the substrate is rotated by an angle $\gamma^A - \gamma^B$ about a suiTable axis parallel to z axis. Then the vapor flux is switched back on at an angle which is changed by $\tau^A + \tau^B$ and the sinusoidal rocking motion is resumed.

Figure 7.12 presents solutions q of the dispersion equation (3.87) at $\lambda_0 = 633$ nm [183]. Both SNTFs are made of titanium oxide and are described by Eqs. (6.16), (7.10), and (7.11), along with data from Table 6.1, $\Omega^A = \Omega^B = 200$ nm, $\tilde{\chi}_v^A = \tilde{\chi}_v^B = 45°$, $\delta_v^A = \delta_v^B = 30°$, $\tau^A = 0°$, and $\tau^B = 180°$. Twofold symmetry exists for Dyakonov-Tamm-wave propagation in relation to the angle γ^A: for every $\gamma^A \in [0°, 180°]$, the propagation characteristics are the same as for $\gamma^A + 180°$, provided that

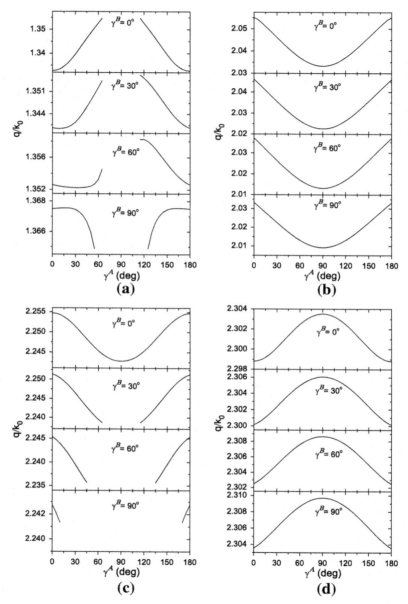

Figure 7.12 Four sets of solutions q of the canonical boundary-value problem illustrated in Figure 7.11 as functions of $\gamma^{\mathcal{A}}$, when $\psi = 0°$ and $\lambda_0 = 633$ nm and $\gamma^{\mathcal{B}} \in \{0°, 30°, 60°, 90°\}$. The Dyakonov-Tamm waves are guided by the interface of two periodically nonhomogeneous titanium-oxide SNTFs described by Eqs. (6.16), (7.10), and (7.11), along with data from Table 6.1, $\Omega^{\mathcal{A}} = \Omega^{\mathcal{B}} = 200$ nm, $\tilde{\chi}_v^{\mathcal{A}} = \tilde{\chi}_v^{\mathcal{B}} = 45°$, $\delta_v^{\mathcal{A}} = \delta_v^{\mathcal{B}} = 30°$, $\tau^{\mathcal{A}} = 0°$, and $\tau^{\mathcal{B}} = 180°$ [183]. (a) First set, (b) second set, (c) third set, and (d) fourth set.

γ^B is fixed. Analogous twofold symmetry also exists with respect to γ^B. Furthermore, solutions for $\{\gamma^A, \gamma^B\}$ are equivalent to those for $\{180° - \gamma^A, 180° - \gamma^B\}$.

The dispersion relation (3.87) yields four sets of solutions (i.e. sets of values of wavenumber q) for $\gamma^A \in [0°, 180°]$ when γ^B is held fixed in the range $[0°, 90°]$. Figure 7.12 contains a representative selection of results wherein $\gamma^A \in [0°, 180°]$ and $\gamma^B \in \{0°, 30°, 60°, 90°\}$. Solutions do not exist in the first and the third sets for all combinations $\{\gamma^A, \gamma^B\}$ considered, but that is not true for the second and fourth sets. Thus, multiple Dyakonov-Tamm-wave modes are possible for any direction.

The solution sets are organized in Figure 7.12 in increasing magnitudes of q and therefore decreasing magnitudes of v_p. The phase speeds of all Dyakonov-Tamm-wave modes in this figure are less than c_0.

The extent to which a Dyakonov-Tamm-wave mode is bound in material $\mathcal{A}$ to the plane $z = 0+$ is determined by the decay constants β_1^A and β_2^A defined in Eqs. (7.6). Likewise, the decay constants

$$\left. \begin{array}{l} \beta_3^B = \exp\left(2\Omega^B \text{Im}\left\{\alpha_3^B\right\}\right) \in (0, 1) \\ \beta_4^B = \exp\left(2\Omega^B \text{Im}\left\{\alpha_4^B\right\}\right) \in (0, 1) \end{array} \right\} \tag{7.14}$$

determine the extent to which the Dyakonov-Tamm-wave mode is bound in material $\mathcal{B}$ to the plane $z = 0-$. Smaller decay constants imply stronger binding.

Figure 7.13 presents all four decay constants for the four sets of solutions for Dyakonov-Tamm-wave propagation identified in Figure 7.12. For the first set of solutions, the decay constants vary between 0.7 and 0.99, and indicate that the Dyakonov-Tamm-wave modes are loosely bound to the interface in both partnering materials. Stronger binding is indicated for the second set of solutions, because the decay constants range from 0.1 to 0.45. The variation in the decay constants for the third set of solutions is considerable, ranging from almost 0 to 0.9. The decay constants range from almost 0 to 0.12 for the fourth set of solutions, indicating thereby that the Dyakonov-Tamm-wave modes in this set of solutions are the most tightly bound to the interface. The Dyakonov-Tamm-wave modes in the fourth set also have the smallest phase speeds.

Spatial profiles of the fields confirm the conclusions offered by the decay constants. For instance, plots of the magnitudes of the Cartesian components of the field phasors $\underline{e}(z)$ and $\underline{h}(z)$, as well as the time-averaged Poynting vector $\underline{P}(x, y, z)$, in the vicinity of the interface $z = 0$ are provided in Figure 7.14 as functions of z for a Dyakonov-Tamm-wave mode belonging to the first set in Figure 7.12. All three quantities decay more rapidly in the half-space $z > 0$ than in the half-space $z < 0$. Indeed, the magnitudes in the half-space $z < 0$ are significant even at distances greater than 30 periods from the interface. These observations are consistent with the relative magnitudes of the decay constants available from Figure 7.13: $\beta_1^A = 0.94$ and $\beta_2^A = 0.74$ for $z > 0$, while $\beta_3^B = 0.99$ and $\beta_4^B = 0.84$ for $z < 0$. Asymmetry of the spatial profiles and the decay constants with respect to the interface $z = 0$ stems from the asymmetric arrangement, about the propagation direction, of the morphologically significant planes of the SNTF on each side of $z = 0$.

The spatial profiles of $\underline{e}(z)$, $\underline{h}(z)$, and $\underline{P}(x, y, z)$ for a solution from the second set are presented in Figure 7.15. Clearly, the Dyakonov-Tamm-wave mode is largely

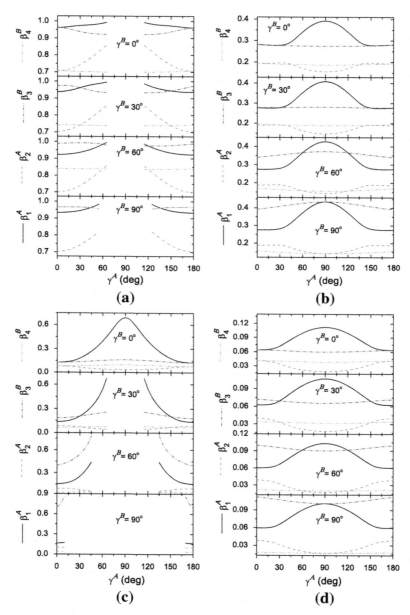

Figure 7.13 Same as Figure 7.12, except that the decay constants β_1^A, β_2^A, β_3^B, and β_4^B are plotted [183]. (a) First set, (b) second set, (c) third set, and (d) fourth set.

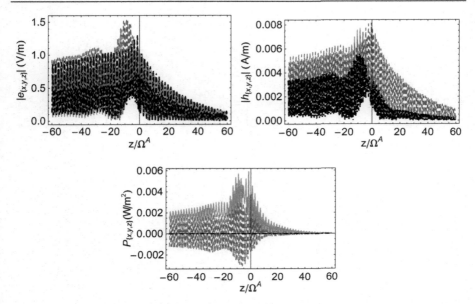

Figure 7.14 Magnitudes of the Cartesian components of $\underline{e}(z)$, $\underline{h}(z)$, and $\underline{P}(x, y, z)$ plotted against $z/\Omega^{\mathcal{A}}$ for a Dyakonov-Tamm-wave mode guided by the interface $z = 0$ of two SNTFs, when $\psi = 0°$ and $\lambda_0 = 633$ nm. All parameters of the two SNTFs are specified in Figure 7.12, except that $\gamma^{\mathcal{A}} = 30°$ and $\gamma^{\mathcal{B}} = 60°$. The wavenumber $q = 1.3522k_0$ belongs to the first set of solutions in Figures 7.12 and 7.13. The components parallel to $\underline{u}_x$, $\underline{u}_y$, and $\underline{u}_z$ are represented by red solid, blue dashed, and black chain-dashed lines, respectively. The data were computed by setting $C_1^{\mathcal{A}} = 1$ V m^{-1} [183]. (For interpretation of the references to color in this figure legend, the reader is referred to the web version of this book.)

confined within one period on either side of the interface. The spatial profiles for $z < 0$ approximately mirror those for $z > 0$, but a small degree of asymmetry is apparent because the morphologically significant planes of the two SNTFs are oriented asymmetrically with respect to the direction of propagation.

7.2.2.2 Interface of Two Chiral Sculptured Thin Films

In a similar vein to the SNTF/SNTF interface of Section 7.2.2.1, multiple Dyakonov-Tamm-wave modes may be guided by the interface of two chiral STFs. A variety of different configurations have been contemplated: the chiral STFs occupying the half-spaces $z > 0$ and $z < 0$ could be dissimilar in (i) orientation about the helical axis [50,51], (ii) structural handedness [52], (iii) period, (iv) vapor flux angle [355], (v) material, or (vi) or any combination of (i)–(v) [53]. As is true for the SNTF/SNTF interface in Section 7.2.2.1, the multiple Dyakonov-Tamm-wave modes guided by any chiral-STF/chiral-STF interface differ in phase speed, spatial profiles, and the degree of binding to the interface.

A notable difference between Dyakonov-Tamm-wave modes guided by the chiral-STF/chiral-STF interface as compared with the SNTF/SNTF interface concerns the

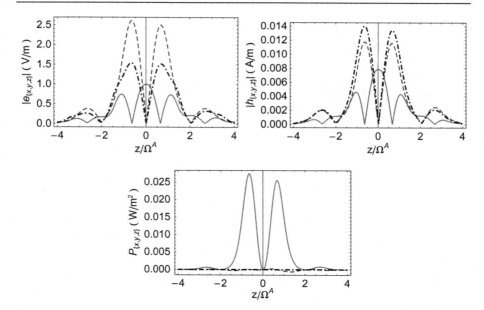

Figure 7.15 Same as Figure 7.14 but for the wavenumber $q = 2.02646k_0$ belonging to the second set of solutions in Figures 7.12 and Figure 7.13 [183].

localization to the interface. If one compares data for the SNTF/SNTF interface presented in Section 7.2.2.1 [183] with the data for a comparable chiral-STF/chiral-STF interface [50,51], then one finds that the most tightly bound Dyakonov-Tamm-wave modes are confined to within one period on either side of the SNTF/SNTF interface, whereas the Dyakonov-Tamm-wave modes are at best confined within two periods on either side of the chiral-STF/chiral-STF interface. This difference may be attributed to the different morphological dimensionality of the two types of STFs—chiral STFs have a three-dimensional morphology, but the morphology of SNTFs is essentially two-dimensional.

7.3 Practical Configurations

The canonical boundary-value problem exemplified in Section 7.2 is appropriate for determining possible wavenumbers for Dyakonov-Tamm waves. However, it is unimplemenTable because both partnering materials occupy half-spaces. In practical configurations, both materials have to be finite in extent in the direction normal to their planar interface.

7.3.1 Prism-Coupled Configuration

Being quite straightforward to implement, the prism-coupled configuration of Figure 5.5 has been employed for experiments with Tamm waves [174–177]. Whereas

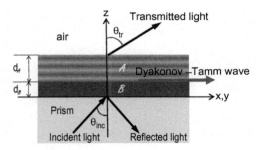

Figure 7.16 Schematic of the prism-coupled configuration for Dyakonov-Tamm waves guided by the planar interface of a periodically nonhomogeneous and anisotropic material $\mathcal{A}$ and a homogeneous isotropic material $\mathcal{B}$.

material $\mathcal{A}$ is isotropic and periodically nonhomogeneous, material $\mathcal{B}$ is isotropic and homogeneous. If material $\mathcal{A}$ is made anisotropic, as depicted in Figure 7.16, the prism-coupled configuration can provide the simplest experimental method to excite Dyakonov-Tamm waves.

The thicknesses of materials $\mathcal{A}$ and $\mathcal{B}$ are $d_{\mathcal{A}}$ and $d_{\mathcal{B}}$, respectively. The layer of material $\mathcal{B}$ is in contact with a prism of refractive index $n_{\text{prism}} > 1$, whereas the layer of material $\mathcal{A}$ is in contact with air ($n_{\text{air}} = 1$). For analytical purposes, a linearly polarized plane wave is taken to be incident on the interface of the prism and material $\mathcal{B}$ at an angle $\theta_{\text{inc}} \in [0°, 90°)$ with respect to the z axis, as shown in Figure 7.16, and at an angle $\psi \in [0°, 360°)$ to the x axis in the xy plane. The plane wave transmitted into air has a wave vector that is inclined at an angle θ_{tr} to the z axis and at the angle $\psi \in [0°, 360°)$ to the x axis in the xy plane. The angle θ_{tr} could be complex valued. The theory underpinning this prism-coupled excitation of surface waves is presented in detail in Section 3.7.

While ψ is kept constant, θ_{inc} is varied from $0°$ towards $90°$, and the reflectance back into the prism is computed as a function of that angle. Even though both partnering materials for Dyakonov-Tamm-wave propagation are supposed to have negligible dissipation, $n^{\mathcal{B}}$ must be endowed with a small positive imaginary part [356] so that some reflectance dips for $\theta_{\text{inc}} > \sin^{-1}(1/n_{\text{prism}})$ indicate the excitation of Dyakonov-Tamm waves, as is illustrated for Tamm waves in Section 5.3.1. The reflectance dips for $\theta_{\text{inc}} > \sin^{-1}(1/n_{\text{prism}})$ also manifest themselves as absorptance peaks.

The correspondence between the prism-coupled configuration and the underlying canonical boundary-value problem is illustrated in Figure 7.17. For this figure, material $\mathcal{A}$ is a titanium-oxide SNTF described by

$$
\left.
\begin{aligned}
\underline{\underline{\epsilon}}^{\mathcal{A}}(z) &= \underline{\underline{S}}_y \left[\chi^{\mathcal{A}}(z - d_{\mathcal{B}})\right] \cdot \left[\epsilon_a^{\mathcal{A}}(z - d_{\mathcal{B}})\underline{u}_z\underline{u}_z \right. \\
&\quad + \epsilon_b^{\mathcal{A}}(z - d_{\mathcal{B}})\underline{u}_x\underline{u}_x + \epsilon_c^{\mathcal{A}}(z - d_{\mathcal{B}})\underline{u}_y\underline{u}_y\Big] \\
&\quad \cdot \underline{\underline{S}}_y^{-1}\left[\chi^{\mathcal{A}}(z - d_{\mathcal{B}})\right] \\
\chi_v^{\mathcal{A}}(z) &= \tilde{\chi}_v^{\mathcal{A}} + \delta_v^{\mathcal{A}}\sin\left(\frac{\pi z}{\Omega^{\mathcal{A}}}\right)
\end{aligned}
\right\}, \quad z > d_{\mathcal{B}}, \tag{7.15}
$$

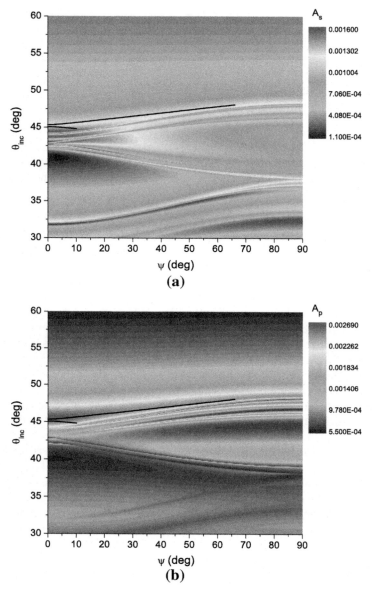

Figure 7.17 Density plots of linear absorptances (a) A_s and (b) A_p as functions of θ_{inc} and ψ in the prism-coupled configuration of Figure 7.16, when $\lambda_0 = 633$ nm. Material $\mathcal{A}$ of thickness $d_{\mathcal{A}} = 6\Omega^{\mathcal{A}}$ is a periodically nonhomogeneous titanium-oxide SNTF described by Eqs. (6.16) and (7.15), along with data from Table 6.1, $\Omega^{\mathcal{A}} = 197$ nm, $\tilde{\chi}_v^{\mathcal{A}} = 19.1°$, and $\delta_v^{\mathcal{A}} = 16.2°$. The refractive index of the homogeneous partnering material $\mathcal{B}$ is $n^{\mathcal{B}} = 1.8\,(1 + 10^{-4}i)$ and its thickness $d_{\mathcal{B}} = 75$ nm. The prism material is zinc selenide ($n_{\text{prism}} = 2.58$) [356]. The black overlaid curves represent values of $\sin^{-1}(q/k_0 n_{\text{prism}})$ obtained for each ψ from solutions of the underlying canonical boundary-value problem; comparison with Figure 7.6 is recommended.

along with Eqs. (6.16), data from Table 6.1, $\Omega^{\mathcal{A}} = 197$ nm, $\tilde{\chi}_v^{\mathcal{A}} = 19.1°$, and $\delta_v^{\mathcal{A}} = 16.2°$, whereas the homogeneous material $\mathcal{B}$ has refractive index $n^{\mathcal{B}} = 1.8\,(1 + 10^{-4}i)$. With the region $z < 0$ filled with a medium of refractive index the same as zinc selenide and the half-space $z > d_{\mathcal{A}} + d_{\mathcal{B}}$ filled with air, the absorptances A_s and A_p at $\lambda_0 = 633$ nm were calculated as functions of θ_{inc} and ψ for sufficiently large thicknesses of both partnering materials [356]. Density plots of A_s and A_p with respect to both angles are provided in Figure 7.17.

The thickness of a partnering material that is periodically nonhomogeneous should be increased in increments of a period, for the identification of surface waves. Absorptance peaks whose angular locations do not depend upon the thicknesses of either material $\mathcal{A}$ or material $\mathcal{B}$, beyond certain threshold thicknesses, indicate the possible excitation of Dyakonov-Tamm-wave modes. In contrast, absorptance peaks whose angular locations depend upon the thicknesses of the partnering materials are attributed to waveguide modes. A narrow band of high absorptance is evident in Figure 7.17, starting at $\{\theta_{\text{inc}} = 45.5°, \psi = 0°\}$ and ending at $\{\theta_{\text{inc}} = 48.6°, \psi = 90°\}$. The high-absorptance band is widest at $\psi = 0°$ and decreases in width as ψ increases.

Overlaid on the density plots are two black curves which represent the locus of $\sin^{-1}(q/k_0 n_{\text{prism}})$ as a function of ψ obtained by solving the underlying canonical boundary-value problem discussed in Section 7.2.1.2. Each curve represents a branch of solutions. Both branches lie entirely within the high-absorptance band. One branch extends over $\psi \in [0°, 66°]$ while the other extends over $\psi \in [0°, 11°]$. As the separation in terms of θ_{inc} for the two branches over $\psi \in [0°, 11°]$ is very small, separate high-absorptance bands, corresponding to the two solutions to the canonical boundary-value problem, cannot be discerned in the density plots.

Since solutions of the canonical boundary-value problem exist only for $\psi \in [0°, 66°]$, it may be inferred that the high-absorptance bands for $\psi \in [66°, 90°]$ do not indicate the excitation of Dyakonov-Tamm waves. But Dyakonov-Tamm-wave excitation definitely arises for $\psi \in [0°, 66°]$, for incident plane waves of both s- and p-polarization states.

7.3.2   Grating-Coupled Configuration

Grating coupling provides another practical configuration for the excitation of Dyakonov-Tamm waves. The general theory describing this configuration is available in Section 3.8.

7.4   Outlook

Two canonical boundary-value problems have been solved to investigate the characteristics of Dyakonov-Tamm waves. Only one of the two partnering materials is anisotropic and periodically nonhomogeneous in one canonical boundary-value problem, while both partnering materials are anisotropic and periodically nonhomogeneous in the other one. Nevertheless, a comprehensive understanding has not yet been developed due to the plethora of constitutive parameters involved when both partnering materials can be bianisotropic and periodically nonhomogeneous in the direction normal to the interface plane. Neither has practical configurations required to excite these surface waves been extensively investigated.

With one partnering material both isotropic and homogeneous, experimental observation of Dyakonov-Tamm waves is not significantly more arduous than that of Tamm waves—which had been observed a few decades ago. Implementation of the prism-coupled configurations is straightforward and the AEDs of Dyakonov-Tamm waves are much more favorable toward experimental observation than those of Dyakonov waves.

The prism-coupled configurations can be exploited for optical sensing and may eventually lead to the development of Dyakonov-Tamm-wave sensors. Application for optical sensing is a significant incentive for researchers to bestow attention on these surface waves.

Appendix A

Dyadics

A brief introduction to dyadics and their properties is provided in this Appendix. For further details about dyadics, especially in the context of electromagnetics, the reader is referred elsewhere [78,357].

A dyad is composed of two vectors:

$$\underline{d} = \underline{a}\,\underline{b}. \tag{A.1}$$

While $\underline{d} \cdot \underline{p} = \underline{a}(\underline{b} \cdot \underline{p})$ and $\underline{p} \cdot \underline{d} = (\underline{p} \cdot \underline{a})\underline{b}$ are vectors, $\underline{d} \times \underline{p} = \underline{a}(\underline{b} \times \underline{p})$ and $\underline{p} \times \underline{d} = (\underline{p} \times \underline{a})\underline{b}$ are dyads. The transpose of a dyad $\underline{a}\,\underline{b}$ is the dyad $\underline{b}\,\underline{a}$. Sometimes, dyads are called bivectors. Some authors prefer to write the dyad $\underline{a}\,\underline{b}$ as $\underline{b} \otimes \underline{a}$.

A dyadic is a linear mapping from one n-vector to another. For 3-vectors, the mapping can be straightforwardly described in terms of Cartesian vectors. Thus, a 3×3 dyadic $\underline{\underline{M}}$ is a linear mapping from a Cartesian 3-vector $\underline{p}$ to Cartesian 3-vector $\underline{q}$:

$$\underline{q} = \underline{\underline{M}} \cdot \underline{p}. \tag{A.2}$$

The identity dyadic $\underline{\underline{I}}$ is such that $\underline{p} \cdot \underline{\underline{I}} = \underline{\underline{I}} \cdot \underline{p} = \underline{p}$; likewise, the null dyadic $\underline{\underline{0}}$ is defined so that $\underline{p} \cdot \underline{\underline{0}} = \underline{\underline{0}} \cdot \underline{p}$ equals the null vector $\underline{0}$.

A dyadic is not necessarily a dyad. The general representation of a dyadic is as a sum of dyads; i.e.

$$\underline{\underline{M}} = \sum_{\ell=1,2,\dots} \underline{\underline{d}}_\ell = \sum_{\ell=1,2,\dots} M_\ell \underline{u}_\ell \underline{v}_\ell, \tag{A.3}$$

where M_ℓ are some scalar coefficients, while $\underline{u}_\ell$ and $\underline{v}_\ell$ are vectors of unit magnitude.

All vectors can be written using matrix notation. Thus, the vector $\underline{a} = \underline{u}_x a_x + \underline{u}_y a_y + \underline{u}_z a_z$ in a Cartesian coordinate system is equivalent to the column 3-vector

$$\underline{a} \equiv \begin{bmatrix} a_x \\ a_y \\ a_z \end{bmatrix} \tag{A.4}$$

and the dyad $\underline{d} = \underline{a}\,\underline{b}$ is equivalent to the 3×3 matrix

$$\underline{\underline{d}} \equiv \begin{bmatrix} a_x b_x & a_x b_y & a_x b_z \\ a_y b_x & a_y b_y & a_y b_z \\ a_z b_x & a_z b_y & a_z b_z \end{bmatrix}. \tag{A.5}$$

Electromagnetic Surface Waves. http://dx.doi.org/10.1016/B978-0-12-397024-4.00008-6

Hence, dyadics in electromagnetics can be written as 3×3 matrixes. The identity dyadic $\underline{\underline{I}}$ is equivalent to the 3×3 identity matrix, and the null dyadic $\underline{\underline{0}}$ to the 3×3 null matrix. The usual algebra of matrixes can thus be used for dyadics as well.

The trace of a dyadic is the sum of the diagonal elements in its matrix representation. Likewise, the determinant of a dyadic is the same as the determinant of its equivalent matrix. A dyadic can be transposed in the same way as a matrix. If it is nonsingular, a dyadic can be inverted.

The anti-symmetric dyadic

$$\underline{a} \times \underline{\underline{I}} = \underline{\underline{I}} \times \underline{a} \tag{A.6}$$

$$= a_x(\underline{u}_z\underline{u}_y - \underline{u}_y\underline{u}_z) + a_y(\underline{u}_x\underline{u}_z - \underline{u}_z\underline{u}_x) + a_z(\underline{u}_y\underline{u}_x - \underline{u}_x\underline{u}_y) \tag{A.7}$$

$$\equiv \begin{bmatrix} 0 & -a_z & a_y \\ a_z & 0 & -a_x \\ -a_y & a_x & 0 \end{bmatrix} \tag{A.8}$$

is often useful to denote gyrotropic electromagnetic properties that are characteristic of ferrites and plasmas. The simplest anti-symmetric dyadic is $\underline{u} \times \underline{\underline{I}}$, where $\underline{u}$ is any vector of unit magnitude. The trace of any anti-symmetric dyadic is zero.

Even vector differential operators can be thought of as dyadics. Thus, the *curl* operator is written as $\nabla \times \underline{\underline{I}}$ and the *divergence* operator as $\nabla \cdot \underline{\underline{I}}$, with

$$\nabla = \underline{u}_x \frac{\partial}{\partial x} + \underline{u}_y \frac{\partial}{\partial y} + \underline{u}_z \frac{\partial}{\partial z} \tag{A.9}$$

written as a vector in a Cartesian coordinate system.

Appendix B

Biaxial Permittivity Dyadic

Two different ways are commonly used to express the permittivity dyadic $\underline{\underline{\epsilon}}$ of a homogeneous biaxial dielectric material. The more transparent way is

$$\underline{\underline{\epsilon}} = \epsilon_1 \underline{u}_1 \underline{u}_1 + \epsilon_2 \underline{u}_2 \underline{u}_2 + \epsilon_3 \underline{u}_3 \underline{u}_3, \tag{B.1}$$

where the unit vectors $\underline{u}_1, \underline{u}_2$, and $\underline{u}_3$ are the eigenvectors of $\underline{\underline{\epsilon}}$ while the principal permittivity scalars ϵ_1, ϵ_2, and ϵ_3 are the corresponding eigenvalues of $\underline{\underline{\epsilon}}$. The three eigenvectors form a right-handed triad:

$$\underline{u}_1 \times \underline{u}_2 = \underline{u}_3, \quad \underline{u}_3 \times \underline{u}_1 = \underline{u}_2, \quad \underline{u}_2 \times \underline{u}_3 = \underline{u}_1. \tag{B.2}$$

In the remainder of this Appendix, dissipation is ignored and all three eigenvalues are assumed to be real valued and positive.

For a biaxial material, all three eigenvalues of $\underline{\underline{\epsilon}}$ are distinct. Without loss of generality, let us assume that

$$\epsilon_1 > \epsilon_2 > \epsilon_3 > 0. \tag{B.3}$$

From these quantities, the three principal refractive indexes of the biaxial dielectric material can be obtained as

$$n_1 = \sqrt{\epsilon_1/\epsilon_0}, \quad n_2 = \sqrt{\epsilon_2/\epsilon_0}, \quad n_3 = \sqrt{\epsilon_3/\epsilon_0}. \tag{B.4}$$

The *birefringence* of the material is defined as

$$\Delta n = n_1 - n_3. \tag{B.5}$$

The *biaxiality* of the material is defined in terms of the angle

$$\delta = \cos^{-1} \sqrt{\frac{\epsilon_1 - \epsilon_2}{\epsilon_1 - \epsilon_3}}. \tag{B.6}$$

Positive biaxiality requires that $\delta \in (0°, 45°)$, which implies that

$$\frac{\epsilon_1 - \epsilon_2}{\epsilon_1 - \epsilon_3} \in \left(\frac{1}{2}, 1\right). \tag{B.7}$$

Electromagnetic Surface Waves. http://dx.doi.org/10.1016/B978-0-12-397024-4.00009-8

Negative biaxiality requires that $\delta \in (45°, 90°)$, which implies that

$$\frac{\epsilon_1 - \epsilon_2}{\epsilon_1 - \epsilon_3} \in \left(0, \frac{1}{2}\right). \tag{B.8}$$

When ϵ_2 is the average of ϵ_1 and ϵ_3, then $\delta = 45°$ and the material is neither positive biaxial nor negative biaxial.

Equation (B.1) can be rewritten as

$$\underline{\underline{\epsilon}} = \epsilon_2 \underline{\underline{I}} + \frac{\epsilon_1 - \epsilon_3}{2}(\underline{a}_+\underline{a}_- + \underline{a}_-\underline{a}_+), \tag{B.9}$$

which is the other common way of expressing $\underline{\underline{\epsilon}}$. Here the unit vectors

$$\underline{a}_+ = \underline{u}_1 \sqrt{\frac{\epsilon_1 - \epsilon_2}{\epsilon_1 - \epsilon_3}} + \underline{u}_3 \sqrt{\frac{\epsilon_2 - \epsilon_3}{\epsilon_1 - \epsilon_3}} \tag{B.10}$$

and

$$\underline{a}_- = \underline{u}_1 \sqrt{\frac{\epsilon_1 - \epsilon_2}{\epsilon_1 - \epsilon_3}} - \underline{u}_3 \sqrt{\frac{\epsilon_2 - \epsilon_3}{\epsilon_1 - \epsilon_3}} \tag{B.11}$$

are parallel to the two optic ray axes of the biaxial dielectric material. The relationship

$$\underline{a}_+ \cdot \underline{a}_- = \cos(2\delta) \tag{B.12}$$

is often useful. When the material is neither positive biaxial nor negative biaxial, then the two optic ray axes are mutually orthogonal.

The inverse of the permittivity dyadic can be expressed as

$$\underline{\underline{\epsilon}}^{-1} = \frac{1}{\epsilon_2}\underline{\underline{I}} + \frac{\frac{1}{\epsilon_3} - \frac{1}{\epsilon_1}}{2}(\underline{b}_+\underline{b}_- + \underline{b}_-\underline{b}_+). \tag{B.13}$$

The unit vectors

$$\underline{b}_+ = \underline{u}_1 \sqrt{\frac{(\epsilon_1 - \epsilon_2)\epsilon_3}{(\epsilon_1 - \epsilon_3)\epsilon_2}} + \underline{u}_3 \sqrt{\frac{(\epsilon_2 - \epsilon_3)\epsilon_1}{(\epsilon_1 - \epsilon_3)\epsilon_2}} \tag{B.14}$$

and

$$\underline{b}_- = -\underline{u}_1 \sqrt{\frac{(\epsilon_1 - \epsilon_2)\epsilon_3}{(\epsilon_1 - \epsilon_3)\epsilon_2}} + \underline{u}_3 \sqrt{\frac{(\epsilon_2 - \epsilon_3)\epsilon_1}{(\epsilon_1 - \epsilon_3)\epsilon_2}} \tag{B.15}$$

are parallel to the two optic axes of the biaxial dielectric material.

A biaxial dielectric material becomes uniaxial, if two of the eigenvalues of its permittivity dyadic are equal and different from the third eigenvalue. Positive uniaxiality requires that $\delta = 0°$, which implies that $\epsilon_1 > \epsilon_2 = \epsilon_3$. Negative uniaxiality requires that $\delta = 90°$; hence, $\epsilon_1 = \epsilon_2 > \epsilon_3$. Regardless of the type of uniaxiality, $\underline{a}_+$ and $\underline{a}_-$ are either co-parallel or anti-parallel, so that a uniaxial dielectric material has just one optic ray axis. Furthermore, there is just one optic axis because $\underline{b}_+$ and $\underline{b}_-$ are either co-parallel or anti-parallel. Finally, the sole optic ray axis of a uniaxial dielectric material coincides with its sole optic axis.

Appendix C

Zenneck Wave

Analytical treatments of the simple SPP wave and the Zenneck wave [6] are identical. Similarly to the canonical boundary-value problem for the simple SPP wave described in Section 2.2, let the half-space $z < 0$ be filled with a homogeneous, isotropic, dielectric material of permittivity $\epsilon_{\ell d}$, and the half-space $z > 0$ be filled with another homogeneous, isotropic, dielectric material of permittivity ϵ_{diel}. Whereas $\epsilon_{\ell d}$ is complex valued with $\text{Re}\{\epsilon_{\ell d}\} > 0$ and $\text{Im}\{\epsilon_{\ell d}\} > 0$, $\epsilon_{\text{diel}} > 0$ is purely real. Unlike the situation for an SPP wave, the material with relative permittivity $\epsilon_{\ell d}$ is not a metal but a dissipative dielectric material.

The fields of a Zenneck wave are p polarized, just as those of a simple SPP wave. In the two half-spaces, the electric and magnetic phasors of a Zenneck wave are given by

$$\underline{E}(\underline{r}) = \begin{cases} \left(\frac{\alpha_{\ell d}\underline{u}_{\text{prop}}+q\underline{u}_z}{k_0 n_{\ell d}} \right) \exp[i(q\underline{u}_{\text{prop}} - \alpha_{\ell d}\underline{u}_z) \cdot \underline{r}], & z < 0, \\ c_p \left(\frac{-\alpha_{\text{diel}}\underline{u}_{\text{prop}}+q\underline{u}_z}{k_0 n_{\text{diel}}} \right) \exp[i(q\underline{u}_{\text{prop}} + \alpha_{\text{diel}}\underline{u}_z) \cdot \underline{r}], & z > 0, \end{cases} \tag{C.1}$$

and

$$\underline{H}(\underline{r}) = \begin{cases} -\frac{n_{\ell d}}{\eta_0}(\underline{u}_z \times \underline{u}_{\text{prop}}) \exp[i(q\underline{u}_{\text{prop}} - \alpha_{\ell d}\underline{u}_z) \cdot \underline{r}], & z < 0, \\ -c_p \frac{n_{\text{diel}}}{\eta_0}(\underline{u}_z \times \underline{u}_{\text{prop}}) \exp[i(q\underline{u}_{\text{prop}} + \alpha_{\text{diel}}\underline{u}_z) \cdot \underline{r}], & z > 0, \end{cases} \tag{C.2}$$

where the unit vector $\underline{u}_{\text{prop}} = \underline{u}_x \cos \psi + \underline{u}_y \sin \psi$ is parallel to the direction of propagation in the xy plane, the refractive index $n_{\text{diel}} = \sqrt{\epsilon_{\text{diel}}/\epsilon_0} > 0$, the refractive index $n_{\ell d} = \sqrt{\epsilon_{\ell d}/\epsilon_0}$ has a non-negative imaginary part, and c_p is a coefficient to be determined.

The variations of the field phasors along the z axis in the two partnering materials are described through the complex-valued wavenumbers

$$\left. \begin{aligned} \alpha_{\ell d} &= \sqrt{\omega^2 \mu_0 \epsilon_{\ell d} - q^2} \\ \alpha_{\text{diel}} &= \sqrt{\omega^2 \mu_0 \epsilon_{\text{diel}} - q^2} \end{aligned} \right\} \tag{C.3}$$

that must satisfy the appropriate restrictions

$$\left. \begin{aligned} \text{Im}\{\alpha_{\ell d}\} &> 0 \\ \text{Im}\{\alpha_{\text{diel}}\} &> 0 \end{aligned} \right\} \tag{C.4}$$

for a surface wave.

Electromagnetic Surface Waves. http://dx.doi.org/10.1016/B978-0-12-397024-4.00010-4

Application of the standard boundary conditions (2.12) yields

$$\left.\begin{array}{r}\alpha_{\ell d}n_{\text{diel}} = -c_p\alpha_{\text{diel}}n_{\ell d} \\ n_{\ell d} = c_p n_{\text{diel}}\end{array}\right\}. \tag{C.5}$$

Hence, the coefficient

$$c_p = n_{\ell d}/n_{\text{diel}} \tag{C.6}$$

is the ratio of the two refractive indexes and the dispersion equation

$$\alpha_{\text{diel}}\epsilon_{\ell d} = -\alpha_{\ell d}\epsilon_{\text{diel}} \tag{C.7}$$

for the Zenneck wave has the same form as the dispersion equation (2.26) of the simple SPP wave. Analysis also shows that

$$\text{Re}\{\alpha_{\text{diel}}\} < 0, \tag{C.8}$$

which is the same as Eq. (2.42) for the simple SPP wave. But the restriction

$$\text{Re}\{\alpha_{\ell d}\} > 0 \tag{C.9}$$

on the Zenneck wave is not the same as Eq. (2.44) for the simple SPP wave.

The similarities and differences between the Zenneck wave and the simple SPP wave are numerically exemplified by a comparison of the last columns of Tables 2.1 and C.1.

Table C.1 Wavenumbers of Zenneck waves, for comparison with Table 2.1.

$\epsilon_{\text{diel}}/\epsilon_0$	$\epsilon_{\ell d}/\epsilon_0$	q/k_0	α_{diel}/k_0	$\alpha_{\ell d}/k_0$
4	$15 + 6i$	$1.800 + 0.067i$	$-0.886 + 0.137i$	$3.526 + 0.817i$
	$7.0405 + 2.5991i$	$1.624 + 0.100i$	$-1.180 + 0.137i$	$2.166 + 0.525i$
	$3 + i$	$1.332 + 0.120i$	$-1.500 + 0.107i$	$1.152 + 0.295i$

Appendix D

Floquet Theory

Floquet theory [358] deals with first-order matrix ordinary differential equations of the kind

$$[\underline{f}(z)]' = i\left[\underline{P}(z)\right] \cdot [\underline{f}(z)],\tag{D.1}$$

where the prime denotes differentiation with respect to the argument, $[\underline{f}(z)]$ is a column vector with n components and the $n \times n$ matrix $\left[\underline{P}(z)\right]$ is a piecewise continuous function of z with period 2Ω, i.e.

$$\left[\underline{P}(z + 2\Omega)\right] = \left[\underline{P}(z)\right].\tag{D.2}$$

The $n \times n$ matrix $\left[\underline{\Phi}(z)\right]$, defined by the functional relationship

$$[\underline{f}(z)] = \left[\underline{\Phi}(z)\right] \cdot [\underline{f}(0)],\tag{D.3}$$

is called the matrizant. It satisfies the matrix ordinary differential equation

$$\left[\underline{\Phi}(z)\right]' = i\left[\underline{P}(z)\right] \cdot \left[\underline{\Phi}(z)\right],\tag{D.4}$$

with initial value $\left[\underline{\Phi}(0)\right] = \left[\underline{I}\right]$, the identity matrix.

By virtue of a change of the independent variable, then [262, Chapter 5]

$$\begin{aligned}\left[\underline{\Phi}(z + 2\Omega)\right]' &= i\left[\underline{P}(z + 2\Omega)\right] \cdot \left[\underline{\Phi}(z + 2\Omega)\right]\\&= i\left[\underline{P}(z)\right] \cdot \left[\underline{\Phi}(z + 2\Omega)\right].\end{aligned}\tag{D.5}$$

Hence, $\left[\underline{\Phi}(z + 2\Omega)\right]$ is also a matrizant, but with initial value $\left[\underline{\Phi}(2\Omega)\right]$. Therefore,

$$\left[\underline{\Phi}(z + 2\Omega)\right] = \left[\underline{\Phi}(z)\right] \cdot \left[\underline{\Phi}(2\Omega)\right].\tag{D.6}$$

As a constant matrix $\left[\underline{A}\right]$ can be defined such that $\left[\underline{\Phi}(2\Omega)\right] = \exp\left\{i2\Omega\left[\underline{A}\right]\right\}$, Eq. (D.6) may be rewritten as

$$\left[\underline{\Phi}(z + 2\Omega)\right] = \left[\underline{\Phi}(z)\right] \cdot \exp\left\{i2\Omega\left[\underline{A}\right]\right\}.\tag{D.7}$$

Equation (D.7) suggests the ansatz

$$\left[\underline{\Phi}(z)\right] = \left[\underline{F}(z)\right] \cdot \exp\left\{i\left[\underline{A}\right]z\right\},\tag{D.8}$$

Electromagnetic Surface Waves. http://dx.doi.org/10.1016/B978-0-12-397024-4.00011-6

where the $n \times n$ matrix $\left[\underline{\underline{F}}(z)\right]$ is not known; accordingly,

$$\left[\underline{\underline{\Phi}}(z+2\Omega)\right] = \left[\underline{\underline{F}}(z+2\Omega)\right] \bullet \exp\left\{i \left[\underline{\underline{A}}\right] z\right\} \bullet \exp\left\{i2\Omega \left[\underline{\underline{A}}\right]\right\}. \tag{D.9}$$

If the periodicity constraint

$$\left[\underline{\underline{F}}(z+2\Omega)\right] = \left[\underline{\underline{F}}(z)\right] \tag{D.10}$$

is imposed, Eq. (D.9) simplifies to

$$\begin{aligned}
\left[\underline{\underline{\Phi}}(z+2\Omega)\right] &= [\underline{\underline{F}}(z)] \bullet \exp\left\{i \left[\underline{\underline{A}}\right] z\right\} \bullet \exp\left\{i2\Omega \left[\underline{\underline{A}}\right]\right\} \\
&= \left[\underline{\underline{\Phi}}(z)\right] \bullet \exp\left\{i2\Omega \left[\underline{\underline{A}}\right]\right\},
\end{aligned} \tag{D.11}$$

which is the same as Eq. (D.7).

Thus, Floquet theory yields the solution of Eq. (D.1) as

$$[\underline{f}(z)] = \left[\underline{\underline{F}}(z)\right] \bullet \exp\left\{i \left[\underline{\underline{A}}\right] z\right\} \bullet [\underline{f}(0)], \tag{D.12}$$

where the matrix $\left[\underline{\underline{F}}(z)\right]$ has an initial value

$$\left[\underline{\underline{F}}(0)\right] = \left[\underline{\underline{I}}\right] \tag{D.13}$$

and is also a periodic function of z. Unfortunately, this theory does not deliver actual expressions for $\left[\underline{\underline{F}}(z)\right]$ and $\left[\underline{\underline{A}}\right]$.

Appendix E

Forward and Inverse Bruggeman Formalisms

Suppose that identical spherical particles made of a certain linear material are randomly dispersed in a linear host material. Let the particle radius be less than a tenth of the minimum wavelength in the two materials. Then the particle is electrically small [89] and the particulate composite material can be considered as an *effectively* homogeneous material.

The mathematical procedure to predict the effective constitutive dyadics of the composite material from the constitutive dyadics of its constituent materials is called homogenization [89]. Many homogenization formalisms exist, of which the Bruggeman formalism is perhaps the most widely used. This formalism has been extended to the homogenization of bianisotropic ellipsoidal particles dispersed in a bianisotropic host material [17,359], and has been applied for CTFs [308] as well as chiral STFs [347].

E.1 Forward Bruggeman Formalism for CTFs

Each column of a CTF may be regarded as a set of elongated ellipsoidal particles strung together end-to-end, as shown in Figure 1.1. All particles have the same orientation and shape. The surface of each ellipsoidal particle is parameterized as

$$\underline{r}_{\text{surf}}(\theta, \phi) = R_{\text{surf}}\underline{\underline{U}} \cdot [(\underline{u}_x \cos \phi + \underline{u}_y \sin \phi) \sin \theta + \underline{u}_z \cos \theta], \tag{E.1}$$

θ and ϕ are spherical polar coordinates in a coordinate system located at the centroid of the particle and $R_{\text{surf}} > 0$ is a measure of the particle's linear dimensions. The shape dyadic

$$\underline{\underline{U}} = \underline{u}_n\underline{u}_n + \gamma_\tau\underline{u}_\tau\underline{u}_\tau + \gamma_b\underline{u}_b\underline{u}_b \tag{E.2}$$

has positive eigenvalues and unit determinant. The unit vectors

$$\left.\begin{array}{l} \underline{u}_n = -\underline{u}_x \sin \chi + \underline{u}_z \cos \chi \\ \underline{u}_\tau = \underline{u}_x \cos \chi + \underline{u}_z \sin \chi \\ \underline{u}_b = -\underline{u}_y \end{array}\right\} \tag{E.3}$$

are defined in terms of the tilt angle $\chi \in (0°, 90°]$ of the columns. Since the columnar morphology is highly aciculate, the shape parameters $\gamma_b \gtrsim 1$ and $\gamma_\tau \gg 1$, but these

Electromagnetic Surface Waves. http://dx.doi.org/10.1016/B978-0-12-397024-4.00012-8

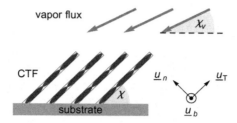

Figure 1.1 The columns of a CTF are represented as a set of elongated ellipsoidal particles, strung together end-to-end. The columns grow tilted at angle χ, along the direction of the unit vector $\underline{u}_\tau$, due to a vapor flux incident at an angle $\chi_v \leq \chi$ on a substrate. When implementing the Bruggeman formalism, the air (approximated by vacuum) between the columns is modeled as an Apollonian distribution [360] of electrically small spheres.

parameters need not be restricted thus for the application of the forward Bruggeman formalism for other composite materials.

As discussed in Section 1.4.3.1, at length scales far greater than the nanoscale, a CTF is effectively a dielectric continuum which may be characterized by the frequency-domain constitutive relation [42,327]

$$\underline{D} = \underline{\underline{\epsilon}}_{CTF} \cdot \underline{E},\tag{E.4}$$

where the permittivity dyadic

$$\underline{\underline{\epsilon}}_{CTF} = \underline{\underline{S}}_y(\chi) \cdot (\epsilon_a \underline{u}_z \underline{u}_z + \epsilon_b \underline{u}_x \underline{u}_x + \epsilon_c \underline{u}_y \underline{u}_y) \cdot \underline{\underline{S}}_y^{-1}(\chi)\tag{E.5}$$

contains the dyadic

$$\underline{\underline{S}}_y(\chi) = \underline{u}_y \underline{u}_y + (\underline{u}_x \underline{u}_x + \underline{u}_z \underline{u}_z)\cos\chi + (\underline{u}_z \underline{u}_x - \underline{u}_x \underline{u}_z)\sin\chi.\tag{E.6}$$

Let the ellipsoidal particles be made of an isotropic dielectric material of refractive index n_s. Let $v_s \in (0, 1)$ be the volume fraction occupied by the ellipsoidal particles, while the void region is filled with air (or vacuum). Thus, the porosity of the CTF equals $1 - v_s$.

The dyadic

$$\underline{\underline{b}}(\epsilon_a, \epsilon_b, \epsilon_c) = v_s \underline{\underline{a}}_s + (1 - v_s)\underline{\underline{a}}_f\tag{E.7}$$

is the volume-fraction-weighted sum of the two polarizability density dyadics

$$\underline{\underline{a}}_s = \left[n_s^2 \epsilon_0 \underline{\underline{I}} - (\epsilon_a \underline{u}_z \underline{u}_z + \epsilon_b \underline{u}_x \underline{u}_x + \epsilon_c \underline{u}_y \underline{u}_y)\right]$$
$$\cdot \left\{\underline{\underline{I}} + i\omega \underline{\underline{D}}_s \cdot \left[n_s^2 \epsilon_0 \underline{\underline{I}} - (\epsilon_a \underline{u}_z \underline{u}_z + \epsilon_b \underline{u}_x \underline{u}_x + \epsilon_c \underline{u}_y \underline{u}_y)\right]\right\}^{-1}\tag{E.8}$$

and

$$\underline{\underline{a}}_f = \left[\epsilon_0 \underline{\underline{I}} - (\epsilon_a \underline{u}_z \underline{u}_z + \epsilon_b \underline{u}_x \underline{u}_x + \epsilon_c \underline{u}_y \underline{u}_y)\right]$$
$$\cdot \left\{\underline{\underline{I}} + i\omega \underline{\underline{D}}_f \cdot \left[\epsilon_0 \underline{\underline{I}} - (\epsilon_a \underline{u}_z \underline{u}_z + \epsilon_b \underline{u}_x \underline{u}_x + \epsilon_c \underline{u}_y \underline{u}_y)\right]\right\}^{-1}.\tag{E.9}$$

Here, $\underline{\underline{a}}_s$ is the polarizability density dyadic which characterizes the deposited material, whereas $\underline{\underline{a}}_f$ characterizes the void region. The depolarization dyadics

$$
\underline{\underline{D}}_s = \frac{1}{i\omega} \frac{2}{\pi} \int_{\phi=0}^{\pi/2} d\phi \int_{\theta=0}^{\pi/2} d\theta
$$
$$
\left[\sin\theta \frac{\frac{\cos^2\theta}{\gamma_\tau^2} \underline{u}_x \underline{u}_x + \sin^2\theta \left(\cos^2\phi\, \underline{u}_z \underline{u}_z + \frac{\sin^2\phi}{\gamma_b^2} \underline{u}_y \underline{u}_y \right)}{\epsilon_b \frac{\cos^2\theta}{\gamma_\tau^2} + \sin^2\theta \left(\epsilon_a \cos^2\phi + \epsilon_c \frac{\sin^2\phi}{\gamma_b^2} \right)} \right]
$$

(E.10)

and

$$
\underline{\underline{D}}_f = \frac{1}{i\omega} \frac{2}{\pi} \int_{\phi=0}^{\pi/2} d\phi \int_{\theta=0}^{\pi/2} d\theta
$$
$$
\left[\sin\theta \frac{\cos^2\theta\, \underline{u}_x \underline{u}_x + \sin^2\theta \left(\cos^2\phi\, \underline{u}_z \underline{u}_z + \sin^2\phi\, \underline{u}_y \underline{u}_y \right)}{\epsilon_b \cos^2\theta + \sin^2\theta \left(\epsilon_a \cos^2\phi + \epsilon_c \sin^2\phi \right)} \right]
$$

(E.11)

usually have to be evaluated using numerical techniques such as Gauss-Legendre quadrature [261].

The forward Bruggeman formalism requires solution of the equation

$$
\underline{b}(\epsilon_a, \epsilon_b, \epsilon_c) = \underline{0}.
$$

(E.12)

In other words, the volume-fraction-weighted sum of $\underline{\underline{a}}_s$ and $\underline{\underline{a}}_f$ must be zero [361]. The quantities n_s, v_s, γ_b, and γ_τ are assumed to be known. As the unknowns ϵ_a, ϵ_b, and ϵ_c appear in the definitions of the depolarization dyadics $\underline{\underline{D}}_s$ and $\underline{\underline{D}}_f$ [362], Eq. (E.12) has to be solved using an indirect numerical technique. The Jacobi iterative technique [261] has been successfully applied to numerically solve Eq. (E.12) [363].

E.2 Inverse Bruggeman Formalism for CTFs

Knowledge of the nanoscale parameters n_s, v_s, γ_b, and γ_τ is not easy to get for a CTF for at least three reasons. One reason is the variability that exists due to differences in deposition conditions [364, 365]. For instance, the bulk material that is evaporated may be quite different from the material that is actually deposited as a thin film. Therefore, while the refractive index of the bulk material is easily known prior to evaporation, the refractive index n_s of the deposited material may well be different, depending on, whether the deposition occurred in an oxidizing or reducing atmosphere, whether trace amounts of water vapor were present, and the temperature. Second, direct determination of porosity $(1 - v_s)$ through a gas-adsorption technique [366–368], although accurate, is very time-consuming. Therefore, porosity is usually measured indirectly through measurement of mass density, which has its own sources of inaccuracy [364]. Third, even when $\gamma_\tau > 10$ is fixed, γ_b cannot be measured as it is a notional quantity rather than a physical quantity.

Provided that ϵ_a, ϵ_b, and ϵ_c of a CTF have been measured by suitable optical experiments [94,345,346] and $\gamma_\tau > 10$ has been fixed [347], an *inverse* homogenization procedure [369] can yield n_s, v_s, and γ_b [308]. For that purpose, the left side of Eq. (E.7) is interpreted as

$$\underline{\underline{b}} \equiv b_x(n_s, v_s, \gamma_b)\underline{u}_x\underline{u}_x + b_y(n_s, v_s, \gamma_b)\underline{u}_y\underline{u}_y + b_z(n_s, v_s, \gamma_b)\underline{u}_z\underline{u}_z \tag{E.13}$$

and the three equations

$$\left.\begin{array}{l} b_x(n_s, v_s, \gamma_b) = 0 \\ b_y(n_s, v_s, \gamma_b) = 0 \\ b_z(n_s, v_s, \gamma_b) = 0 \end{array}\right\} \tag{E.14}$$

have to be solved with ϵ_a, ϵ_b, and ϵ_c being known.

Solutions of Eqs. (E.14) may be computed using a modified Newton-Raphson technique [261,370]. In this recursive technique, the estimated solutions at step $k+1$, namely $\left\{n_s^{(k+1)}, v_s^{(k+1)}, \gamma_b^{(k+1)}\right\}$, are derived from those at step k, namely $\left\{n_s^{(k)}, v_s^{(k)}, \gamma_b^{(k)}\right\}$, as follows:

$$\left.\begin{array}{l} n_s^{(k+1)} = n_s^{(k)} - \dfrac{b_x\left(n_s^{(k)}, v_s^{(k)}, \gamma_b^{(k)}\right)}{\frac{\partial}{\partial n_s} b_x\left(n_s^{(k)}, v_s^{(k)}, \gamma_b^{(k)}\right)} \\[2em] v_s^{(k+1)} = v_s^{(k)} - \dfrac{b_y\left(n_s^{(k+1)}, v_s^{(k)}, \gamma_b^{(k)}\right)}{\frac{\partial}{\partial v_s} b_y\left(n_s^{(k+1)}, v_s^{(k)}, \gamma_b^{(k)}\right)} \\[2em] \gamma_b^{(k+1)} = \gamma_b^{(k)} - \dfrac{b_z\left(n_s^{(k+1)}, v_s^{(k+1)}, \gamma_b^{(k)}\right)}{\frac{\partial}{\partial \gamma_b} b_z\left(n_s^{(k+1)}, v_s^{(k+1)}, \gamma_b^{(k)}\right)} \end{array}\right\} . \tag{E.15}$$

In order for the scheme (E.15) to converge, it is crucial that the initial estimate $\left\{n_s^{(0)}, v_s^{(0)}, \gamma_b^{(0)}\right\}$ be sufficiently close to the true solution. A suitable initial estimate may be found by exploiting the forward Bruggeman formalism as follows [308]:

Let $\epsilon_{A,B,C}$ denote estimates of the CTF permittivity parameters $\epsilon_{a,b,c}$, computed using the forward Bruggeman formalism for physically reasonable ranges of the parameters n_s, v_s, and γ_b—namely, $n_s \in \left(n_s^L, n_s^U\right)$, $v_s \in \left(v_s^L, v_s^U\right)$, and $\gamma_b \in \left(\gamma_b^L, \gamma_b^U\right)$. Then:

i. Fix $n_s = \left(n_s^L + n_s^U\right)/2$ and $\gamma_b = \left(\gamma_b^L + \gamma_b^U\right)/2$. For all values of $v_s \in \left(v_s^L, v_s^U\right)$, identify the value $v_s^\sharp$ for which the quantity

$$\Delta = \sqrt{(\epsilon_a - \epsilon_A)^2 + (\epsilon_b - \epsilon_B)^2 + (\epsilon_c - \epsilon_C)^2} \tag{E.16}$$

is minimized.

ii. Fix $v_s = v_s^\sharp$ and $\gamma_b = \left(\gamma_b^L + \gamma_b^U\right)/2$. For all values of $n_s \in \left(n_s^L, n_s^U\right)$, identify the value $n_s^\sharp$ for which Δ is minimized.

iii. Fix $v_s = v_s^\sharp$ and $n_s = n_s^\sharp$. For all values of $\gamma_b \in \left(\gamma_b^L, \gamma_b^U\right)$, identify the value $\gamma_b^\sharp$ for which Δ is minimized.

Steps (i)–(iii) are repeated, using $n_s^\sharp$ and $\gamma_b^\sharp$ as the fixed values of n_s and γ_b in step (i), and $\gamma_b^\sharp$ as the fixed value of γ_b in step (ii), until Δ becomes sufficiently small. Numerical experiments [308] indicate that $n_s^\sharp$, $v_s^\sharp$, and $\gamma_b^\sharp$ are suitable initial estimates for the modified Newton-Raphson technique when Δ decreases to less than 0.01.

E.3 Forward Bruggeman Formalism for Fluid-Infiltrated CTFs

Suppose that $\{n_s, v_s, \gamma_b\}$ have been determined for a CTF. It is then completely infiltrated with a fluid of refractive index $n_{f\ell}$. The permittivity dyadic of the fluid-infiltrated CTF shall be

$$\underline{\underline{\epsilon}}_{CTF}^{f\ell} = \underline{\underline{S}}_y(\chi) \bullet \left(\epsilon_a^{f\ell} \underline{u}_z \underline{u}_z + \epsilon_b^{f\ell} \underline{u}_x \underline{u}_x + \epsilon_c^{f\ell} \underline{u}_y \underline{u}_y \right) \bullet \underline{\underline{S}}_y^{-1}(\chi), \tag{E.17}$$

instead of $\underline{\underline{\epsilon}}_{CTF}$ defined in Eq. (E.5).

In order to predict $\epsilon_a^{f\ell}$, $\epsilon_b^{f\ell}$, and $\epsilon_c^{f\ell}$, the forward Bruggeman formalism can again be employed to yield the dyadic equation

$$v_s \underline{\underline{a}}_s^{f\ell} + (1 - v_s) \underline{\underline{a}}_\ell^{f\ell} = \underline{\underline{0}}, \tag{E.18}$$

where the polarizability density dyadics

$$\underline{\underline{a}}_s^{f\ell} = \left[n_s^2 \epsilon_0 \underline{\underline{I}} - \left(\epsilon_a^{f\ell} \underline{u}_z \underline{u}_z + \epsilon_b^{f\ell} \underline{u}_x \underline{u}_x + \epsilon_c^{f\ell} \underline{u}_y \underline{u}_y \right) \right]$$
$$\bullet \left\{ \underline{\underline{I}} + i\omega \underline{\underline{D}}_s^{f\ell} \bullet \left[n_s^2 \epsilon_0 \underline{\underline{I}} - \left(\epsilon_a^{f\ell} \underline{u}_z \underline{u}_z + \epsilon_b^{f\ell} \underline{u}_x \underline{u}_x + \epsilon_c^{f\ell} \underline{u}_y \underline{u}_y \right) \right] \right\}^{-1} \tag{E.19}$$

and

$$\underline{\underline{a}}_\ell^{f\ell} = \left[n_{f\ell}^2 \epsilon_0 \underline{\underline{I}} - \left(\epsilon_a^{f\ell} \underline{u}_z \underline{u}_z + \epsilon_b^{f\ell} \underline{u}_x \underline{u}_x + \epsilon_c^{f\ell} \underline{u}_y \underline{u}_y \right) \right]$$
$$\bullet \left\{ \underline{\underline{I}} + i\omega \underline{\underline{D}}_\ell^{f\ell} \bullet \left[n_{f\ell}^2 \epsilon_0 \underline{\underline{I}} - \left(\epsilon_a^{f\ell} \underline{u}_z \underline{u}_z + \epsilon_b^{f\ell} \underline{u}_x \underline{u}_x + \epsilon_c^{f\ell} \underline{u}_y \underline{u}_y \right) \right] \right\}^{-1}. \tag{E.20}$$

The depolarization dyadics are given by

$$\underline{\underline{D}}_s^{f\ell} = \frac{1}{i\omega} \frac{2}{\pi} \int_{\phi=0}^{\pi/2} d\phi \int_{\theta=0}^{\pi/2} d\theta$$
$$\times \left[\sin\theta \, \frac{\frac{\cos^2\theta}{\gamma_\tau^2} \underline{u}_x \underline{u}_x + \sin^2\theta \left(\cos^2\phi \, \underline{u}_z \underline{u}_z + \frac{\sin^2\phi}{\gamma_b^2} \underline{u}_y \underline{u}_y \right)}{\epsilon_b^{f\ell} \frac{\cos^2\theta}{\gamma_\tau^2} + \sin^2\theta \left(\epsilon_a^{f\ell} \cos^2\phi + \epsilon_c^{f\ell} \frac{\sin^2\phi}{\gamma_b^2} \right)} \right] \tag{E.21}$$

and

$$
\underline{\underline{D}}_\ell^{f\ell} = \frac{1}{i\omega} \frac{2}{\pi} \int_{\phi=0}^{\pi/2} d\phi \int_{\theta=0}^{\pi/2} d\theta
$$
$$
\times \left[\sin\theta \, \frac{\cos^2\theta \, \underline{u}_x\underline{u}_x + \sin^2\theta \left(\cos^2\phi \, \underline{u}_z\underline{u}_z + \sin^2\phi \, \underline{u}_y\underline{u}_y \right)}{\epsilon_b^{f\ell}\cos^2\theta + \sin^2\theta \left(\epsilon_a^{f\ell} \cos^2\phi + \epsilon_c^{f\ell} \sin^2\phi \right)} \right]. \qquad \text{(E.22)}
$$

The Jacobi iterative technique [261] has been applied to numerically solve Eq. (E.18) for $\epsilon_a^{f\ell}$, $\epsilon_b^{f\ell}$, and $\epsilon_c^{f\ell}$ [98,347].

E.4   Cautionary Remarks

No homogenization formalism should ever be viewed as *the truth*. Each has at least one underlying assumption that all particles are electrically small [89,91]. Each formalism has inherent limitations. For instance, both the forward [371–373] and the inverse [374] Bruggeman formalisms should be applied for metal/dielectric composite materials with great caution, because they can yield spurious results. Each formalism should yield reasonable conclusions, e.g. if the relative permittivities of both constituent materials of a composite material are doubled, then the relative permittivity of the homogenized composite material is also doubled [375]. Finally, there can be significant disparities between the predictions of a homogenization formalism and experimentally gathered data, because homogenization formalisms are theories of averages and do not yield standard deviations [376].

Bibliography

[1] F. de Fornel, Evanescent Waves: From Newtonian Optics to Atomic Optics, Springer, Heidelberg, Germany, 2001.

[2] B. Hecht, B. Sick, U.P. Wild, V. Deckert, R. Zenobi, O.J.F. Martin, D.W. Pohl, Scanning near-field optical microscopy with aperture probes: fundamentals and applications, J. Chem. Phys. 112 (2000) 7761–7774, http://dx.doi.org/10.1063/1.481382.

[3] J. Homola (Ed.), Surface Plasmon Resonance Based Sensors, Springer, Heidelberg, Germany, 2006, http://dx.doi.org/10.1007/b100321.

[4] I. Abdulhalim, M. Zourob, A. Lakhtakia, Surface plasmon resonance for biosensing: a mini-review, Electromagnetics 28 (2008) 214–242, http://dx.doi.org/10.1080/02726340 801921650.

[5] J. Zenneck, Über die Fortpflanzung ebener elektromagnetischer Wellen längs einer ebenen Lieterfläche und ihre Beziehung zur drahtlosen Telegraphie, Ann. Phys. Lpz. 23 (1907) 846–866, http://dx.doi.org/10.1002/andp.19073281003.

[6] D.A. Hill, J.R. Wait, On the excitation of the Zenneck surface wave over the ground at 10 MHz, Ann. Télécommun. 35 (1980) 179–182, http://dx.doi.org/10.1007/BF03000465.

[7] A. Sommerfeld, Über die Ausbreitung der Wellen in der drahtlosen Telegraphie, Ann. Phys. Lpz. 28 (1909) 665–736, http://dx.doi.org/10.1002/andp.19093330402.

[8] C.J. Bouwkamp, On Sommerfeld's surface wave, Phys. Rev. 80 (1950) 294, http://dx.doi.org/10.1103/PhysRev.80.294.

[9] V.N. Datsko, A.A. Kopylov, On surface electromagnetic waves, Phys. Usp. 51 (2008) 101–102, http://dx.doi.org/10.1070/PU2008v051n01ABEH006208.

[10] A.V. Kukushkin, On the existence and physical meaning of the Zenneck wave, Phys. Usp. 52 (2009) 755–756, http://dx.doi.org/10.3367/UFNe.0179.200907j.0801.

[11] D. Bohm, E.P. Gross, Theory of plasma oscillations. A. Origin of medium-like behavior, Phys. Rev. 75 (1949) 1851–1864, http://dx.doi.org/10.1103/PhysRev.75.1851.

[12] D. Bohm, E.P. Gross, Theory of plasma oscillations. B. Excitation and damping of oscillations, Phys. Rev. 75 (1949) 1864–1876, http://dx.doi.org/10.1103/PhysRev.75.1864.

[13] R. Kronig, A collective description of electron interactions, Phys. Rev. 86 (1952) 795, http://dx.doi.org/10.1103/PhysRev.86.795.

[14] J.R. Pierce, A note on plasma oscillation, Phys. Rev. 76 (1949) 565, http://dx.doi.org/10.1103/PhysRev.76.565.2.

[15] D. Pines, Collective energy losses in solids, Rev. Mod. Phys. 28 (1956) 184–198, http://dx.doi.org/10.1103/RevModPhys.28.184.

[16] R.H. Ritchie, Plasma losses by fast electrons in thin films, Phys. Rev. 106 (1957) 874–881, http://dx.doi.org/10.1103/PhysRev.106.874.

[17] T.G. Mackay, A. Lakhtakia, Electromagnetic Anisotropy and Bianisotropy: A Field Guide, World Scientific, Singapore, 2010.

Electromagnetic Surface Waves. http://dx.doi.org/10.1016/B978-0-12-397024-4.00020-7
© 2013 Elsevier Inc. All rights reserved.

[18] J.J. Hopfield, Theory of the contribution of excitons to the complex dielectric constant of crystals, Phys. Rev. 112 (1958) 1555–1567, http://dx.doi.org/10.1103/PhysRev.112.1555.

[19] U. Fano, The theory of anomalous diffraction gratings and of quasi-stationary waves on metallic surfaces (Sommerfeld's waves), J. Opt. Soc. Am. 31 (1941) 213–222, http://dx.doi.org/10.1364/JOSA.31.000213.

[20] T. Turbadar, Complete absorption of light by thin metal films, Proc. Phys. Soc. 73 (1959) 40–44, http://dx.doi.org/10.1088/0370-1328/73/1/307.

[21] T. Turbadar, Complete absorption of plane polarized light by thin metal films, Opt. Acta 11 (1964) 207–210, http://dx.doi.org/10.1080/713817875.

[22] A. Otto, Excitation of nonradiative surface plasma waves in silver by the method of frustrated total reflection, Z. Phys. 216 (1968) 398–410, http://dx.doi.org/10.1007/BF01391532.

[23] E. Kretschmann, H. Raether, Radiative decay of nonradiative surface plasmons excited by light, Z. Naturforsch. A 23 (1968) 2135–2136.

[24] J. Fahrenfort, Attenuated total reflection: a new principle for the production of useful infra-red reflection spectra of organic compounds, Spectrochim. Acta 17 (1961) 698–709, http://dx.doi.org/10.1016/0371-1951(61)80136-7.

[25] R.W. Wood, On a remarkable case of uneven distribution of light in a diffraction grating spectrum, Proc. Phys. Soc. Lond. 18 (1902) 269–275, http://dx.doi.org/10.1088/1478-7814/18/1/325. This paper was also published in another journal [26].

[26] R.W. Wood, On a remarkable case of uneven distribution of light in a diffraction grating spectrum, Phil. Mag. 4 (1902) 396–402, http://dx.doi.org/10.1080/14786440209462857. This paper was also published in another journal [25].

[27] C. Nylander, B. Liedberg, T. Lind, Gas detection by means of surface plasmon resonance, Sens. Actuat. 3 (1982) 79–88, http://dx.doi.org/10.1016/0250-6874(82)80008-5.

[28] G.J. Sprokel, R. Santo, J.D. Swalen, Determination of the surface tilt angle by attenuated total reflection, Mol. Cryst. Liq. Cryst. 68 (1981) 29–38, http://dx.doi.org/10.1080/00268948108073550.

[29] G.J. Sprokel, The reflectivity of a liquid crystal cell in a surface plasmon experiment, Mol. Cryst. Liq. Cryst. 68 (1981) 39–45, http://dx.doi.org/10.1080/00268948108073551.

[30] K.R. Welford, J.R. Sambles, M.G. Clark, Guided modes and surface plasmon-polaritons observed with a nematic liquid crystal using attenuated total reflection, Liq. Cryst. 2 (1987) 91–105, http://dx.doi.org/10.1080/02678298708086641.

[31] S.J. Elston, J.R. Sambles, Surface plasmon-polaritons on an anisotropic substrate, J. Mod. Opt. 37 (1990) 1895–1902, http://dx.doi.org/10.1080/09500349014552101.

[32] M.G. Blaber, M.D. Arnold, M.J. Ford, Designing materials for plasmonic systems: the alkali-noble intermetallics, J. Phys.: Condens. Matter 22 (2010) 095501, http://dx.doi.org/10.1088/0953-8984/22/9/095501.

[33] P.R. West, S. Ishii, G.V. Naik, N.K. Emani, V.M. Shalaev, A. Boltasseva, Searching for better plasmonic materials, Laser Photon. Rev. 4 (2010) 795–808, http://dx.doi.org/10.1002/lpor.200900055.

[34] R.F. Wallis, J.J. Brion, E. Burstein, A. Hartstein, Theory of surface polaritons in anisotropic dielectric media with application to surface magnetoplasmons in semiconductors, Phys. Rev. B 9 (1974) 3424–3437, http://dx.doi.org/10.1103/PhysRevB.9.3424.

[35] M.I. D'yakonov, New type of electromagnetic wave propagating at an interface, Sov. Phys. JETP 67 (1988) 714–716.

[36] O. Takayama, L.-C. Crasovan, S.K. Johansen, D. Mihalache, D. Artigas, L. Torner, Dyakonov surface waves: a review, Electromagnetics 28 (2008) 126–145, http://dx.doi.org/10.1080/02726340801921403.

[37] S. He, Electromagnetic surface waves for some artificial bianisotropic media, J. Electromag. Waves Appl. 12 (1998) 449–466, http://dx.doi.org/10.1163/156939398X00881.

[38] V.M. Galynsky, A.N. Furs, L.M. Barkovsky, Integral formalism for surface electromagnetic waves in bianisotropic media, J. Phys. A: Math. Gen. 37 (2004) 5083–5096, http://dx.doi.org/10.1088/0305-4470/37/18/012.

[39] R.H. Tarkhanyan, D.G. Niarchos, Nonradiative surface electromagnetic waves at the interface of uniaxially bianisotropic enantiomeric media, Phys. Status Solidi B 248 (2011) 1499–1504, http://dx.doi.org/10.1002/pssb.201046374.

[40] T.G. Mackay, A. Lakhtakia, Surface-plasmon-polariton waves guided by the uniformly moving planar interface of a metal film and dielectric slab, Optik 123 (2012) 49–57, http://dx.doi.org/10.1016/j.ijleo.2011.03.003.

[41] O. Takayama, L. Crasovan, D. Artigas, L. Torner, Observation of Dyakonov surface waves, Phys. Rev. Lett. 102 (2009) 043903, http://dx.doi.org/10.1103/PhysRevLett.102.043903.

[42] A. Lakhtakia, R. Messier, Sculptured Thin Films: Nanoengineered Morphology and Optics, SPIE Press, Bellingham, WA, USA, 2005.

[43] J.A. Polo Jr., S.R. Nelatury, A. Lakhtakia, Propagation of surface waves at the planar interface of a columnar thin film and an isotropic substrate, J. Nanophoton. 1 (2007) 013501, http://dx.doi.org/10.1117/1.2515618.

[44] A. Lakhtakia, J.A. Polo Jr., Morphological influence on surface-wave propagation at the planar interface of a metal and a columnar thin film, Asian J. Phys. 17 (2008) 185–191.

[45] J.A. Polo Jr., A. Lakhtakia, Morphological effects on surface-plasmon-polariton waves at the planar interface of a metal and a columnar thin film, Opt. Commun. 281 (2008) 5453–5457, http://dx.doi.org/10.1016/j.optcom.2008.07.051.

[46] A. Lakhtakia, J.A. Polo Jr., Dyakonov-Tamm wave at the planar interface of a chiral sculptured thin film and an isotropic dielectric material, J. Eur. Opt. Soc.–Rapid Pub. 2 (2007) 07021, http://dx.doi.org/10.2971/jeos.2007.07021.

[47] I. Tamm, Über eine mögliche Art der Elektronenbindung an Kristalloberflächen, Z. Phys. A 76 (1932) 849–850, http://dx.doi.org/10.1007/BF01341581.

[48] K. Agarwal, J.A. Polo Jr., A. Lakhtakia, Theory of Dyakonov-Tamm waves at the planar interface of a sculptured nematic thin film and an isotropic dielectric material, J. Opt. A: Pure Appl. Opt. 11 (2009) 074003, http://dx.doi.org/10.1088/1464-4258/11/7/074003.

[49] J.A. Polo Jr., A. Lakhtakia, Dyakonov-Tamm waves guided by the planar interface of an isotropic dielectric material and an electro-optic ambichiral Reusch Pile, J. Opt. Soc. Am. B 28 (2011) 567–576, http://dx.doi.org/10.1364/JOSAB.28.000567.

[50] J. Gao, A. Lakhtakia, J.A. Polo Jr., M. Lei, Dyakonov-Tamm wave guided by a twist defect in a structurally chiral material, J. Opt. Soc. Am. A 26 (2009) 1615–1621, http://dx.doi.org/10.1364/JOSAA.26.001615. Corrections: 26 (2009) 2399, http://dx.doi.org/10.1364/JOSAA.26.002399.

[51] J. Gao, A. Lakhtakia, M. Lei, On Dyakonov-Tamm waves localized to a central twist defect in a structurally chiral material, J. Opt. Soc. Am. B 26, (2009) B74–B82, http://dx.doi.org/10.1364/JOSAB.26.000B74.

[52] J. Gao, A. Lakhtakia, M. Lei, Dyakonov-Tamm waves guided by the interface between two structurally chiral materials that differ only in handedness, Phys. Rev. A 81 (2010) 013801, http://dx.doi.org/10.1103/PhysRevA.81.013801.

[53] J. Gao, A. Lakhtakia, M. Lei, Synoptic view of Dyakonov-Tamm waves localized to the planar interface of two chiral sculptured thin films, J. Nanophoton. 5 (2011) 051502, http://dx.doi.org/10.1117/1.3543814.

[54] M.A. Motyka, A. Lakhtakia, Multiple trains of same-color surface plasmon-polaritons guided by the planar interface of a metal and a sculptured nematic thin film, J. Nanophoton. 2 (2008) 021910, http://dx.doi.org/10.1117/1.3033757.

[55] M.A. Motyka, A. Lakhtakia, Multiple trains of same-color surface plasmon-polaritons guided by the planar interface of a metal and a sculptured nematic thin film. Part II: Arbitrary incidence, J. Nanophoton. 3 (2009) 033502, http://dx.doi.org/10.1117/1.3147876.

[56] J.A. Polo Jr., A. Lakhtakia, On the surface plasmon polariton wave at the planar interface of a metal and a chiral sculptured thin film, Proc. R. Soc. Lond. A 465 (2009) 87–107, http://dx. doi.org/10.1098/rspa.2008.0211.

[57] A. Lakhtakia, Y.-J. Jen, C.-F. Lin, Multiple trains of same-color surface plasmon-polaritons guided by the planar interface of a metal and a sculptured nematic thin film. Part III: Experimental evidence, J. Nanophoton. 3 (2009) 033506, http://dx.doi.org/10.1117/1.3249629.

[58] Devender, D.P. Pulsifer, A. Lakhtakia, Multiple surface plasmon polariton waves, Electron. Lett. 45 (2009) 1137–1138, http://dx.doi.org/10.1049/el.2009.2049.

[59] T.H. Gilani, N. Dushkina, W.L. Freeman, M.Z. Numan, D.N. Talwar, D.P. Pulsifer, Surface plasmon resonance due to the interface of a metal and a chiral sculptured thin film, Opt. Eng. 49 (2010) 120503, http://dx.doi.org/10.1117/1.3525282.

[60] P.W. Baumeister, Optical Coating Technology, SPIE Press, Bellingham, WA, USA, 2004.

[61] M. Faryad, H. Maab, A. Lakhtakia, Rugate-filter-guided propagation of multiple Fano waves, J. Opt. (United Kingdom) 13 (2011) 075101, http://dx.doi.org/10.1088/2040-8978/13/7/075101.

[62] H. Maab, M. Faryad, A. Lakhtakia, Surface electromagnetic waves supported by the interface of two semi-infinite rugate filters with sinusoidal refractive-index profiles, J. Opt. Soc. Am. B 28 (2011) 1204–1212, http://dx.doi.org/10.1364/JOSAB.28.001204.

[63] M. Faryad, A. Lakhtakia, On surface plasmon-polariton waves guided by the interface of a metal and a rugate filter with a sinusoidal refractive-index profile, J. Opt. Soc. Am. B 27 (2010) 2218–2223, http://dx.doi.org/10.1364/JOSAB.27.002218.

[64] E. Reusch, Untersuchung über Glimmercombinationen, Ann. Phys. Chem. Lpz. 138 (1869) 628–638, http://dx.doi.org/10.1002/andp.18692141211.

[65] I.J. Hodgkinson, A. Lakhtakia, Q.h. Wu, L. De Silva, M.W. McCall, Ambichiral, equichiral and finely chiral layered structures, Opt. Commun. 239 (2004) 353–358, http://dx.doi.org/10.1016/j.optcom.2004.06.005.

[66] S.K. Morrison, Y.S. Kivshar, Tamm states and nonlinear surface modes in photonic crystals, Opt. Commun. 266 (2006) 323–326, http://dx.doi.org/10.1016/j.optcom.2006.04.068.

[67] R.D. Meade, K.D. Brommer, A.M. Rappe, J.D. Joannopoulos, Electromagnetic Bloch waves at the surface of a photonic crystal, Phys. Rev. B 44 (1991) 10961–10964, http://dx.doi.org/10.1103/PhysRevB.44.10961.

[68] S.A. Ramakrishna, T.M. Grzegorczyk, Physics and Applications of Negative Refractive Index Materials, SPIE Press, Bellingham, WA, USA, 2009.

[69] L.-C. Crasovan, O. Takayama, D. Artigas, S.K. Johansen, D. Mihalache, L. Torner, Enhanced localization of Dyakonov-like surface waves in left-handed materials, Phys. Rev. B 74 (2006) 155120, http://dx.doi.org/10.1103/PhysRevB.74.155120.

[70] Y.-Y. Jiang, H.-Y. Shi, Y.-Q. Zhang, C.-F. Hou, X.-D. Sun, Characteristics of surface waves in anisotropic left-handed materials, Chin. Phys. 16 (2007) 1959–1962, http://dx.doi.org/10.1088/1009-1963/16/7/026.

[71] H.J. Lezec, J.A. Dionne, H.A. Atwater, Negative refraction at visible frequencies, Science 316 (2007) 430–432, http://dx.doi.org/10.1126/science.1139266.

[72] N.M. Litchinitser, V.M. Shalaev, Metamaterials: transforming theory into reality, J. Opt. Soc. Am. B 26, (2009) B161–B169, http://dx.doi.org/10.1364/JOSAB.26.00B161.

[73] T.G. Mackay, Effective constitutive parameters of linear nanocomposites in the long-wavelength regime, J. Nanophoton. 5 (2011) 051001, http://dx.doi.org/10.1117/1.3626857.

[74] R. Ruppin, Surface polaritons of a left-handed medium, Phys. Lett. A 277 (2000) 61–64, http://dx.doi.org/10.1016/S0375-9601(00)00694-0.

[75] S.A. Darmanyan, M. Nevière, A.A. Zakhidov, Surface modes at the interface of conventional and left-handed media, Opt. Commun. 225 (2003) 233–240, http://dx.doi.org/10.1016/j.optcom.2003.07.047.

[76] M. Zeller, M. Cuevas, R.A. Depine, Surface plasmon polaritons in attenuated total reflection systems with metamaterials: homogeneous problem, J. Opt. Soc. Am. B 28 (2011) 2042–2047, http://dx.doi.org/10.1364/JOSAB.28.002042.

[77] W.L. Barnes, Surface plasmon-polariton length scales: a route to sub-wavelength optics, J. Opt. A Pure Appl. Opt. 8 (2006) S87–S93, http://dx.doi.org/10.1088/1464-4258/8/4/S06.

[78] H.C. Chen, Theory of Electromagnetic Waves. A Co-ordinate Free Approach, McGraw-Hill, New York, NY, USA, 1983.

[79] D. Mihalache, D. Mazilu, F. Lederer, Nonlinear TE-polarized surface plasmon polaritons guided by metal films, Opt. Commun. 59 (1986) 391–394, http://dx.doi.org/10.1016/0030-4018(86)90364-0.

[80] A. Lakhtakia, V.K. Varadan, V.V. Varadan, Scattering by a partially illuminated, doubly periodic, doubly infinite surface, J. Acoust. Soc. Am. 77 (1985) 1999–2004, http://dx.doi.org/10.1121/1.391771.

[81] P.C. Waterman, Scattering by periodic surfaces, J. Acoust. Soc. Am. 57 (1975) 791–802, http://dx.doi.org/10.1121/1.380521.

[82] V.E. Ferry, L.A. Sweatlock, D. Pacifici, H.A. Atwater, Plasmonic nanostructure design for efficient light coupling into solar cells, Nano Lett. 8 (2008) 4391–4397, http://dx.doi.org/10.1021/nl8022548.

[83] J. Chen, Q. Wang, H. Li, Microstructured design of metallic diffraction gratings for light trapping in thin-film silicon solar cells, Opt. Commun. 283 (2010) 5236–5244, http://dx.doi.org/10.1016/j.optcom.2010.07.052.

[84] M. Faryad, A. Lakhtakia, Enhanced absorption of light due to multiple surface-plasmon-polariton waves, Proc. SPIE 8110 (2011) 81100F, http://dx.doi.org/10.1117/12.893492.

[85] I. Dolev, M. Volodarsky, G. Porat, A. Arie, Multiple coupling of surface plasmons in quasi-periodic gratings, Opt. Lett. 36 (2011) 1584–1586, http://dx.doi.org/10.1364/OL.36.001584.

[86] M. Faryad, A. Lakhtakia, Excitation of multiple surface-plasmon-polariton waves using a compound surface-relief grating, J. Nanophoton. 6 (2012) 061701, http://dx.doi.org/10.1117/1.JNP.6.061701.

[87] G.I. Stegeman, R.F. Wallis, A.A. Maradudin, Excitation of surface polaritons by end-fire coupling, Opt. Lett. 8 (1983) 386–389, http://dx.doi.org/10.1364/OL.8.000386.

[88] J.F. Nye, Physical Properties of Crystals, Oxford University Press, Oxford, United Kingdom, 1985.

[89] A. Lakhtakia (Ed.), Selected Papers on Linear Optical Composite Materials, SPIE Press, Bellingham, WA, USA, 1996.

[90] L. Lewin, The electrical constants of a material loaded with spherical particles, J. Inst. Elect. Eng. Pt. III: Radio Commun. Eng. 94 (1947) 65–68, http://dx.doi.org/10.1049/ji-3-2.1947.0013.

[91] T.G. Mackay, Lewin's homogenization formula revisited for nanocomposite materials, J. Nanophoton. 2 (2008) 029503, http://dx.doi.org/10.1117/1.3028260.

[92] H.C. van de Hulst, Light Scattering by Small Particles, Dover, New York, NY, USA, 1981.

[93] B. Michel, Recent developments in the homogenization of linear bianisotropic composite materials, in: O.N. Singh, A. Lakhtakia (Eds.), Electromagnetic Fields in Unconventional Materials and Structures, Wiley, New York, NY, USA, 2000, pp. 39–82.

[94] I. Hodgkinson, Q.h. Wu, J. Hazel, Empirical equations for the principal refractive indices and column angle of obliquely deposited films of tantalum oxide, titanium oxide, and zirconium oxide, Appl. Opt. 37 (1998) 2653–2659, http://dx.doi.org/10.1364/AO.37.002653.

[95] L. Wei, P. Parhi, E.A. Vogler, T.M. Ritty, A. Lakhtakia, Thickness-controlled hydrophobicity of fibrous Parylene-C films, Mater. Lett. 64 (2010) 1063–1065, http://dx.doi.org/10.1016/j.matlet.2010.02.009.

[96] A. Lakhtakia, J.B. Geddes III, Thin-film metamaterials called sculptured thin films, in: A. Aldea, V. Bârsan (Eds.), Trends in Nanophysics, Springer, Heidelberg, Germany, 2010, pp. 59–71, http://dx.doi.org/doi:10.1007/978-3-642-12070-1_3.

[97] V.C. Venugopal, A. Lakhtakia, R. Messier, J.-P. Kucera, Low-permittivity nanocomposite materials using sculptured thin film technology, J. Vac. Sci. Technol. B 18 (2000) 32–36, http://dx.doi.org/doi:10.1116/1.591146.

[98] T.G. Mackay, A. Lakhtakia, Modeling columnar thin films as platforms for surface-plasmonic-polaritonic optical sensing, Photon. Nanostruct. Fundam. Appl. 8 (2010) 140–149, http://dx.doi.org/doi:10.1016/j.photonics.2010.02.003.

[99] R. Messier, T. Gehrke, C. Frankel, V.C. Venugopal, W. Otaño, A. Lakhtakia, Engineered sculptured nematic thin films, J. Vac. Sci. Technol. A 15 (1997) 2148–2152, http://dx.doi.org/doi:10.1116/1.580621.

[100] N.O. Young, J. Kowal, Optically active fluorite films, Nature 183 (1959) 104–105, http://dx.doi.org/doi:10.1038/183104a0.

[101] K. Robbie, M.J. Brett, A. Lakhtakia, Chiral sculptured thin films, Nature 384 (1996) 616, http://dx.doi.org/doi:10.1038/384616a0.

[102] Q. Wu, I.J. Hodgkinson, A. Lakhtakia, Circular polarization filters made of chiral sculptured thin films: experimental and simulation results, Opt. Eng. 39 (2000) 1863–1868, http://dx.doi.org/doi:10.1117/1.602570.

[103] I.J. Hodgkinson, Q.H. Wu, K.E. Thorn, A. Lakhtakia, M.W. McCall, Spacerless circular-polarization spectral-hole filters using chiral sculptured thin films: theory and experiment, Opt. Commun. 184 (2000) 57–66, http://dx.doi.org/doi:10.1016/S0030-4018(00)00935-4.

[104] M. Suzuki, Y. Taga, Integrated sculptured thin films, Jap. J. Appl. Phys. Part 2 40, (2001) L358–L359, http://dx.doi.org/doi:10.1143/JJAP.40.L358.

[105] C. Zhou, D. Gall, Two-component nanorod arrays by glancing-angle deposition, Small 4 (2008) 1351–1354, http://dx.doi.org/10.1002/smll.200701289.

[106] J.D. Joannopoulos, S.G. Johnson, J.N. Winn, R.D. Meade, Photonic Crystals: Molding the Flow of Light, second ed., Princeton University Press, Princeton, NJ, USA, 2008.

[107] R.H. Lipson, C. Lu, Photonic crystals: a unique partnership between light and matter, Eur. J. Phys. 30 (2009) S33–S48, http://dx.doi.org/10.1088/0143-0807/30/4/S04.

[108] Lord Rayleigh, On the maintenance of vibrations by forces of double frequency, and on the propagation of waves through a medium endowed with a periodic structure, Phil. Mag. 24 (1887) 145–159, http://dx.doi.org/10.1017/CBO9780511703980.002.

[109] K. Ohtaka, Energy band of photons and low-energy photon diffraction, Phys. Rev. B 19 (1979) 5057–5067, http://dx.doi.org/10.1103/PhysRevB.19.5057.

[110] G. Shambat, M.S. Mirotznik, G. Euliss, V.O. Smolski, E.G. Johnson, R.A. Athale, Photonic crystal filters for multi-band optical filtering on a monolithic substrate, J. Nanophoton. 3 (2009) 031506, http://dx.doi.org/10.1117/1.3110223.

[111] R.J. Martín-Palma, A. Lakhtakia, Nanotechnology: A Crash Course, SPIE Press, Bellingham, WA, USA, 2010.

[112] S.Y. Lin, J.G. Fleming, D.L. Hetherington, B.K. Smith, R. Biswas, K.M. Ho, M.M. Sigalas, W. Zubrzycki, S.R. Kurtz, J. Bur, A three-dimensional photonic crystal operating at infrared wavelengths, Nature 394 (1998) 251–253, http://dx.doi.org/10.1038/28343.

[113] A. Chelnokov, K. Wang, S. Rowson, P. Garoche, J.M. Lourtioz, Near-infrared Yablonovite-like photonic crystals by focused-ion-beam etching of macroporous silicon, Appl. Phys. Lett. 77 (2000) 2943–2945, http://dx.doi.org/10.1063/1.1322630.

[114] F. García-Santamaría, H.T. Miyazaki, A. Urquía, M. Ibisate, M. Belmonte, N. Shinya, F. Meseguer, C. López, Nanorobotic manipulation of microspheres for on-chip diamond architectures, Adv. Mater. 14 (2002) 1144–1147, http://dx.doi.org/10.1002/1521-4095(20020816)14:16<1144::AID-ADMA1144>3.0.CO;2-I.

[115] J.F. Galisteo-López, M. Ibisate, R. Sapienza, L. Froufe-Pérez, Á. Blanco, C. López, Self-assembled photonic structures, Adv. Mater. 23 (2011) 30–69, http://dx.doi.org/10.1002/adma.201000356.

[116] M. Maldovan, E.L. Thomas, Periodic Materials and Interference Lithography for Photonics, Phononics and Mechanics, Wiley-VCH, Weinheim, Germany, 2009.

[117] K. Aoki, H.T. Miyazaki, H. Hirayama, K. Inoshita, T. Baba, K. Sakoda, N. Shinya, Y. Aoyagi, Microassembly of semiconductor three-dimensional photonic crystals, Nature Mater. 2 (2003) 117–121, http://dx.doi.org/10.1038/nmat802.

[118] M. Maldovan, C.K. Ullal, W.C. Carter, E.L. Thomas, Exploring for 3D photonic bandgap structures in the 11 f.c.c. space groups, Nature Mater. 2 (2003) 664–667, http://dx.doi.org/10.1038/nmat979.

[119] B.G. Bovard, Rugate filter theory: an overview, Appl. Opt. 32 (1993) 5427–5442, http://dx.doi.org/10.1364/AO.32.005427.

[120] F. Villa, R. Machorro, A. Martínez, Rugate absorbing thin films and the 2×2 inhomogeneous matrix, Appl. Opt. 34 (1995) 3711–3714, http://dx.doi.org/10.1364/AO.34.003711.

[121] W.J. Gunning, R.L. Hall, F.J. Woodberry, W.H. Southwell, N.S. Gluck, Codeposition of continuous composition rugate filters, Appl. Opt. 28 (1989) 2945–2948, http://dx.doi.org/10.1364/AO.28.002945.

[122] A.F. Jankowski, L.R. Schrawyer, P.L. Perry, Reactive sputtering of molybdenum-oxide gradient-index filters, J. Vac. Sci. Technol. A 9 (1991) 1184–1187, http://dx.doi.org/10.1116/1.577599.

[123] P.L. Swart, P.V. Bulkin, B.M. Lacquet, Rugate filter manufacturing by electron cyclotron resonance plasma-enhanced chemical vapor deposition of SiN_x, Opt. Eng. 36 (1997) 1214–1219, http://dx.doi.org/10.1117/1.601281.

[124] E. Lorenzo, C.J. Oton, N.E. Capuj, M. Ghulinyan, D. Navarro-Urrios, Z. Gaburro, L. Pavesi, Fabrication and optimization of rugate filters based on porous silicon, phys. stat. sol. (c) 2 (2005) 3227–3231, http://dx.doi.org/10.1002/pssc.200461125.

[125] A.J. McPhun, Q.H. Wu, I.J. Hodgkinson, Birefringent rugate filters, Electron. Lett. 34 (1998) 360–361, http://dx.doi.org/10.1049/el:19980309.

[126] S. Chandrasekhar, Liquid Crystals, second ed., Cambridge University Press, Cambridge, United Kingdom, 1992.

[127] P.G. de Gennes, J.A. Prost, The Physics of Liquid Crystals, second ed., Clarendon Press, Oxford, United Kingdom, 1993.

[128] G.R. Luckhurst, Biaxial nematic liquid crystals: fact or fiction? Thin Solid Films 393 (2001) 40–52, http://dx.doi.org/10.1016/S0040-6090(01)01091-4.

[129] L.A. Madsen, T.J. Dingemans, M. Nakata, E.T. Samulski, Thermotropic biaxial nematic liquid crystals, Phys. Rev. Lett. 92 (2004) 145505, http://dx.doi.org/10.1103/PhysRevLett.92.145505.

[130] S. Chandrasekhar, B.K. Sadashiva, K.A. Suresh, Liquid crystals of disc-like molecules, Pramāna 9 (1977) 471–480, http://dx.doi.org/10.1007/BF02846252.

[131] S. Chandrasekhar, G.S. Ranganath, Discotic liquid crystals, Rep. Prog. Phys. 53 (1990) 57–84, http://dx.doi.org/10.1088/0034-4885/53/1/002.

[132] G. Pelzl, S. Diele, W. Weissflog, Banana-shaped compounds—a new field of liquid crystals, Adv. Mater. 11 (1999) 707–724, http://dx.doi.org/10.1002/(SICI)1521-4095 (199906)11:9<707::AID-ADMA707>3.0.CO;2-D.

[133] F. Reinitzer, Beiträge zur Kenntniss des Cholesterins, Monatsh. Chem. 9 (1888) 421–441.

[134] T.J. Sluckin, D.A. Dunmur, H. Stegemeyer, Crystals That Flow: Classic Papers From the History of Liquid Crystals, Taylor and Francis, London, United Kingdom, 2004.

[135] G. Joly, J. Billard, Quelques champs électromagnétiques dans les piles de Reusch. II. Piles éclairées sous l'incidence normale par des ondes monochromatiques planes et uniformes, J. Opt. (Paris) 13 (1982) 227–238, http://dx.doi.org/10.1088/0150-536X/13/4/008.

[136] R.A. Depine, A. Lakhtakia, A new condition to identify isotropic dielectric-magnetic materials displaying negative phase velocity, Microw. Opt. Technol. Lett. 41 (2004) 315–316, http://dx.doi.org/10.1002/mop.20127.

[137] A.D. Boardman, N. King, L. Velasco, Negative refraction in perspective, Electromagnetics 25 (2005) 365–389, http://dx.doi.org/10.1080/02726340590957371.

[138] E. Shamonina, L. Solymar, Metamaterials: how the subject started, Metamaterials 1 (2007) 12–18, http://dx.doi.org/10.1016/j.metmat.2007.02.001.

[139] R.A. Shelby, D.R. Smith, S. Schultz, Experimental verification of negative index of refraction, Science 292 (2001) 77–79, http://dx.doi.org/10.1126/science.1058847.

[140] T.G. Mackay, A. Lakhtakia, Negative refraction, negative phase velocity, and counterposition in bianisotropic materials and metamaterials, Phys. Rev. B 79 (2009) 235121, http://dx.doi.org/10.1103/PhysRevB.79.235121.

[141] Q.A. Naqvi, T.G. Mackay, A. Lakhtakia, Optical refraction in silver: counterposition, negative phase velocity and orthogonal phase velocity, Eur. J. Phys. 32 (2011) 883–893, http://dx.doi.org/10.1088/0143-0807/32/4/004.

[142] Y.-J. Jen, A. Lakhtakia, C.-W. Yu, C.-T. Lin, Negative refraction in a uniaxial absorbent dielectric material, Eur. J. Phys. 30 (2009) 1381–1390, http://dx.doi.org/10.1088/0143-0807/30/6/016.

[143] R. Marqués, F. Medina, R. Rafii-El-Idrissi, Role of bianisotropy in negative permeability and left-handed metamaterials, Phys. Rev. B 65 (2002) 144440, http://dx.doi.org/10.1103/PhysRevB.65.144440.

[144] T.G. Mackay, A. Lakhtakia, Plane waves with negative phase velocity in Faraday chiral mediums, Phys. Rev. E 69 (2004) 026602, http://dx.doi.org/10.1103/PhysRevE.69.026602.

[145] A. Pimenov, A. Loidl, K. Gehrke, V. Moshnyaga, K. Samwer, Negative refraction observed in a metallic ferromagnet in the gigahertz frequency range, Phys. Rev. Lett. 98 (2007) 197401, http://dx.doi.org/10.1103/PhysRevLett.98.197401.

[146] C.G. Parazzoli, R.B. Greegor, K. Li, B.E.C. Koltenbah, M. Tanielian, Experimental verification and simulation of negative index of refraction using Snell's law, Phys. Rev. Lett. 90 (2003) 107401, http://dx.doi.org/10.1103/PhysRevLett.90.107401.

[147] A.A. Houck, J.B. Brock, I.L. Chuang, Experimental observations of a left-handed material that obeys Snell's law, Phys. Rev. Lett. 90 (2003) 137401, http://dx.doi.org/10.1103/PhysRevLett.90.137401.

[148] B.A. Munk, Metamaterials: Critique and Alternatives, Wiley, Hoboken, NJ, USA, 2009.

[149] R.A. Depine, M.L. Gigli, Resonant excitation of surface modes at a single flat uniaxial-metal interface, J. Opt. Soc. Am. A 14 (1997) 510–519, http://dx.doi.org/10.1364/JOSAA.14.000510.

[150] R. Li, C. Cheng, F.-F. Ren, J. Chen, Y.-X. Fan, J. Ding, H.-T. Wang, Hybridized surface plasmon polaritons at an interface between a metal and a uniaxial crystal, Appl. Phys. Lett. 92 (2008) 141115, http://dx.doi.org/10.1063/1.2908920.

[151] Y.-J. Jen, C.-W. Yu, Optical configuration for unpolarized ultra-long-range surface-plasmon-polariton waves, Appl. Opt. 50 (2011) C154–C158, http://dx.doi.org/10.1364/AO.50.00C154.

[152] Y.-J. Jen, Arbitrarily polarized long-range surface-plasmon-polariton waves, J. Nanophoton. 5 (2011) 050304, http://dx.doi.org/10.1117/1.3634056.

[153] M. Faryad, A. Lakhtakia, Grating-coupled excitation of multiple surface plasmon-polariton waves, Phys. Rev. A 84 (2011) 033852, http://dx.doi.org/10.1103/PhysRevA.84.033852.

[154] J.A. Gaspar-Armenta, F. Villa, Photonic surface-wave excitation: photonic crystal–metal interface, J. Opt. Soc. Am. B 20 (2003) 2349–2354, http://dx.doi.org/10.1364/JOSAB.20.002349.

[155] R. Das, R. Jha, On the modal characteristics of surface plasmon polaritons at a metal–Bragg interface at optical frequencies, Appl. Opt. 48 (2009) 4904–4908, http://dx.doi.org/10.1364/AO.48.004904.

[156] V.N. Konopsky, Plasmon-polariton waves in nanofilms on one-dimensional photonic crystal surfaces, New J. Phys. 12 (2010) 093006, http://dx.doi.org/10.1088/1367-2630/12/9/093006.

[157] M. Kaliteevski, I. Iorsh, S. Brand, R.A. Abram, J.M. Chamberlain, A.V. Kavokin, I.A. Shelykh, Tamm plasmon-polaritons: possible electromagnetic states at the interface of a metal and a dielectric Bragg mirror, Phys. Rev. B 76 (2007) 165415, http://dx.doi.org/10.1103/PhysRevB.76.165415.

[158] M.E. Sasin, R.P. Seisyan, M.A. Kalitteevski, S. Brand, R.A. Abram, J.M. Chamberlain, A.Yu. Egorov, A.P. Vasil'ev, V.S. Mikhrin, A.V. Kavokin, Tamm plasmon polaritons: slow and spatially compact light, Appl. Phys. Lett. 92 (2008) 251112, http://dx.doi.org/10.1063/1.2952486.

[159] M.R.M. Atalla, M. Faryad, A. Lakhtakia, On surface-plasmon-polariton waves guided by the interface of a metal and a rugate filter with a sinusoidal refractive-index profile. Part II: High-phase-speed solutions, J. Opt. Soc. Am. B 29 (2012) 3078–3086, http://dx.doi.org/10.1364/JOSAB.29.003078.

[160] M.E. Sasin, R.P. Seisyan, M.A. Kalitteevski, S. Brand, R.A. Abram, J.M. Chamberlain, I.V. Iorsh, I.A. Shelykh, A.Yu. Egorov, A.P. Vasil'ev, V.S. Mikhrin, A.V. Kavokin, Tamm plasmon polaritons: first experimental observation, Superlattice Microstruct. 47 (2010) 40–44, http://dx.doi.org/10.1016/j.spmi.2009.09.003.

[161] J.A. Polo Jr., A. Lakhtakia, Energy flux in a surface-plasmon-polariton wave bound to the planar interface of a metal and a structurally chiral material, J. Opt. Soc. Am. A 26 (2009) 1696–1703, http://dx.doi.org/10.1364/JOSAA.26.001696.

[162] D.B. Walker, E.N. Glytsis, T.K. Gaylord, Surface mode at isotropic-uniaxial and isotropic-biaxial interfaces, J. Opt. Soc. Am. A 15 (1998) 248–260, http://dx.doi.org/10.1364/JOSAA.15.000248.

[163] A.N. Furs, V.M. Galynsky, L.M. Barkovsky, Surface polaritons in symmetry planes of biaxial crystals, J. Phys. A: Math. Gen. 38 (2005) 8083–8101, http://dx.doi.org/10.1088/0305-4470/38/37/010.

[164] A.N. Darinskiǐ, Dispersionless polaritons on a twist boundary in optically uniaxial crystals, Crystallogr. Rep. 46 (2001) 842–844, http://dx.doi.org/10.1134/1.1405874.

[165] J.A. Polo Jr., S. Nelatury, A. Lakhtakia, Surface electromagnetic wave at a tilted uniaxial bicrystalline interface, Electromagnetics 26 (2006) 629–642, http://dx.doi.org/10.1080/02726340600978364.

[166] S.R. Nelatury, J.A. Polo Jr., A. Lakhtakia, Surface waves with simple exponential transverse decay at a biaxial bicrystalline interface, J. Opt. Soc. Am. A 24 (2007)

856–865, http://dx.doi.org/10.1364/JOSAA.24.000856. Corrections: 24 (2007) 2102, http://dx.doi.org/10.1364/JOSAA.24.002102.

[167] J.A. Polo Jr., S.R. Nelatury, A. Lakhtakia, Surface waves at a biaxial bicrystalline interface, J. Opt. Soc. Am. A 24 (2007) 2974–2979, http://dx.doi.org/10.1364/JOSAA.24.002974.

[168] D. Artigas, L. Torner, Dyakonov surface waves in photonic metamaterials, Phys. Rev. Lett. 94 (2005) 013901, http://dx.doi.org/10.1103/PhysRevLett.94.013901.

[169] L.-C. Crasovan, D. Artigas, D. Mihalache, L. Torner, Optical Dyakonov surface waves at magnetic interfaces, Opt. Lett. 30 (2005) 3075–3077, http://dx.doi.org/10.1364/OL.30.003075.

[170] A. Boardman, N. King, Y. Rapoport, L. Velasco, Gyrotropic impact upon negatively refracting surfaces, New J. Phys. 7 (2005) 191, http://dx.doi.org/10.1088/1367-2630/7/1/191.

[171] M.A. Noginov, Compensation of surface plasmon loss by gain in dielectric medium, J. Nanophoton. 2 (2008) 021855, http://dx.doi.org/10.1117/1.3073670.

[172] A.N. Furs, L.M. Barkovsky, A new type of surface polaritons at the interface of the magnetic gyrotropic media, J. Phys. A: Math. Theor. 40 (2007) 309–328, http://dx.doi.org/10.1088/1751-8113/40/2/010.

[173] P. Yeh, A. Yariv, C.-S. Hong, Electromagnetic propagation in periodic stratified media. I. General theory, J. Opt. Soc. Am. 67 (1977) 423–438, http://dx.doi.org/10.1364/JOSA.67.000423.

[174] P. Yeh, A. Yariv, A.Y. Cho, Optical surface waves in periodic layered media, Appl. Phys. Lett. 32 (1978) 104–105, http://dx.doi.org/10.1063/1.89953.

[175] W.M. Robertson, M.S. May, Surface electromagnetic wave excitation on one-dimensional photonic band-gap arrays, Appl. Phys. Lett. 74 (1999) 1800–1802, http://dx.doi.org/10.1063/1.123090.

[176] W.M. Robertson, Experimental measurement of the effect of termination on surface electromagnetic waves in one-dimensional photonic bandgap arrays, J. Lightwave Technol. 17 (1999) 2013–2017, http://dx.doi.org/10.1109/50.802988.

[177] V.N. Konopsky, E.V. Alieva, Photonic crystal surface waves for optical biosensors, Anal. Chem. 79 (2007) 4729–4735, http://dx.doi.org/10.1021/ac070275y.

[178] J. Martorell, D.W.L. Sprung, G.V. Morozov, Surface TE waves on 1D photonic crystals, J. Opt. A: Pure Appl. Opt. 8 (2006) 630–638, http://dx.doi.org/10.1088/1464-4258/8/8/003.

[179] F. Villa-Villa, J.A. Gaspar-Armenta, A. Mendoza-Suárez, Surface modes in one dimensional photonic crystals that include left handed materials, J. Electromagn. Waves Appl. 21 (2007) 485–499, http://dx.doi.org/10.1163/156939307779367323.

[180] D.P. Pulsifer, M. Faryad, A. Lakhtakia, Grating-coupled excitation of Tamm waves, J. Opt. Soc. Am. B 29 (2012) 2260–2269, http://dx.doi.org/10.1364/JOSAB.29.002260. Corrections: 30 (2013) 177, http://dx.doi.org/10.1364/JOSAB.30.000177.

[181] A. Namdar, I.V. Shadrivov, Y.S. Kivshar, Backward Tamm states in left-handed metamaterials, Appl. Phys. Lett. 89 (2006) 114104, http://dx.doi.org/10.1063/1.2352794.

[182] M. Faryad, A. Lakhtakia, Propagation of surface waves and waveguide modes guided by a dielectric slab inserted in a sculptured nematic thin film, Phys. Rev. A 83 (2011) 013814, http://dx.doi.org/10.1103/PhysRevA.83.013814.

[183] M. Faryad, A. Lakhtakia, Dyakonov-Tamm waves guided by a phase-twist combination defect in a sculptured nematic thin film, Opt. Commun. 284 (2011) 160–168, http://dx.doi.org/10.1016/j.optcom.2010.08.042.

[184] D.R. Smith, D. Schurig, Electromagnetic wave propagation in media with indefinite permittivity and permeability tensors, Phys. Rev. Lett. 90 (2003) 077405, http://dx. doi.org/10.1103/PhysRevLett.90.077405Lett.90.077405.

[185] T.G. Mackay, A. Lakhtakia, R.A. Depine, Uniaxial dielectric media with hyperbolic dispersion relations, Microw. Opt. Technol. Lett. 48 (2006) 363–367, http://dx.doi.org/10. 1002/mop.21350.

[186] V.M. Galynsky, A.N. Furs, L.M. Barkovsky, Surface polaritons at the interface of gyrotropic and nonlinear isotropic media, Nonlin. Phenom. Complex Syst. 8 (2005) 406–408.

[187] L. Torner, J.P. Torres, F. Lederer, D. Mihalache, D.M. Baboiu, M. Ciumac, Nonlinear hybrid waves guided by birefringent interfaces, Electron. Lett. 29 (1993) 1186–1188, http://dx.doi.org/10.1049/el:19930793.

[188] S.R. Nelatury, J.A. Polo Jr., A. Lakhtakia, Electrical control of surface-wave propagation at the planar interface of a linear electro-optic material and an isotropic dielectric material, Electromagnetics 28 (2008) 162–174, http://dx.doi.org/10.1080/02726340801921486.

[189] S.R. Nelatury, J.A. Polo Jr., A. Lakhtakia, On widening the angular existence domain for Dyakonov surface waves using the Pockels effect, Microw. Opt. Technol. Lett. 50 (2008) 2360–2362, http://dx.doi.org/10.1002/mop.23698.

[190] E.G. Matveeva, Z. Gryczynski, J. Malicka, J. Lukomska, S. Makowiec, K. Berndt, J. Lakowicz, I. Gryczynski, Directional surface plasmon-coupled emission: application for an immunoassay in whole blood, Anal. Biochem. 344 (2005) 161–167, http://dx.doi.org/ 10.1016/j.ab.2005.07.005.

[191] M.J. Linman, Q.J. Cheng, Surface plasmon resonance: new biointerface designs and high-throughput affinity screening, in: M. Zourob, A. Lakhtakia (Eds.), Optical Guided-Wave Chemical and Biosensors I, Springer, Berlin, Germany, 2010, http://dx.doi.org/10. 1007/978-3-540-88242-8_5.

[192] A. Ramakrishnan, Y. Tan, A. Sadana, A kinetic study of analyte-receptor binding and dissociation for surface plasmon resonance biosensors applications, IEEE Sensors J. 5 (2005) 356–364, http://dx.doi.org/10.1109/JSEN.2004.839594.

[193] X.D. Hoa, A.G. Kirk, M. Tabrizian, Towards integrated and sensitive surface plasmon resonance biosensors: a review of recent progress, Biosens. Bioelectron. 23 (2007) 151–160, http://dx.doi.org/10.1016/j.bios.2007.07.001.

[194] F. Romanato, K.H. Lee, H.K. Kang, G. Ruffato, C.C. Wong, Sensitivity enhancement in grating coupled surface plasmon resonance by azimuthal control, Opt. Exp. 17 (2009) 12145–12154, http://dx.doi.org/10.1364/OE.17.012145.

[195] W.-K. Kuo, C.-H. Chang, C.-J. Wu, H.-H. Yu, Phase-detection-sensitivity enhancement of grating-coupled surface plasmon resonance sensor with light incident at nonzero azimuth angle, J. Nanophoton. 6 (2012) 063524, http://dx.doi.org/10.1117/1.JNP.6.063524.

[196] E. Yeatman, E.A. Ash, Surface plasmon microscopy, Electron. Lett. 23 (1987) 1091–1092, http://dx.doi.org/10.1049/el:19870762.

[197] B. Rothenhäusler, W. Knoll, Surface-plasmon microscopy, Nature 332 (1988) 615–617, http://dx.doi.org/10.1038/332615a0.

[198] A.V. Zayats, I.I. Smolyaninov, A.A. Maradudin, Nano-optics of surface plasmon polaritons, Phys. Rep. 408 (2005) 131–314, http://dx.doi.org/10.1016/j.physrep.2004.11.001.

[199] V. Coello, Near-field optical microscopy of surface plasmon polaritons nano-optics, in: A. Méndez-Vilas, J. Díaz (Eds.), Modern Research and Educational Topics in Microscopy, Formatex, Badajoz, Spain, 2007, pp. 828–839.

[200] A. Drezet, D. Koller, A. Hohenau, A. Leitner, F.R. Aussenegg, J.R. Krenn, Surface plasmon polariton microscope with parabolic reflectors, Opt. Lett. 32 (2007) 2414–2416, http://dx.doi.org/10.1364/OL.32.002414.

[201] M.G. Somekh, Surface plasmon and surface wave microscopy, in: P. Török, F.-J. Kao (Eds.), Optical Imaging and Microscopy, Springer, Berlin, Germany, 2007, pp. 347–399, http://dx.doi.org/10.1007/978-3-540-69565-3_14.

[202] L. Berguiga, T. Roland, K. Monier, J. Elezgaray, F. Argoul, Amplitude and phase images of cellular structures with a scanning surface plasmon microscope, Opt. Exp. 19 (2011) 6571–6586, http://dx.doi.org/10.1364/OE.19.006571.

[203] G. Stabler, M.G. Somekh, C.W. See, High-resolution wide-field surface plasmon microscopy, J. Microsc. 214 (2004) 328–333, http://dx.doi.org/10.1111/j.0022-2720.2004.01309.x.

[204] S.P. Frisbie, C.F. Chesnutt, M.E. Holtz, A. Krishnan, L. Grave de Peralta, A.A. Bernussi, Image formation in wide-field microscopes based on leakage of surface plasmon-coupled fluorescence, IEEE Photon. J. 1 (2009) 153–162, http://dx.doi.org/10.1109/JPHOT.2009.2028307.

[205] A. Gombert, A. Luque, Photonics in photovoltaic systems, phys. stat. sol. (a) 205 (2008) 2757–2765, http://dx.doi.org/10.1002/pssa.200880459.

[206] H.A. Atwater, A. Polman, Plasmonics for improved photovoltaic devices, Nature Mater. 9 (2010) 205–213, http://dx.doi.org/10.1038/nmat2629.

[207] P. Sheng, A.N. Bloch, R.S. Stepleman, Wavelength-selective absorption enhancement in thin-film solar cells, Appl. Phys. Lett. 43 (1983) 579–581, http://dx.doi.org/10.1063/1.94432.

[208] C. Heine, R.H. Morf, Submicrometer gratings for solar energy applications, Appl. Opt. 34 (1995) 2476–2482, http://dx.doi.org/10.1364/AO.34.002476.

[209] V.E. Ferry, M.A. Verschuuren, H.B.T. Li, E. Verhagen, R.J. Walters, R.E.I. Schropp, H.A. Atwater, A. Polman, Light trapping in ultrathin plasmonic solar cells, Opt. Exp. 18 (2010) A237–A245, http://dx.doi.org/10.1364/OE.18.00A237.

[210] W.J. Nam, L. Ji, V.V. Varadan, S.J. Fonash, Exploration of nano-element array architectures for substrate solar cells using an a-Si:H absorber, J. Appl. Phys. 111 (2012) 123103, http://dx.doi.org/10.1063/1.4729539.

[211] W.L. Barnes, A. Dereux, T.W. Ebbesen, Surface plasmon subwavelength optics, Nature 424 (2003) 824–830, http://dx.doi.org/10.1038/nature01937.

[212] J.S. Sekhon, S.S. Verma, Plasmonics: the future wave of communication, Curr. Sci. India 101 (2011) 484–488.

[213] J.T. Kim, J.J. Ju, S. Park, M.-s. Kim, S.K. Park, M.-H. Lee, Chip-to-chip optical interconnect using gold long-range surface plasmon polariton waveguides, Opt. Exp. 16 (2008) 13133–13138, http://dx.doi.org/10.1364/OE.16.013133.

[214] K.F. MacDonald, N.I. Zheludev, Active plasmonics: current status, Laser Photon. Rev. 4 (2010) 562–567, http://dx.doi.org/10.1002/lpor.200900035.

[215] R.J. Walters, R.V.A. van Loon, I. Brunets, J. Schmitz, A. Polman, A silicon-based electrical source of surface plasmon polaritons, Nature Mater. 9 (2010) 21–25, http://dx.doi.org/10.1038/NMAT2595.

[216] P. Berini, Long-range surface plasmon polaritons, Adv. Opt. Photon. 1 (2009) 484–588, http://dx.doi.org/10.1364/AOP.1.000484.

[217] F. Flory, L. Escoubas, G. Berginc, Optical properties of nanostructured materials: a review, J. Nanophoton. 5 (2011) 052502, http://dx.doi.org/10.1117/1.3609266.

[218] D. Mihalache, Recent trends in micro- and nanophotonics: a personal selection, J. Optoelectron. Adv. Mater. 13 (2011) 1055–1066.

[219] P.B. Johnson, R.W. Christy, Optical constants of the noble metals, Phys. Rev. B 6 (1972) 4370–4379, http://dx.doi.org/10.1103/PhysRevB.6.4370.

[220] I.I. Smolyaninov, C.C. Davis, J. Elliott, A.V. Zayats, Resolution enhancement of a surface immersion microscope near the plasmon resonance, Opt. Lett. 30 (2005) 382–384, http://dx.doi.org/10.1364/OL.30.000382.

[221] P. Tobiška, O. Hugon, A. Trouillet, H. Gagnaire, An integrated optic hydrogen sensor based on SPR on palladium, Sens. Actuat. B: Chem. 74 (2001) 168–172, http://dx.doi.org/10.1016/S0925-4005(00)00728-0.

[222] S. Baldelli, A.S. Eppler, E. Anderson, Y.-R. Shen, G.A. Somorjai, Surface enhanced sum frequency generation of carbon monoxide adsorbed on platinum nanoparticle arrays, J. Chem. Phys. 113 (2000) 5432–5438, http://dx.doi.org/10.1063/1.1290024.

[223] M.G. Blaber, M.D. Arnold, N. Harris, M.J. Ford, M.B. Cortie, Plasmon absorption in nanospheres: a comparison of sodium, potassium, aluminium, silver and gold, Physica B 394 (2007) 184–187, http://dx.doi.org/10.1016/j.physb.2006.12.011.

[224] D.A. Bobb, G. Zhu, M. Mayy, A.V. Gavrilenko, P. Mead, V.I. Gavrilenko, M.A. Noginov, Engineering of low-loss metal for nanoplasmonic and metamaterials applications, Appl. Phys. Lett. 95 (2009) 151102, http://dx.doi.org/10.1063/1.3237179.

[225] R.F. Wallis, Optical properties associated with surface excitations of semiconductors, in: M. Balkanski (Ed.), Handbook on Semiconductors, vol. 2, Elsevier Science, Amsterdam, The Netherlands, 1994, pp. 1–31.

[226] M. Jablan, H. Buljan, M. Soljačić, Plasmonics in graphene at infrared frequencies, Phys. Rev. B 80 (2009) 245435, http://dx.doi.org/10.1103/PhysRevB.80.245435.

[227] P. Tassin, T. Koschny, M. Kafesaki, C.M. Soukoulis, A comparison of graphene, superconductors and metals as conductors for metamaterials and plasmonics, Nature Photon. 6 (2012) 259–264, http://dx.doi.org/10.1038/NPHOTON.2012.27.

[228] A. Lakhtakia, Surface-plasmon wave at the planar interface of a metal film and a structurally chiral medium, Opt. Commun. 279 (2007) 291–297, http://dx.doi.org/10.1016/j.optcom.2007.07.026.

[229] L.G. Parratt, Surface studies of solids by total reflection of X-rays, Phys. Rev. 95 (1954) 359–369, http://dx.doi.org/10.1103/PhysRev.95.359.

[230] R. Rashed, A pioneer in anaclastics, Ibn Sahl on burning mirrors and lenses, Isis 81 (1990) 464–491.

[231] R.C. Jorgenson, S.S. Yee, A fiber-optic chemical sensor based on surface plasmon resonance, Sens. Actuat. B: Chem. 12 (1993) 213–220, http://dx.doi.org/10.1016/0925-4005(93)80021-3.

[232] A. Trouillet, C. Ronot-Trioli, C. Veillas, H. Gagnaire, Chemical sensing by surface plasmon resonance in a multimode optical fibre, Pure Appl. Opt. 5 (1996) 227–237, http://dx.doi.org/10.1088/0963-9659/5/2/006.

[233] A.J.C. Tubb, F.P. Payne, R.B. Millington, C.R. Lowe, Single-mode optical fibre surface plasma wave chemical sensor, Sens. Actuat. B: Chem. 41 (1997) 71–79, http://dx.doi.org/10.1016/S0925-4005(97)80279-1.

[234] A. Díez, M.V. Andrés, J.L. Cruz, In-line fiber-optic sensors based on the excitation of surface plasma modes in metal-coated tapered fibers, Sens. Actuat. B: Chem. 73 (2001) 95–99, http://dx.doi.org/10.1016/S0925-4005(00)00649-3.

[235] R. Charbonneau, E. Lisicka-Shrzek, P. Berini, Broadside coupling to long-range surface plasmons using an angle-cleaved optical fiber, Appl. Phys. Lett. 92 (2008) 101102, http://dx.doi.org/10.1063/1.2892678.

[236] J. Homola, Optical fiber sensor based on surface plasmon excitation, Sens. Actuat. B: Chem. 29 (1995) 401–405, http://dx.doi.org/10.1016/0925-4005(95)01714-3.

[237] D.K. Kambhampati, W. Knoll, Surface-plasmon optical techniques, Curr. Opin. Colloid Interf. Sci. 4 (1999) 273–280, http://dx.doi.org/10.1016/S1359-0294(99)90008-0.

[238] D. Sarid, Long-range surface-plasma waves on very thin metal films, Phys. Rev. Lett. 47 (1981) 1927–1930, http://dx.doi.org/10.1103/PhysRevLett.47.1927.

[239] D. Sarid, W. Challener, Modern Introduction to Surface Plasmons: Theory, Mathematica Modeling and Applications, Cambridge University Press, New York, NY, USA, 2010.

[240] L. Wendler, R. Haupt, Long-range surface plasmon-polaritons in asymmetric layer structures, J. Appl. Phys. 59 (1986) 3289–3291, http://dx.doi.org/10.1063/1.336884.

[241] L. Wendler, R. Haupt, An improved virtual mode theory of ATR experiments on surface polaritons: application to long-range surface plasmon-polaritons in asymmetric layer structures, phys. stat. sol. (b) 143 (1987) 131–148, http://dx.doi.org/10.1002/pssb.2221430115.

[242] J. Guo, R. Adato, Extended long range plasmon waves in finite thickness metal film and layered dielectric materials, Opt. Exp. 14 (2006) 12409–12418, http://dx.doi.org/10.1364/OE.14.012409.

[243] R. Adato, J. Guo, Characteristics of ultra-long range surface plasmon waves at optical frequencies, Opt. Exp. 15 (2007) 5008–5017, http://dx.doi.org/10.1364/OE.15.005008.

[244] Y.-J. Jen, A. Lakhtakia, C.-W. Yu, T.-Y. Chan, Multilayered structures for p- and s-polarized long-range surface-plasmon-polariton propagation, J. Opt. Soc. Am. A 26 (2009) 2600–2606, http://dx.doi.org/10.1364/JOSAA.26.002600.

[245] J.T. Hastings, J. Guo, P.D. Keathley, P.B. Kumaresh, Y. Wei, S. Law, L.G. Bachas, Optimal self-referenced sensing using long- and short-range surface plasmons, Opt. Exp. 15 (2007) 17661–17672, http://dx.doi.org/10.1364/OE.15.017661.

[246] R.F. Oulton, V.J. Sorger, D.A. Genov, D.F.P. Pile, X. Zhang, A hybrid plasmonic waveguide for subwavelength confinement and long-range propagation, Nature Photon. 2 (2008) 496–500, http://dx.doi.org/10.1038/nphoton.2008.131.

[247] D. Dai, S. He, A silicon-based hybrid plasmonic waveguide with a metal cap for a nanoscale light confinement, Opt. Exp. 17 (2009) 16646–16653, http://dx.doi.org/10.1364/OE.17.016646.

[248] Y. Song, J. Wang, Q. Li, M. Yan, M. Qiu, Broadband coupler between silicon waveguide and hybrid plasmonic waveguide, Opt. Exp. 18 (2010) 13173–13179, . http://dx.doi.org/10.1364/OE.18.013173

[249] Z. Chen, I.R. Hooper, J.R. Sambles, Strongly coupled surface plasmons on thin shallow metallic gratings, Phys. Rev. B 77 (2008) 161405(R), http://dx.doi.org/10.1103/PhysRevB.77.161405.

[250] E.N. Glytsis, T.K. Gaylord, Rigorous three-dimensional coupled-wave diffraction analysis of single and cascaded anisotropic gratings, J. Opt. Soc. Am. A 4 (1987) 2061–2080, http://dx.doi.org/10.1364/JOSAA.4.002061.

[251] L. Li, Multilayer modal method for diffraction gratings of arbitrary profile, depth, and permittivity, J. Opt. Soc. Am. A 10 (1993) 2581–2591, http://dx.doi.org/10.1364/JOSAA.10.002581.

[252] L. Li, Multilayer modal method for diffraction gratings of arbitrary profile, depth, and permittivity: addendum, J. Opt. Soc. Am. A 11 (1994) 1685–1685, http://dx.doi.org/10.1364/JOSAA.11.001685.

[253] A. Lakhtakia, V.K. Varadan, V.V. Varadan, On filling up the grooves of a perfectly—conducting grating with a dielectric material, J. Mod. Opt. 38 (1991) 659–669, http://dx.doi.org/10.1080/09500349114550661.

[254] J. Chandezon, Les equations de Maxwell sous forme covariante. Application à l'étude de la propagation dans les guides periodques et à la diffraction par les reseaux, Ph.D. Thesis, Clermont-Ferrand University, Aubiere, France, 1979.

[255] J. Chandezon, M.T. Dupuis, G. Cornet, D. Maystre, Multicoated gratings: a differential formalism applicable in the entire optical region, J. Opt. Soc. Am. 72 (1982) 839–846, http://dx.doi.org/10.1364/JOSA.72.000839.

[256] N. Chateau, J.-P. Hugonin, Algorithm for the rigorous coupled-wave analysis of grating diffraction, J. Opt. Soc. Am. A 11 (1994) 1321–1331, http://dx.doi.org/10.1364/JOSAA. 11.001321.

[257] M.G. Moharam, E.B. Grann, D.A. Pommet, T.K. Gaylord, Formulation for stable and efficient implementation of the rigorous coupled-wave analysis of binary gratings, J. Opt. Soc. Am. A 12 (1995) 1068–1076, http://dx.doi.org/10.1364/JOSAA.12.001068.

[258] F. Wang, M.W. Horn, A. Lakhtakia, Rigorous electromagnetic modeling of near-field phase-shifting contact lithography, Microelectron. Eng. 71 (2004) 34–53, http://dx.doi.org/10.1016/j.mee.2003.09.003.

[259] M. Faryad, A. Lakhtakia, Multiple trains of same-color surface plasmon-polaritons guided by the planar interface of a metal and a sculptured nematic thin film. Part V: Grating-coupled excitation, J. Nanophoton. 5 (2011) 053527, http://dx.doi.org/10.1117/1.3663210.

[260] D. Maystre (Ed.), Selected Papers on Diffraction Gratings, SPIE Press, Bellingham, WA, USA, 1993.

[261] Y. Jaluria, Computer Methods for Engineering, Taylor and Francis, Washington, DC, USA, 1996.

[262] H. Hochstadt, Differential Equations: A Modern Approach, Dover Press, New York, NY, USA, 1975.

[263] D. Marcuse, Wave propagation along a dielectric interface, J. Opt. Soc. Am. 64 (1974) 794–797, http://dx.doi.org/10.1364/JOSA.64.000794.

[264] K.S. Kunz, R.J. Luebbers, The Finite Difference Time Domain Method for Electromagnetics, CRC Press, Boca Raton, FL, USA, 1993.

[265] M. Hochberg, T. Baehr-Jones, C. Walker, A. Scherer, Integrated plasmon and dielectric waveguides, Opt. Exp. 12 (2004) 5481–5486, http://dx.doi.org/10.1364/OPEX.12.005481.

[266] F. Liu, Y. Rao, Y. Huang, W. Zhang, J. Peng, Coupling between long range surface plasmon polariton mode and dielectric waveguide mode, Appl. Phys. Lett. 90 (2007) 141101, http://dx.doi.org/10.1063/1.2719169.

[267] A. Degiron, S.-Y. Cho, T. Tyler, N.M. Jokerst, D.R. Smith, Directional coupling between dielectric and long-range plasmon waveguides, New J. Phys. 11 (2009) 015002, http://dx.doi.org/10.1088/1367-2630/11/1/015002.

[268] F. Liu, R. Wan, Y. Li, Y. Huang, Y. Miura, D. Ohnishi, J. Peng, Extremely high efficient coupling between long range surface plasmon polariton and dielectric waveguide mode, Appl. Phys. Lett. 95 (2009) 091104, http://dx.doi.org/10.1063/1.3212145.

[269] H.S. Won, K.C. Kim, S.H. Song, C.-H. Oh, P.S. Kim, S. Park, S.I. Kim, Vertical coupling of long-range surface plasmon polaritons, Appl. Phys. Lett. 88 (2006) 011110, http://dx.doi.org/10.1063/1.2159558.

[270] R. Wan, F. Liu, X. Tang, Y. Huang, J. Peng, Vertical coupling between short range surface plasmon polariton mode and dielectric waveguide mode, Appl. Phys. Lett. 94 (2009) 141104, http://dx.doi.org/10.1063/1.3111001.

[271] R.W. Boyd, Nonlinear Optics, third ed., Academic Press, London, United Kingdom, 2008.

[272] Y.J. Chen, G.M. Carter, Attenuated total reflection calculations for nonlinear surface plasmon dispersion, Solid State Commun., 45 (1983) 277–280, http://dx.doi.org/10.1016/0038-1098(83)90480-5.

[273] G.I. Stegeman, C.T. Seaton, Nonlinear surface plasmons guided by thin metal films, Opt. Lett. 9 (1984) 235–237, http://dx.doi.org/10.1364/OL.9.000235.

[274] G.I. Stegeman, J.D. Valera, C.T. Seaton, J. Sipe, A.A. Maradudin, Nonlinear s-polarized surface plasmon polaritons, Solid State Commun. 52 (1984) 293–297, http://dx.doi.org/10.1016/0038-1098(84)90829-9.

[275] J. Ariyasu, C.T. Seaton, G.I. Stegeman, A.A. Maradudin, R.F. Wallis, Nonlinear surface polaritons guided by metal films, J. Appl. Phys. 58 (1985) 2460–2466, http://dx.doi.org/10.1063/1.335921.

[276] D. Mihalache, G.I. Stegeman, C.T. Seaton, E.M. Wright, R. Zanoni, A.D. Boardman, T. Twardowski, Exact dispersion relations for transverse magnetic polarized guided waves at a nonlinear interface, Opt. Lett. 12 (1987) 187–189, http://dx.doi.org/10.1364/OL.12.000187.

[277] P.D. Maker, R.W. Terhune, C.M. Savage, Intensity-dependent changes in the refractive index of liquids, Phys. Rev. Lett. 12 (1964) 507–509, http://dx.doi.org/10.1103/PhysRevLett.12.507. Erratum: 16 (1966) 832, http://dx.doi.org/10.1103/PhysRevLett.16.832.

[278] G.M. Carter, Y.J. Chen, S.K. Tripathy, Intensity-dependent index of refraction in multilayers of polydiacetylene, Appl. Phys. Lett. 43 (1983) 891–893, http://dx.doi.org/10.1063/1.94195.

[279] T.S. Moss, Theory of intensity dependence of refractive index, phys. stat. sol. (b) 101 (1980) 555–561, http://dx.doi.org/10.1002/pssb.2221010214.

[280] A.E. Kaplan, Theory of hysteresis reflection and refraction of light by a boundary of a nonlinear medium, Sov. Phys. JETP 45 (1977) 896–905.

[281] R.K. Hickernell, D. Sarid, Optical bistability using prism-coupled, long-range surface plasmons, J. Opt. Soc. Am. B 3 (1986) 1059–1069, http://dx.doi.org/10.1364/JOSAB.3.001059.

[282] G.I. Stegeman, G. Assanto, R. Zanoni, C.T. Seaton, E. Garmire, A.A. Maradudin, R. Reinisch, G. Vitrant, Bistability and switching in nonlinear prism coupling, Appl. Phys. Lett. 52 (1988) 869–871, http://dx.doi.org/10.1063/1.99257.

[283] T.H. O'Dell, The Electrodynamics of Magneto-Electric Media, North-Holland, Amsterdam, The Netherlands, 1970.

[284] H. Schmid, Magnetoelectric effects in insulating magnetic materials, in: W.S. Weiglhofer, A. Lakhtakia (Eds.), Introduction to Complex Mediums for Optics and Electromagnetics, SPIE Press, Bellingham, WA, USA, 2003, pp. 167–195.

[285] N. Miura, Physics of Semiconductors in High Magnetic Fields, Oxford University Press, Oxford, United Kingdom, 2008.

[286] B.A. Auld, Acoustic Fields and Waves in Solids, Krieger, Malabar, FL, USA, 1990.

[287] X. Zhao, Z. Suo, Electrostriction in elastic dielectrics undergoing large deformation, J. Appl. Phys. 104 (2008) 123530, http://dx.doi.org/10.1063/1.3031483.

[288] F. Wang, A. Lakhtakia, R. Messier, On piezoelectric control of the optical response of sculptured thin films, J. Mod. Opt. 50 (2003) 239–249, http://dx.doi.org/10.1080/09500340308235173.

[289] A. Lakhtakia, Electrically switchable exhibition of circular Bragg phenomenon by an isotropic slab, Microw. Opt. Technol. Lett. 48 (2006) 2148–2153, http://dx.doi.org/10.1002/mop.21941.

[290] P. Haefner, E. Luck, E. Mohler, Magnetooptical properties of surface plasma waves on copper, silver, gold, and aluminum, phys. stat. sol. (b) 185 (1994) 289–299, http://dx.doi.org/10.1002/pssb.2221850125.

[291] D. Martín-Becerra, J.B. González-Díaz, V.V. Temnov, A. Cebollada, G. Armelles, T. Thomay, A. Leitenstorfer, R. Bratschitsch, A. García-Martín, M.U. González, Enhancement of the magnetic modulation of surface plasmon polaritons in Au/Co/Au films, Appl. Phys. Lett. 97 (2010) 183114, http://dx.doi.org/10.1063/1.3512874.

[292] J.D. Jackson, Classical Electrodynamics, third ed., Wiley, New York, NY, USA, 1999.

[293] T.G. Mackay, A. Lakhtakia, Electromagnetic fields in linear bianisotropic mediums, Prog. Opt. 51 (2008) 121–209, http://dx.doi.org/10.1016/S0079-6638(07)51003-6.

[294] E. Kreyszig, Advanced Engineering Mathematics, sixth ed., Wiley, New York, NY, USA, 1988.

[295] A. Lakhtakia, Reflection of an obliquely incident plane wave by a half space filled by a helicoidal bianisotropic medium, Phys. Lett. A 374 (2010) 3887–3894, http://dx.doi.org/10.1016/j.physleta.2010.07.047.

[296] A. Lakhtakia, Reflection from a semi-infinite rugate filter, J. Mod. Opt. 58 (2011) 562–565, http://dx.doi.org/10.1080/09500340.2011.554897.

[297] M. Faryad, A. Lakhtakia, Multiple surface-plasmon-polariton waves localized to a metallic defect layer in a sculptured nematic thin film, Phys. Status Sol. RRL 4 (2010) 265–267, http://dx.doi.org/10.1002/pssr.201004049.

[298] M. Faryad, A. Lakhtakia, Surface plasmon-polariton wave propagation guided by a metal slab in a sculptured nematic thin film, J. Opt. (United Kingdom) 12 (2010) 085102, http://dx.doi.org/10.1088/2040-8978/12/8/085102.

[299] N.S. Kapany, J.J. Burke, Optical Waveguides, Academic Press, New York, NY, USA, 1972.

[300] M.G. Moharam, T.K. Gaylord, Rigorous coupled-wave analysis of planar-grating diffraction, J. Opt. Soc. Am. 71 (1981) 811–818, http://dx.doi.org/10.1364/JOSA.71.000811.

[301] F. Wang, A. Lakhtakia, Lateral shifts of optical beams on reflection by slanted chiral sculptured thin films, Opt. Commun. 235 (2004) 107–132, http://dx.doi.org/10.1016/j.optcom.2004.02.050.

[302] M. Onishi, K. Crabtree, R.A. Chipman, Formulation of rigorous coupled-wave theory for gratings in bianisotropic media, J. Opt. Soc. Am. A 28 (2011) 1747–1758, http://dx.doi.org/10.1364/JOSAA.28.001747.

[303] F.N. Marchevskiĭ, V.L. Strizhevskiĭ, S.V. Strizhevskiĭ, Singular electromagnetic waves in bounded anisotropic media, Sov. Phys. Solid State 26 (1984) 911–912.

[304] V.I. Alshits, V.N. Lyubimov, Dispersionless surface polaritons in the vicinity of different sections of optically uniaxial crystals, Phys. Solid State 44 (2002) 386–390, http://dx.doi.org/10.1134/1.1451033.

[305] V.I. Alshits, V.N. Lyubimov, Dispersionless polaritons at symmetrically oriented surfaces of biaxial crystals, Phys. Solid State 44 (2002) 1988–1992, http://dx.doi.org/10.1134/1.1514793.

[306] N.S. Averkiev, M.I. Dyakonov, Electromagnetic waves localized at the interface of transparent anisotropic media, Opt. Spectrosc. (USSR) 68 (1990) 653–655.

[307] S. Brugioni, R. Meucci, Refractive indices of the nematic mixture E7 at 1550 nm, Infrared Phys. Technol. 49 (2007) 210–212, http://dx.doi.org/10.1016/j.infrared.2006.06.006.

[308] T.G. Mackay, A. Lakhtakia, Determination of constitutive and morphological parameters of columnar thin films by inverse homogenization, J. Nanophoton. 4 (2010) 041535, http://dx.doi.org/10.1117/1.3332584.

[309] A. Yariv, P. Yeh, Photonics: Optical Electronics in Modern Communications, sixth ed., Oxford University Press, New York, NY, USA, 2007.

[310] A. Lakhtakia, Electrically tunable, ultranarrowband, circular-polarization rejection filters with electro-optic structurally chiral materials. J. Eur. Opt. Soc.–Rapid Pubs. 1 (2006) 06006, http://dx.doi.org/10.2971/jeos.2006.06006.

[311] N. Feth, C. Enkrich, M. Wegener, S. Linden, Large-area magnetic metamaterials via compact interference lithography, Opt. Exp. 15 (2007) 501–507, http://dx.doi.org/10.1364/OE.15.000501.

[312] W. Cai, U.K. Chettiar, H.-K. Yuan, V.C. de Silva, A.V. Kildishev, V.P. Drachev, V.M. Shalaev, Metamagnetics with rainbow colors, Opt. Exp. 15 (2007) 3333–3341, http://dx.doi.org/10.1364/OE.15.003333.

[313] Y.-J. Jen, A. Lakhtakia, C.-W. Yu, Y.-H. Wang, Negative real parts of the equivalent permittivity, permeability, and refractive index of sculptured-nanorod arrays of silver, J. Vac. Sci. Technol. B 28 (2010) 1078–1083, http://dx.doi.org/10.1116/1.3456125.

[314] W.S. Weiglhofer, Constitutive characterization of simple and complex mediums, in: W.S. Weiglhofer, A. Lakhtakia (Eds.), Introduction to Complex Mediums for Optics and Electromagnetics, SPIE Press, Bellingham, WA, USA, 2003, pp. 27–61.

[315] D.R. Smith, P. Kolinko, D. Schurig, Negative refraction in indefinite media, J. Opt. Soc. Am. B 21 (2004) 1032–1043, http://dx.doi.org/10.1364/JOSAB.21.001032.

[316] W. Yan, L. Shen, L. Ran, J.A. Kong, Surface modes at the interfaces between isotropic media and indefinite media, J. Opt. Soc. Am. A 24 (2007) 530–535, http://dx.doi.org/10.1364/JOSAA.24.000530.

[317] A.N. Darinskii, E. Le Clezio, G. Feuillard, Electromagnetic surface wave attenuation caused by acoustic wave radiation, Electromagnetics 28 (2008) 175–185, http://dx.doi.org/10.1080/02726340801921551.

[318] A. Shalabney, A. Lakhtakia, I. Abdulhalim, A. Lahav, C. Patzig, I. Hazek, A. Karabchevsky, B. Rauschenbach, F. Zhang, J. Xu, Surface plasmon resonance from metallic columnar thin films, Photon. Nanostruct. Fundam. Appl. 7 (2009) 176–185, http://dx.doi.org/10.1016/j.photonics.2009.03.003.

[319] A.P. Vinogradov, A.V. Dorofeenko, S.G. Erokhin, M. Inoue, A.A. Lisyansky, A.M. Merzlikin, A.B. Granovsky, Surface state peculiarities in one-dimensional photonic crystal interfaces, Phys. Rev. B 74 (2006) 045128, http://dx.doi.org/10.1103/PhysRevB.74.045128.

[320] F.Y. Kou, T. Tamir, Range extension of surface plasmons by dielectric layers, Opt. Lett. 12 (1987) 367–369, http://dx.doi.org/10.1364/OL.12.000367.

[321] Z. Salamon, H.A. Macleod, G. Tollin, Coupled plasmon-waveguide resonators: a new spectroscopic tool for probing proteolipids film structure and properties, Biophys. J. 73 (1997) 2791–2797, http://dx.doi.org/10.1016/S0006-3495(97)78308-5.

[322] Z. Salamon, H.A. Macleod, G. Tollin, Surface plasmon resonance spectroscopy as a tool for investigating the biochemical and biophysical properties of membrane protein systems. I: Theoretical principles, Biochim. Biophys. Acta 1331 (1997) 117–129, http://dx.doi.org/10.1016/S0304-4157(97)00004-X.

[323] Z. Salamon, H.A. Macleod, G. Tollin, Surface plasmon resonance spectroscopy as a tool for investigating the biochemical and biophysical properties of membrane protein systems. II: Applications to biological systems, Biochim. Biophys. Acta 1331 (1997) 131–152, http://dx.doi.org/10.1016/S0304-4157(97)00003-8.

[324] G. Borstel, H.J. Falge, Surface phonon-polaritons, in: A.D. Boardman (Ed.), Electromagnetic Surface Modes, Wiley, New York, NY, USA, 1982 (Chapter 6).

[325] R.F. Wallis, Surface magnetoplasmons on semiconductors, in: A.D. Boardman (Ed.), Electromagnetic Surface Modes, Wiley, New York, NY, USA, 1982 (Chapter 15).

[326] I. Abdulhalim, Surface plasmon TE and TM waves at the anisotropic film-metal interface, J. Opt. A: Pure Appl. Opt. 11 (2009) 015002, http://dx.doi.org/10.1088/1464-4258/11/1/015002.

[327] I.J. Hodgkinson, Q.-h. Wu, Birefringent Thin Films and Polarizing Elements, World Scientific, Singapore, 1997.

[328] I.J. Hodgkinson, Q.-h. Wu, S. Collet, Dispersion equations for vacuum-deposited tilted-columnar biaxial media, Appl. Opt. 40 (2001) 452–457, http://dx.doi.org/10.1364/AO.40.000452.

[329] J. Gospodyn, J.C. Sit, Characterization of dielectric columnar thin films by variable angle Mueller matrix and spectroscopic ellipsometry, Opt. Mater. 29 (2006) 318–325, http://dx.doi.org/10.1016/j.optmat.2005.10.004.

[330] J.A. Thornton, Influence of apparatus geometry and deposition conditions on the structure and topography of thick sputtered coatings, J. Vac. Sci. Technol. 11 (1974) 666–670, http://dx.doi.org/10.1116/1.1312732.

[331] R. Messier, The nano-world of thin films, J. Nanophoton. 2 (2008) 021995, http://dx.doi.org/10.1117/1.3000671.

[332] J.A. Polo Jr., A. Lakhtakia, Surface electromagnetic waves: a review, Laser Photon. Rev. 5 (2011) 234–246, http://dx.doi.org/10.1002/lpor.200900050.

[333] M. Faryad, A. Lakhtakia, On multiple surface-plasmon-polariton waves guided by the interface of a metal and a rugate filter in the Kretschmann configuration, Opt. Commun. 284 (2011) 5678–5687, http://dx.doi.org/10.1016/j.optcom.2011.08.055.

[334] M. Faryad, J.A. Polo Jr., A. Lakhtakia, Multiple trains of same-color surface plasmon-polaritons guided by the planar interface of a metal and a sculptured nematic thin film. Part IV: Canonical Problem. 4 (2010) 043505, http://dx.doi.org/10.1117/1.3365052.

[335] A. Lakhtakia, J.A. Polo Jr., Engineering the phase speed of surface-plasmon wave at the planar interface of a metal and a chiral sculptured thin film, Microw. Opt. Technol. Lett. 50 (2008) 1966–1970, http://dx.doi.org/10.1002/mop.23511. Corrections: 50 (2008) 3279–3280, http://dx.doi.org/10.1002/mop.23899. More corrections: 51 (2009) 2524, http://dx.doi.org/10.1002/mop.24633.

[336] J.A. Polo Jr., T.G. Mackay, A. Lakhtakia, Mapping multiple surface-plasmon-polariton-wave modes at the interface of a metal and a chiral sculptured thin film, J. Opt. Soc. Am. B 28 (2011) 2656–2666, http://dx.doi.org/10.1364/JOSAB.28.002656.

[337] P.G. de Gennes, J. Prost, The Physics of Liquid Crystals, Clarendon Press, Oxford, United Kingdom, 1993.

[338] T.G. Mackay, A. Lakhtakia, Modeling chiral sculptured thin films as platforms for surface-plasmonic-polaritonic optical sensing, IEEE Sens. J. 12 (2012) 273–280, http://dx.doi.org/10.1109/JSEN.2010.2067448.

[339] A.S. Hall, M. Faryad, G.D. Barber, L. Liu, S. Erten, T.S. Mayer, A. Lakhtakia, T.E. Mallouk, Broadband light absorption with multiple surface plasmon polariton waves excited at the interface of a metallic grating and photonic crystal, ACS Nano 7 (2013) 4995–5007, http://dx.doi.org/10.1021/nn4003488.

[340] S.E. Swiontek, D.P. Pulsifer, J. Xu, A. Lakhtakia, Suppression of circular Bragg phenomenon in chiral sculptured thin films produced with simultaneous rocking and rotation of substrate during serial bideposition, J. Nanophoton. 7 (2013) 073599, http://dx.doi.org/10.1117/1.JNP.7.073599.

[341] J.B. Kim, Y. Zou, Y.D. Kim, J.J. Kim, C.K. Hwangbo, Multiple surface plasmon waves in [prism/Ag/SiO$_2$ helical thin film] Kretschmann configuration, Thin Solid Films 520 (2011) 1451–1453, http://dx.doi.org/10.1016/j.tsf.2011.10.001.

[342] A. Lakhtakia, On determining gas concentrations using dielectric thin-film helicoidal bianisotropic medium bilayers, Sens. Actuat. B: Chem. 52 (1998) 243–250, http://dx.doi.org/10.1016/S0925-4005(98)00245-7.

[343] T.G. Mackay, A. Lakhtakia, S.S. Jamaian, Chiral sculptured thin films as integrated dual-modality optical sensors, Proc. SPIE 8465 (2012) 84650X, http://dx.doi.org/10.1117/12.928981.

[344] S.E. Swiontek, D.P. Pulsifer, A. Lakhtakia, Optical sensing of analytes in aqueous solutions with a multiple surface-plasmon-polariton-wave platform, Sci. Rep. 3 (2013) 1409, http://dx.doi.org/10.1038/srep01409.

[345] H. Jänchen, D. Endelema, N. Kaiser, F. Flory, Determination of the refractive indices of highly biaxial anisotropic coatings using guided modes, Pure Appl. Opt. 5 (1996) 405–415, http://dx.doi.org/10.1088/0963-9659/5/4/007.

[346] N.A. Beaudry, Y. Zhao, R. Chipman, Dielectric tensor measurement from a single Mueller matrix image, J. Opt. Soc. Am. A 24 (2007) 814–824, http://dx.doi.org/10.1364/JOSAA.24.000814.

[347] A. Lakhtakia, Enhancement of optical activity of chiral sculptured thin films by suitable infiltration of void regions, Optik 112 (2001) 145–148, http://dx.doi.org/10.1078/0030-4026-00024. Corrections: 112 (2001) 544, http://dx.doi.org/10.1078/0030-4026-00088.

[348] C.F. Bohren, E.E. Clothiaux, Fundamentals of Atmospheric Radiation, Wiley-VCH, Weinheim, Germany, 2006.

[349] M. Solano, M. Faryad, A.S. Hall, T.E. Mallouk, P.B. Monk, A. Lakhtakia, Optimization of the absorption efficiency of an amorphous-silicon thin-film tandem solar cell backed by a metallic surface-relief grating, Appl. Opt. 52 (2013) 966–979, http://dx.doi.org/10.1364/AO.52.000966.

[350] A. Lakhtakia, Surface multiplasmonics, Proc. SPIE 8104 (2011) 810403, http://dx.doi.org/10.1117/12.893129.

[351] S.V. Shiyanovskii, Theory of surface electromagnetic waves in cholesteric liquid crystals, Mol. Cryst. Liq. Cryst. 179 (1990) 133–138, http://dx.doi.org/10.1080/0026894900 8055363.

[352] M. Ciumac, D.-M. Baboiu, D. Mihalache, Hybrid surface modes in periodic stratified media: transfer matrix technique, Opt. Commun. 111 (1994) 548–555, http://dx.doi.org/10.1016/0030-4018(94)90534-7.

[353] A.M. Merzlikin, A.P. Vinogradov, A.V. Dorofeenko, M. Inoue, M. Levy, A.B. Ganovsky, Controllable Tamm states in magnetophotonic crystal, Physica B 394 (2007) 277–280, http://dx.doi.org/10.1016/j.physb.2006.12.027.

[354] M. Zgonik, R. Schlesser, I. Biaggio, E. Volt, J. Tscherry, P. Günter, Materials constants of KNbO$_3$ relevant for electro- and acousto-optics, J. Appl. Phys. 74 (1993) 1287–1297, http://dx.doi.org/10.1063/1.354934.

[355] J. Gao, A. Lakhtakia, M. Lei, Simultaneous propagation of two Dyakonov-Tamm waves guided by the planar interface created in a chiral sculptured thin film by a sudden change of vapor flux direction, Phys. Lett. A 374 (2010) 3370–3372, http://dx.doi.org/10.1016/j.physleta.2010.06.042.

[356] M. Faryad, A. Lakhtakia, Prism-coupled excitation of Dyakonov-Tamm waves, Opt. Commun. 294 (2013) 192–197, http://dx.doi.org/10.1016/j.optcom.2012.12.072.

[357] J. Van Bladel, Electromagnetic Fields, Hemisphere, Washington, DC, USA, 1985.

[358] G. Floquet, Sur les équations différentielles linéaires à coefficients périodiques, Ann. sci. l'École Norm. Supér. Ser. 2 (12) (1883) 47–88.

[359] W.S. Weiglhofer, A. Lakhtakia, B. Michel, Maxwell Garnett and Bruggeman formalisms for a particulate composite with bianisotropic host medium, Microw. Opt. Technol. Lett. 15 (1997) 263–266, http://dx.doi.org/10.1002/(SICI)1098-2760(199707)15:4<263::AID-MOP19>3.0.CO;2-8; Corrections: 22 (1999) 221, http://dx.doi.org/10.1002/(SICI)1098-2760(19990805)22:3<221::AID-MOP21>3.0.CO;2-R.

[360] T. Aste, Circle, sphere, and drop packings, Phys. Rev. E 53 (1996) 2571–2579, http://dx.doi.org/10.1103/PhysRevE.53.2571.

[361] B.M. Ross, A. Lakhtakia, Bruggeman approach for isotropic chiral mixtures revisited, Microw. Opt. Technol. Lett. 44 (2005) 524–527, http://dx.doi.org/10.1002/(ISSN)1098-2760.

[362] T.G. Mackay, A. Lakhtakia, Bruggeman formalism versus "Bruggeman formalism": particulate composite materials comprising oriented ellipsoidal particles, J. Nanophoton. 6 (2012) 069501, http://dx.doi.org/10.1117/1.JNP.6.069501. Corrections: 6 (2013) 060106, http://dx.doi.org/10.1117/1.JNP.6.060106.

[363] B. Michel, A. Lakhtakia, W.S. Weiglhofer, Homogenization of linear bianisotropic particulate composite media—numerical studies, Int. J. Appl. Electromag. Mech. 9 (1998) 167–178. Corrections: 10 (1999) 537–538.

[364] R. Messier, T. Takamori, R. Roy, Structure-composition variation in rf-sputtered films of Ge caused by process parameter changes, J. Vac. Sci. Technol. 13 (1976) 1060–1065, http://dx.doi.org/10.1116/1.569060.

[365] J.R. Blanco, P.J. McMarr, J.E. Yehoda, K. Vedam, R. Messier, Density of amorphous germanium films by spectroscopic ellipsometry, J. Vac. Sci. Technol. A 4 (1986) 577–582, http://dx.doi.org/10.1116/1.573851.

[366] S. Brunauer, P.H. Emmett, E. Teller, Adsorption of gases in multimolecular layers, J. Am. Chem. Soc. 60 (1938) 309–319, http://dx.doi.org/10.1021/ja01269a023.

[367] G. Bomchil, R. Herino, K. Barla, J.C. Pfister, Pore size distribution in porous silicon studied by adsorption isotherms, J. Electrochem. Soc. 130 (1983) 1611–1614, http://dx.doi.org/10.1149/1.2120044.

[368] J.V. Ryan, M. Horn, A. Lakhtakia, C.G. Pantano, Characterization of sculptured thin films, Proc. SPIE 5593 (2004) 643–649, http://dx.doi.org/10.1117/12.573910.

[369] W.S. Weiglhofer, On the inverse homogenization problem of linear composite materials, Microw. Opt. Technol. Lett. 28 (2001) 421–423, http://dx.doi.org/10.1002/1098-2760 (20010320)28:6<421::AID-MOP1059>3.0.CO;2-1.

[370] R.D. Kampia, A. Lakhtakia, Bruggeman model for chiral particulate composites, J. Phys. D: Appl. Phys. 25 (1992) 1390–1394, http://dx.doi.org/10.1088/0022-3727/25/10/002.

[371] T.G. Mackay, A. Lakhtakia, A limitation of the Bruggeman formalism for homogenization, Opt. Commun. 234 (2004) 35–42, http://dx.doi.org/10.1016/j.optcom.2004.02.007. Corrections: 282 (2009) 4028, http://dx.doi.org/10.1016/j.optcom.2009.07.019.

[372] A.J. Duncan, T.G. Mackay, A. Lakhtakia, On the Bergman-Milton bounds for the homogenization of dielectric composite materials, Opt. Commun. 271 (2007) 470–474, http://dx.doi.org/10.1016/j.optcom.2006.10.056.

[373] T.G. Mackay, On the effective permittivity of silver-insulator nanocomposites, J. Nanophoton. 1 (2007) 019501, http://dx.doi.org/10.1117/1.2472372.

[374] S.S. Jamaian, T.G. Mackay, On limitations of the Bruggeman formalism for inverse homogenization, J. Nanophoton. 4 (2010) 043510, http://dx.doi.org/10.1117/1.3460908.

[375] C.F. Bohren, Do extended effective-medium formulas scale properly? J. Nanophoton. 3 (2009) 039501, http://dx.doi.org/10.1117/1.3157171.

[376] C.F. Bohren, X. Xiao, A. Lakhtakia, The missing ingredient in effective-medium theories: standard deviations, J. Mod. Opt. 59 (2012) 1312–1315, http://dx.doi.org/10.1080/09500340.2012.713521.

Printed in the United States
By Bookmasters